Michael Meißer

Resonant Behaviour of Pulse Generators for the Efficient Drive of Optical Radiation Sources Based on Dielectric Barrier Discharges

Resonant Behaviour of Pulse Generators for the Efficient Drive of Optical Radiation Sources Based on Dielectric Barrier Discharges

by
Michael Meißer

Dissertation, Karlsruher Institut für Technologie (KIT)
Fakultät für Elektrotechnik und Informationstechnik, 2013

Impressum

Karlsruher Institut für Technologie (KIT)
KIT Scientific Publishing
Straße am Forum 2
D-76131 Karlsruhe
www.ksp.kit.edu

KIT – Universität des Landes Baden-Württemberg und
nationales Forschungszentrum in der Helmholtz-Gemeinschaft

KIT Scientific Publishing 2013
Print on Demand

ISBN 978-3-7315-0083-4

Resonant Behaviour of Pulse Generators for the Efficient Drive of Optical Radiation Sources Based on Dielectric Barrier Discharges

Resonanzverhalten von Pulsgeneratoren zum effizienten Betrieb von optischen Strahlungsquellen basierend auf der Dielektrisch Behinderten Entladung

Zur Erlangung des akademischen Grades eines

DOKTOR-INGENIEURS

von der Fakultät für

Elektrotechnik und Informationstechnik

des Karlsruher Instituts für Technologie (KIT)

genehmigte

DISSERTATION

von

Dipl.-Ing. Michael Meißer

geboren in Schwerin

Tag der mündlichen Prüfung: 10.07.2013

Hauptreferent: Prof. Dr.-Ing. Wolfgang Heering

Korreferent: Prof. Dr.-Ing. Michael Braun

“Knowledge is the only possession
that proliferates
when shared.”

Translated statement of Marie Freifrau von Ebner-Eschenbach, Austrian author.

I. ZUSAMMENFASSUNG

Die vorliegende Arbeit umfasst die Untersuchung neuartiger Schaltungstopologien zum gepulsten Betrieb von Excimer-Lampen basierend auf Dielektrisch Behinderten Entladungen (DBE), die zur Erzeugung optischer Strahlung im sichtbaren und ultravioletten Bereich eingesetzt werden. Der Fokus liegt dabei auf der effizienten Erzeugung kurzer Pulse und vorteilhafter Pulsformen um einen hohen Systemwirkungsgrad sicher zu stellen.

Erstmals werden in dieser Arbeit bekannte und neu entwickelte Schaltungstopologien zum Betrieb von DBE grundlegend klassifiziert und durchgängig experimentell validiert. Die neu entwickelten Topologien und Betriebsmodi basieren sowohl auf rein resonantem Betrieb als auch auf der Drosselvorladung, welche die Adaptivität an unterschiedliche Lampen sowie eine höhere Spannungsverstärkung bietet.

Der Fokus dieser Arbeit liegt auf der Untersuchung von transformatorlosen Topologien, die – ausgestattet mit modernen Silizium-Karbid Leistungshalbleitern – die Anregung der DBE-Lampen mit Pulsen kürzer als 1 µs erlauben. Der positive Effekt der kurzen Pulsdauer auf die Lampeneffizienz und -leistungsdichte wurde experimentell nachgewiesen. Eine flache DBE-Lampe mit vergleichsweise großem Energiebedarf (3.18 nF Kapazität und 2 kV Zündspannung) konnte mit Pulsfrequenzen von mehr als 1 MHz betrieben werden, während eine kleinere Koaxial-DBE mit bis zu 3.1 MHz Pulsfrequenz angeregt wurde. Aufgebaute Pulsinverter erreichten Spitzenwirkungsgrade von 92 %.

Anstelle von Simulationen wurden in der Arbeit hauptsächlich Kalkulationen und elektrische Messungen durchgeführt, um Bauteile und Schaltungen zu beschreiben. Die Hauptkomponenten der Hochfrequenz-Pulsinverter, wie Leistungshalbleiter und magnetische Bauelemente sowie die DBD-Lampen wurden bezüglich ihrer parasitären Eigenschaften mittels spektral aufgelöster Impedanzmessungen charakterisiert. Aufbauend auf den Messergebnissen konnte die parasitäre Induktivität von Gate-Treibern und Leistungsstufen signifikant reduziert werden. Darüber hinaus wurden thermische Messungen zum Erfassen von Verlustleistungen sowie statische Messungen des Einschaltwiderstandes von Leistungshalbleitern vorgenommen.

Neben dem Betrieb von Silizium-Karbid MOSFETs und JFETs wurde auch die Verschaltung von Silizium MOSFETs zu einem Matrix-Schalter mit 2,7 kV Betriebsspannung untersucht.

Ausgehend vom recherchierten Stand der Technik werden vorteilhafte Spannungspulsformen zum Betrieb der DBE diskutiert. Darüber hinaus enthält diese Arbeit eine Abhandlung zu den Einflüssen der verschiedenen DBE-Kapazitäten auf die Entladung und die Energie, die vom Pulsgenerator zur Verfügung gestellt werden muss.

II. Abstract

This dissertation presents novel methods of operating dielectric barrier discharge (DBD) excimer lamps with short, high-voltage pulses of favourable wave-shape in order to ensure operation at highest efficiency.

Novel and state-of-the-art power electronic pulse inverter topologies are investigated, experimentally validated and classified in terms of their main properties. The novel topologies introduced are based on pure resonant operation or a combination of resonant operation and inductor pre-charge which offers an improved adaptability to different lamps and higher voltage amplification.

Many topologies are modified towards transformer-less variants which are equipped with SiC power semiconductor switches.

A flat DBD lamp with a capacitance of 3.18 nF and an ignition voltage of 2 kV was operated with a pulse frequency of more than 1 MHz while a smaller experimental coaxial DBD was operated at 3.1 MHz. Peak inverter efficiencies of 92 % were achieved. Experiments verified the positive effect of multi-MHz pulses on DBD lamp efficiency and power density.

Instead of simulations, the development was driven by theoretical calculations and real world measurements. The main pulse inverter components as well as the DBD lamps are characterised by means of frequency-resolved impedance measurements with high-voltage DC bias. Building on the results, the parasitic inductance of power stages and gate drive circuits was significantly reduced. Basing on the requirements for HF operation SiC power semiconductor switches are benchmarked and by means of an alternative approach, a modular Si-MOSFET equipped 2.7 kV matrix switch design is presented.

Based on the state-of-the-art, characteristic electrical parameters of possible pulse voltage wave-shapes are discussed. Beyond that, this work contains a theoretical study on the influence of the different DBD capacitances on the energy support of the transient DBD plasma and on the energy that needs to be provided by the inverter.

Having the focus both on mathematical backgrounds of pulsed resonant circuits and practical implementation of low-inductive power stages, the work at hand marries theory and practice.

III. ACKNOWLEDGEMENT

This work would not have been possible without the continuous support of my colleagues at the LTI. I owe general thanks to the students who supported my research by their work and valuable feedback. I thank especially Kay Messerschmidt, Timon Messner, Fabian Denk, Hannes Ruf, Denis Krämer, Thomas Lüth, Philip Leuger, Manuel Popp, Martin Perner, Qiao Guo, Gennady Benderman, William Truong and Carina Hansmann.

I would like to express my sincere gratitude to my doctorate supervisor Prof. Heering for his valued feedback and in-depth review of this work. I appreciate the commitment of Prof. Braun for taking over the second assessment.

I acknowledge the continued support throughout the course of this work of my supervisor Dr. Rainer Kling.

I am glad to have been colleague of Karsten Haehre and Christoph Simon owing special thanks to them for the review of this work and others, inspiring discussions and constructive comments.

I am grateful to Mark Paravia and Klaus Trampert for their valued suggestions and guidance during my first months at the LTI.

Special thanks go to Prof. Philip Mawby at the University of Warwick, UK, for accepting me as a visiting researcher and Dr. Dean Hamilton for his valuable support. I enjoyed the friendly atmosphere at Warwick that was last but not least contributed to by Dr. Mike Jennings.

I am grateful to my colleagues Christoph Kaiser and Mohan Ögün for their continued support and fruitful discussions. I will definitely miss our talks during lunch, which mostly started at the quality of the meal and ended up in discussions of aspects of global interest.

My studies and research at the LTI were rewarding and enjoyable not least due to the excellent work and comprehensive support by the LTI workshop staff, the IT administrators and the secretaries. I owe thanks especially to Siegfried Kettlitz who was always helpful when I faced software problems. Special thanks go to Felix Geislhöringer, the good spirit of the LTI, for his extensive support.

I cannot thank enough Ellen Hofmann for the extensive proof-reading which (hopefully) makes this work also readable for English native speakers.

I owe thanks to my loved parents, my brother and dear friends for their understanding, intense care and passionate support.

This work was generously supported by the Karlsruhe House of Young Scientists (KHYS).

IV. Table of Contents

1 DBD SYSTEM OVERVIEW AND STRUCTURE OF THIS WORK

Dielectric barrier discharges (DBD) efficiently generate non-thermal plasmas which play a vital role in many industrial applications. The main difference to galvanically coupled thermal plasma generation is the electrical insulation of the electrodes by non-conductive dielectric barriers. They prevent a continuous power flow into the plasma and thereby support non-local thermal equilibrium (non-LTE) plasma conditions. The focus of this work is to break fresh grounds in efficiently driving these incoherently emitting light sources with pulses of optimised wave-shape. The attention is directed towards possible power electronic topologies that are suited to generate these pulses.

A DBD radiation system consisting of the DBD lamp and a tailored inverter is diagrammatically personated in Figure 1. The system is depicted with its pivotal function blocks and associated loss mechanisms.

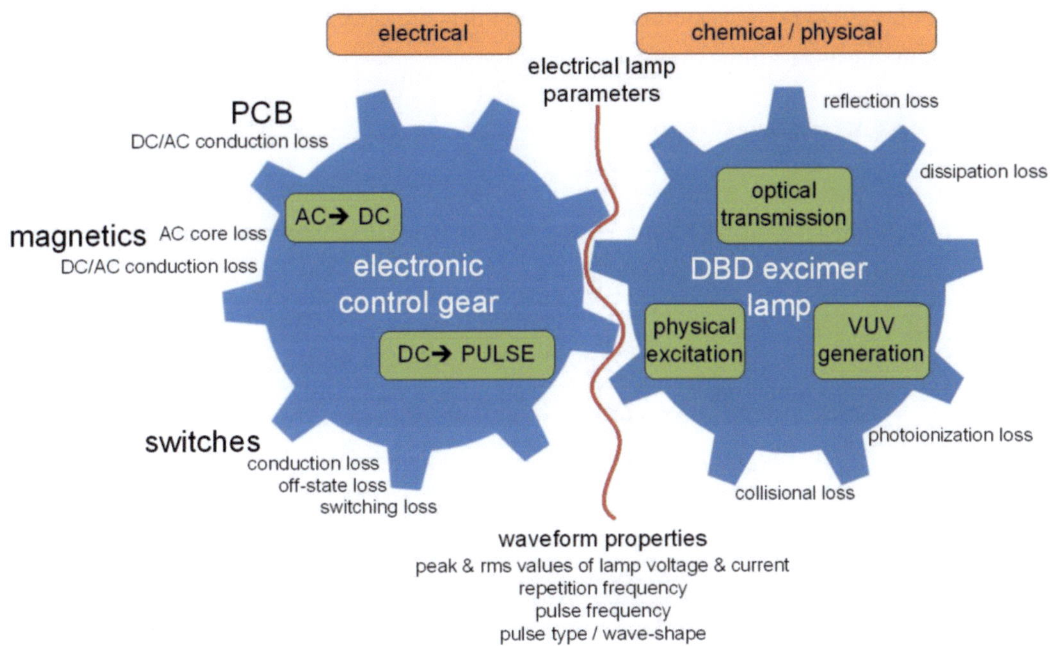

Figure 1: DBD System overview. The main processes including loss processes of the inverter (electronic control gear, ECG) and the DBD excimer lamp are depicted.

As the chart shows, the boundary between the inverter and the DBD excimer lamp is defined by the electrical parameters of the lamp and the electrical properties of the driving waveform. The chart visualises that in order to achieve high system efficiency, both system parts – the inverter and the DBD lamp – have to work hand in hand. Operation modes where either lamp or pulse inverter benefits while the other is disproportionally affected are not expedient.

Beside a stable set of electrical parameters, the lamp has highly non-linear electrical behaviour due to the influence of the plasma. The DBD lamp operation principles and their relevant properties and scaling laws are discussed in Chapter 2. The important aspect of the waveform properties that widely determine efficiency and appearance of the DBD's optical radiation is discussed in the subsequent Chapter 3. The pulse inverter that has to deliver tailored pulses to the DBD lamp, its principles of operation and principal scaling rules are presented in Chapter 4. Here, also the main components such as power semiconductor switches and power magnetics are perused. The main focus of the work at hand is the investigation of novel topologies and operation modes of pulse inverters. For the first time, novel as well as common topologies are classified into three classes. The base of this classification is explained in Chapter 4.5. New and known members of these classes are described in the Chapters 5 to 7. A qualitative comparison of the presented topologies follows in Chapter 8. The whole work is then summarised in Chapter 9. The final Chapter 10 gives an outlook and provides ideas and tasks that would be further steps towards an advanced state of knowledge regarding power electronic systems for pulsed DBD drive.

Topologies and operation schemes published here could also find their application in other fields of pulsed power such as laser pumping, (piezo) actuator driving and the like. Notwithstanding the fact that this dissertation is written in British English, the electrical symbols are according to German DIN standards and all units are according to SI.

This work was significantly funded by the German public. As a matter of course, this thesis will be available to everybody without charge following the open access concept. The author herewith expresses his antipathy with contemporary commercial publishing companies enriching themselves on the authors' publications while simultaneously curtailing their rights.

1.1 Historical Background

The principle of dielectric barrier discharges (DBD) is known since 1857 when Werner von Siemens used a first DBD reactor set-up to produce ozone (Siemens 1857). Following research led to a better understanding of the DBD-specific electrical breakdown processes (Honda 1955). Beside other applications, the efficient generation of deep and even vacuum ultraviolet (VUV) radiation by the relaxation of excited dimers (excimers) is pursued even further since it was first reported by Volkova (1984).

The DBD excimer lamp is a non-coherent source of ultraviolet (UV) radiation. Depending on the gas filling, quasi-monochromatic radiation in the range from near UV ($\lambda = 354$ nm) to VUV ($\lambda = 126$ nm) can be generated (IUPAC 2007). As of this writing, especially VUV DBD sources attained scientific and commercial interest.

At the Light Technology Institute, research on excimer DBD lamps started in 1987 with investigations of Schorpp (1988) focusing on XeCl excimers. Since then, the efficient and appropriate drive of DBD lamps with the aim of maximising system efficiency and homogeneity of the discharge is a major topic of research. Enthusiasm concerning the excimer-DBD as radiation and light source rose in the 1990's (Newman 1995) and continued to the 21st century (Sosnin 2010). Starting with radio-frequency (RF) generators (Fließ 1989) the focus was set on pulse generators (Hassall 1994), especially after the favourable properties of the pulsed excitation scheme were made public by two Osram employees (Vollkommer 1994; Huber 1996). In the 21st century, the interest in new topologies and operation modes was boosted and, besides other approaches, was led to high-voltage current source bridge topologies (Kyrberg 2007b) and complex flyback-derived topologies (El-Deib 2010a). Research and transfer to commercial products will intensify even more since the corresponding intellectual property on the pulsed operation of DBDs hold by Osram will expire soon.

1.2 DBD Products and Applications

As of this writing, the application of DBDs occurred in three main application fields. Beside DBD reactors, DBDs are used for the generation of light – both for short wavelengths and for visible light. Their propagation into the lighting market declined to sparse niche applications after initial prestige products as Osram's Planon® and Saint-Gobain's Planilum®. Nevertheless, their application in the UV and especially in the VUV market is promising. Especially the broad requirement for safe and efficient water treatment solutions fed by the world-wide growth of economies can lead to a renaissance of DBD excimer technology that will by far exceed the current market share. Even though the most important companies in the UV field have DBD excimer sources in their portfolio, their focus is still on low-volume, high-margin applications such as semiconductor wafer treatment or TFT screen surface cleaning. The following sub-chapters give an overview of commercially available products as well as reported research.

1.2.1 DBD Reactors

The energy of electrons accelerated by the DBD can be directed into chemical energy. As early as 1857 Werner von Siemens reported on ozone production ("Umwandlung des Sauerstoffs der Luft in Ozon") and a first reactor set-up using DBDs (Siemens 1857). Broad use of DBD reactors for ozone generation came into play in the early 1900's with first large-scale installations in Paris, France. Secondary information is found in (Hosselet 1971; Eliasson 1987). Being the second strongest oxidizing agent the triatomic ozone is broadly used in water purification processes and for chemical bleaching since then. According to Formula (1.1), ozone is formed by radiation-induced dissociation of O_2 and a subsequent three-body reaction where M = O, O_2, O_3 is a third collision particle:

$$O+O_2+M \rightarrow O_3^*+M \rightarrow O_3+M \tag{1.1}$$

As multiple competing side reactions steal predecessors as well as parts of the end product, this poses an upper limit on the possible relative ozone concentration at 100 ppm. Long-time storage is impossible because ozone is inherently instable. Therefore, ozone production must be on site and controlled to meet the process demand. Ozone reactors such as depicted in Figure 2 consist of cylindrical glass tubes with wall thickness between 1.5 – 2 mm. The gaps where the gas is led through are up to 2 mm in width. The individual tubes arranged in the reactor have a length of up to 2 m. Reactors can comprise of hundreds of those tubes which are mostly fed by a common power supply. The tubes are cooled with water and the DBDs are driven by approximately 20 kV at 50 Hz or thyristor-controlled frequency converters with up to 5 kHz operating frequency.

Beside this, reactors are also used for cleaning of gaseous matter. Chemical processing, print and paint shops as well as semiconductor production sites produce large amounts of toxic gases. DBDs can be used to generate reactive chemical gases such as N_2, O_3 or OH which readily attack organic molecules of volatile organic compounds (VOCs). DBDs are especially efficient when the concentration of pollutants is below 1,000 ppm since active species are generated without noticeably raising the enthalpy of the background gas (Kogelschatz 2008). Nozaki (2002) reported on the simultaneous removal of NO_X, SO_X and fly ash particles from coal combustion exhaust gas while Zrilli (2005) reduced the NO_X content with the help of DBDs. Shoyama (2006) found that NO_X removal from diesel exhaust gases improved when the DBD reactor was operated with short voltage pulses that simultaneously reduced the discharge current density. This is in line with other findings related to DBD excimer lamps that will be discussed in Chapter 3.4.

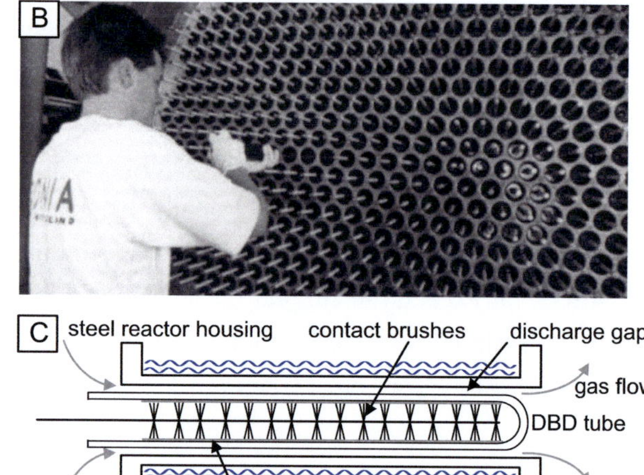

Figure 2: Ozone generator installation (A) and operator inserting DBD tubes into reactor housing (B) (Kogelschatz 1999). C: Reprint of basic reactor structure according to Kogelschatz (2008).

Falkenstein (1998) describes the ozone production by DBD reactors for off-gas treatment. Ozone production and synthesis from air is also discussed by Hosselet (1971) and Nozaki (2002). Simulations are contributed by Eliasson (1987).

A third application of DBD reactors is the treatment of surfaces of solid matter. Beside radicals generated by the discharges, heavy and lightweight particles accelerated towards the surfaces can be directly used to modify surface properties such as adhesion (Chen 2004), wettability (Finantu-Dinu 2007) and surface tension. Here, the DBD converts electrical energy into kinetic energy. Zacher (2004) reports on surface cleaning in advance of layer deposition. DBD surface treatment of next-generation solar cells is presented by Zen (2012). Bente (2004) uses DBDs operated in different gases to create water repellent characteristics on wooden surfaces. In contrast, DBDs operated in air can modify wool (Finantu-Dinu 2007) and nonwoven fabric (Zhan 2007) to achieve hydrophilic properties. The surface properties of plastic foils can be improved by application of DBD treatment, mostly with filamented discharges. The reactors are designed to process foil material with up to 10 m width which is being moved continuously with speeds of up to a few hundred m/min. Beside air, also reactive gases or mixtures of gases are used in open or closed systems (Kogelschatz 2008). Examples of such reactors are depicted in Figure 3. In addition to continuous treatment according to Figure 3A, patterned electrodes allow for a structured surface modification as depicted in Figure 3B. For instance, wettability of foils could be modified with a certain areal distribution in order to pattern the subsequently deposited lacquer. The homogeneous discharge that maximises the efficiency of DBD excimer lamps could also help in surface modification applications, where uniformity of the surface treatment is an important attribute.

Beside surface modification, DBDs can also assist chemical vapour deposition (CVD) used for thin film layer deposition (Nozaki 2002; Kim 2012). Thin metal, insulation or semiconducting films may also be deposited by means of UV or VUV catalyst processes (Wagner 2003) which require respective DBD lamps discussed in the next chapter.

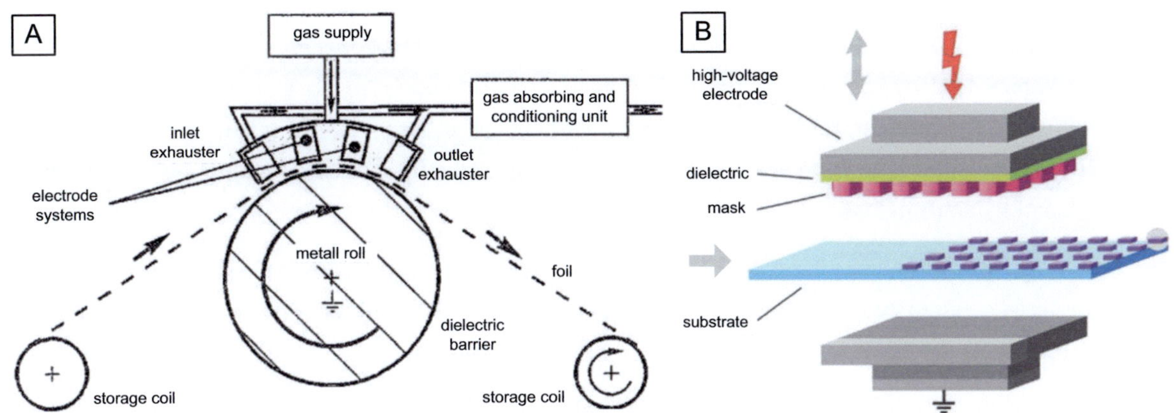

Figure 3A: Continuous DBD treatment of foils, reprint of (Kogelschatz 2008). B: Electro-chemical surface modification and functionalisation by a plasma-stamp device, reprint of (Borris 2010).

1.2.2 Excimer DBD UV-Sources

The nature of DBDs is especially suited to efficiently operate excimer VUV and UV radiation sources. The set-up and operation principle of these lamps is described in Chapter 2.1. Depending on the gas filling, these incoherent high energy optical radiation sources can generate a variety of spectral output characteristics that can be tailored to the respective application. See Figure 4 for overview. The main advantages of these lamps are listed below:

- As excimer generation is almost independent on gas temperature, excimer DBD lamps do not require a warm-up time and may be re-ignited anytime. In contrast to high-pressure gas discharge lamps DBDs are comparatively cold light sources which in most applications do not need extra cooling (Kunze 2001). The DBD's light output is available instantly which allows for their application with repetitive on-off operation. For the same reason, excimer DBDs can be dimmed by means of pulse density modulation (PDM) or pulse package density modulation and operated at very low output power.
- The electrodes are not galvanically connected with the gas discharge. Due to the missing electrode wear-out, excimer DBD lamps theoretically offer a long lifetime of up to 80,000 h (Kogelschatz 2008).
- Excimer DBD lamps may be operated with very high (pulse) power densities that potentially exceed those of mercury low-pressure lamps.
- The most efficient excimer DBD lamps operate with noble-gas or noble-gas halogen fillings. They are mercury-free which makes them especially environmentally friendly.
- Depending on the gas or gas mixture used the VUV/ UV output to electrical input efficiency of Xe excimer DBD lamps can theoretically be as high as 78 % (Vollkommer 1994). Efficiencies of up to 58 % (Dichtl 1998) and 60 % (Beleznai 2006) were already experimentally verified.

- The optical output spectral distribution of excimer DBD lamps is mostly very narrow which is beneficial to wavelength-selective processes. The 172 nm line of the popular Xe excimer has a full-width half-maximum (FWHM) value of 14 nm, for instance.
- The spectral output can be tailored by selecting from a variety of excimer gases and gas mixtures (see Figure 4) or by transferring the radiation spectrum towards longer wavelengths by a phosphor (Jüstel 2011). However, the limited conversion efficiency of the phosphor is not to be neglected. The greater the difference between exciting wavelength and radiated wavelength (stokes shift), the higher the quantum loss.
- Excimer DBD lamps may be constructed in a broad variety of possible geometries. Linear, coaxial and plane configurations are already available as commercial products or are in research state (see Chapter 2.1).

Even though reported to be still in research stage (Bolton 2008) a broad range of excimer-based UV and VUV lamps is commercially available. According to Kogelschatz (2003a) most common excimer complexes used in lamps are Xe_2^* and XeCl*. Although lifetimes of up to 80,000 h were predicted in general (Kogelschatz 2008), the following commercial 172 nm radiation sources only offer up to 2,500 h.

The 20 W to 100 W tubular Xeradex® family of Osram (OSRAM 2012) generates a 172 nm irradiance of 40 mW/cm^2 at a lifetime of 2500 h. Up to 80 mW/cm^2 can be reached at the cost of necessary active cooling and reduced life. Similar lamps are offered by Heraeus Noblelight but with a higher datasheet irradiance of 50 mW/cm^2. Competitor Ushio offers the same irradiance with its comparable tubular compact excimer UV system family. For industrial applications they are combined to large irradiation units with an irradiance of 40 mW/cm^2 provided across an operation width of 220 cm (Ushio 2000). However, compact excimer laboratory units generate only 10 mW/cm^2 at 1,000 h lifetime out of a 100x100 mm window while a head-on system offers higher irradiance of 50 mW/cm^2 out of a round 30 mm diameter window at a lifetime of solely 700 h. The CiMAX lamp (Ushio 2006) generates up to 100 mW/cm^2 irradiance at a lifetime of 1,000 h. The QEX1600 lamp (quark-tec) generates an VUV irradiance of 70 mW/cm^2 at a lifetime of 2500 h. The tubular UVS 172-01® system of UV Solutions Inc. (UV-Solutions 2011b) is a corona discharge tube offering an outstanding radiation power of 700 mW/cm per unit length. The German Company IOT GmbH provides the Excirad radiation system having active widths of 10 to 250 cm and a radiation power of 30 mW per cm unit length (IOT_GmbH 2012). Lomaev (2012) reports on a Xe^*_2 lamp with up to 25 mW/cm^2 irradiance that is housed in a reactor used to reduce the water content of natural gas.

Excimer lamps are used for various kinds of photon-induced chemical reactions due to their relatively low cost compared to excimer lasers. Their application in the semiconductor and TFT screen industry is already vital. Excimer lamps are used for surface cleaning (Falkenstein 2001b) of optical surface (Carman 2004b) or even medical skin treatment (Sosnin 2010). Applications also include UV-induced thin-film metal deposition (Boyd 1997b), rapid silicon low-temperature oxidation (Boyd 1997a; Liaw 2008), deposition and annealing of high-κ Layers (Yu 2004), photo-assisted sol-gel processing and photo-assisted chemical vapour deposition (CVD) (Boyd 2010). 308 nm XeCl excimer lamps are used to cure deposited layers in printing processes on heat sensitive materials (Mehnert 1999). Since the above mentioned UV-induced processes run under low temperature they bring about only little damage on the processed surfaces. They are therefore especially suited for nano-scale samples with reduced thermal persistence (Liaw 2008). Thus they reduce problems such as substrate warpage, dopant redistribution and defect generation that are usually caused by high temperature processing. An application that is presently still dominated by mercury discharge lamps but bears possibilities for a growing market for excimer DBD radiation sources is water treatment (Legrini 1993). As early as 1903 it was discovered that inactivation of bacteria is efficient at wavelengths around 250 nm (Barnard 1903). A more detailed declarative to the spectral efficiency of E-coli bacteria inactivation is found in DIN5031-10 (2000) which is also depicted in Figure 4. Since excimer lamps have been commercially available, researchers were investigating their application in water purification.

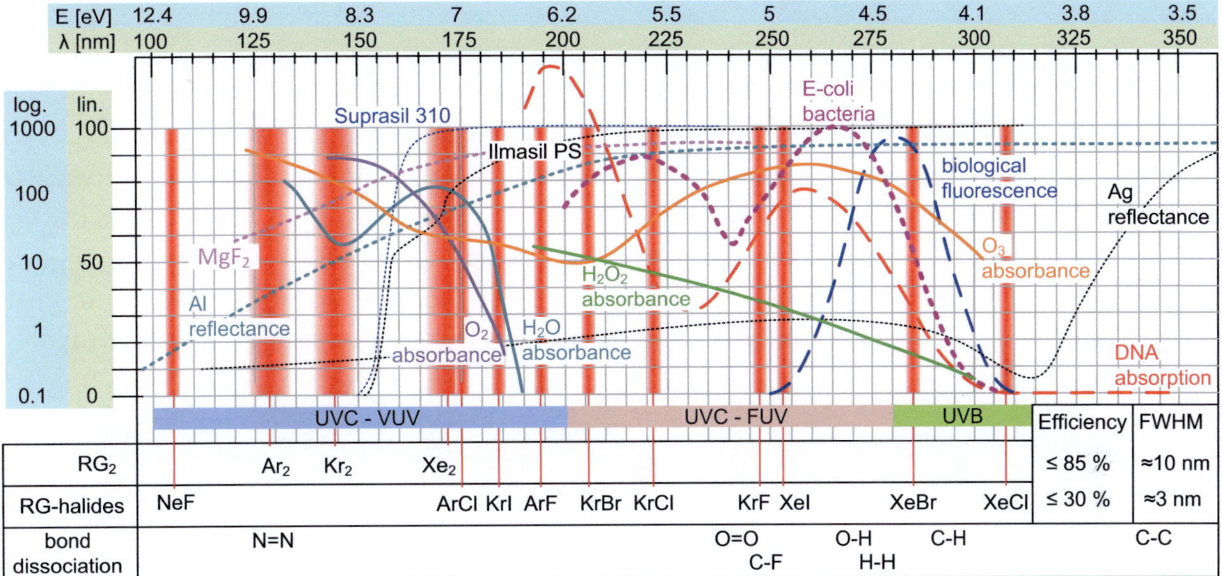

Figure 4: Overview of UV wavelengths in relation to emission spectra of rare-gas (RG) excimers and rare-gas halide excimers based on (UV-Solutions 2011a), linear optical transmission of quartz glasses based on (Qsil 2011; Heraeus 2013) and of 5 mm MgF_2 window glass based on (Crystan), linear optical reflectance of aluminium and silver based on (Oppenländer 2003), linear DNA absorbance [1/(100 cm)] based on (Sonntag 1988), linear spectral inactivation efficiency of E-coli bacteria according to (DIN5031-10 2000) and logarithmic absorbance of important chemicals based on (Oppenländer 2003), chemical bond dissociation energy levels (Liaw 2008). UV classification according to DIN5031-7.

An excimer DBD equipped flatbed reactor for photo-chemical water treatment is presented by Oppenländer (1997). A demonstration-stage excimer system for treatment of conductive fluids had been presented later by Ushio in 2001 (Falkenstein 2001a). Beside DBD reactors, also excimer DBD UV/VUV lamps can be used as a powerful source of ozone. As Figure 4 indicates, O_2 absorbance is significant at wavelengths below 185 nm. Contrarily, ozone destruction peaks around 260 nm. This makes ozone production by low-pressure mercury gas discharge lamps which emit at 185 nm and with higher intensity at 256 nm sparsely efficient since huge portions of the generated ozone are simultaneously destructed. Compared to silent discharge reactors discussed in the previous chapter, 172 nm emitting excimer DBD lamps can achieve twice the ozone saturation concentration and allow for size independent O_3 yields in the order of 200 g/kWh (Salvermoser 2009) since total absorption occurs within 2 mm in dry air (Falkenstein 2001b). Photo-induced production of hydrogen peroxide (H_2O_2) being a strong oxidizing agent is very energy efficient (Bolton 1994). KrCl excimer DBD lamps are well suited for H_2O_2 generation (Oppenländer 1997) due to their radiation peak at 222 nm. Kogelschatz (1991; 1993) reviewed several photo-chemical pollution control techniques as photo-catalysis, photo-oxidation, direct photo-cleavage and photolysis. Modified Linex® lamps were used in air reactors for automotive interior air cleaning (Große-Ophoff 2007). Investigations of engine exhaust gas cleaning were performed at the LTI by Wolf (2000) and Trompeter (2001). An exotic application remains the use of Xe^*_2 DBD flash lamps for laser pumping (Paul 1987).

1.2.3 EXCIMER DBD LIGHT SOURCES

Around the end of the last century, several excimer DBD light sources entered the market. Equipped with a phosphor coating which converts the 172 nm Xe excimer radiation into visible light, these flat lamps were soon recognised as being predestined for applications where dazzle-free bright light was needed. Due to phosphor conversion losses, the electrical input to VIS output efficiency is restricted to values between 50 lm/W (Jüstel 2007) and 80 lm/W (Beleznai 2008a).

Due to the competitive LED technology, the production of Osram's Planon® (Ilmer 2000) is already shut down, as well as the production of Saint-Gobain's Planilum® (Waite 2008) DBD lamp. Examples of both lamps are depicted in Figure 5. However, there are also new approaches as Ushio's XeFL™. The XeFL™ is a tubular T4 diameter lamp with a length of up to 59". Ushio (2009) claims to achieve an efficiency of 55 lm/W at a colour temperature of 5800 K and colour rendering index (CRI) of 90 with a lifetime of up to 60,000 h. Another flat micro-plasma lamp (Herring 2012) is presented by Eden Park offering a projected lifetime exceeding 50,000 h. This Xe excimer lamp with micro-cavities etched into its glass housing and internal phosphor coating is the result of research at the University of Illinois (Park 2007b; Eden 2010).

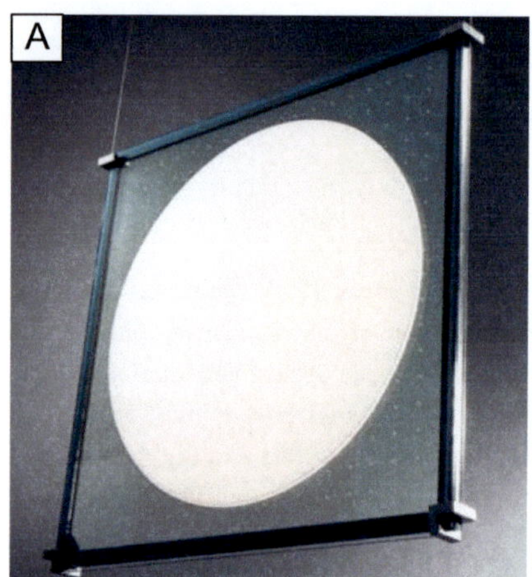

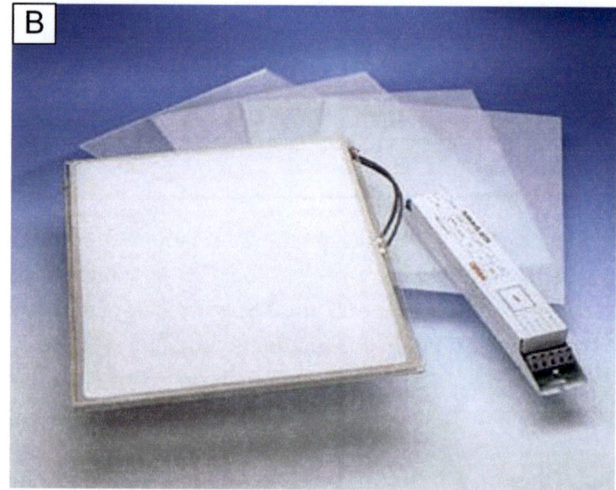

Figure 5A: Planilum® DBD lamp (Waite 2008). B: Planon® flat DBD lamp with ECG (EDC_magazine 2003).

However, due to the evolving LED technology, illumination panels with widths of only 13 mm, an efficiency of 70 lm/W and a CRI of 90 are available (Lunitronix 2010) which already beat the performance of the above introduced commercial DBD light sources. Therefore, excimer DBD-derived light sources ultimately found wide-spread application solely in two mass markets.

The first is plasma display panels (PDP). Despite the advancements in LCD technology, PDPs are still the choice for large flat screens as they provide a wide viewing angle, a high contrast ratio, and a long life span. Due to very small gap distances (refer to Figure 6) PDP cells are operated with high-frequency AC-voltage with amplitudes of about 200 V and peak currents for a large sized display of about 100 A (Chung-Wook 2003). The luminous efficiency is quite poor with up to 2.4 lm/W (Lee 2008a).

Being deployed in the second market - photo copying machines - the linear light source Linex® of Osram (Hitzschke 1998; Osram 2002) benefits from the excimer DBD characteristic of instantly available, temperature independent light. The Linex® is still in production and is also available in a UV-A version.

A lot of secretiveness is made around the so-called e^3 (energy-efficient excitation) light sources of Wammes & Partner which are referred to as being an advancement of the CCFL (Härter 2010). Presumably basing on the key patent of (Wammes 2002), the e³ source contains 2 – 700 mbar of a rare-gas mixture from which exciplexes can be generated that are claimed to emit in the UV, VIS and IR wavelength range (Wammes 2013). The e³ light source bases on a 3 mm diameter glass tube coated with doped ceramics, getter material and phosphor. e³ illumination systems for cash machine backlighting are claimed to exceed an MTTH lifetime of 75.000 hours .

1.3 Objectives of This Work

The dissertation at hand focuses on the efficient and favourable pulsed operation of xenon excimer DBD lamps by means of power electronic inverters. It is preconditioned that pulsed excitation is the superior driving mode both in terms of DBD lamp efficiency and discharge homogeneity.

The objective of this work is summarised by the following key points:

1. Provide an insight into fundamental scaling laws of DBDs as capacitive loads and of the inverter set-up (Chapters 3 and 4).
2. Present an analysis of resonant circuit properties with emphasis on the drive of capacitive loads (Chapter 4).
3. Present an analysis of novel topologies both theoretically and experimentally (Chapters 5 - 7).
4. Give a comprehensive overview and classification scheme of state-of-the-art topologies and own developments (Chapters 4 - 7).
5. Provide investigation results and solutions to manage probable parasitic resonances that have negative effects on efficiency and reliability (Chapters 5.1.3, 5.2.2, 6.3.2 and 7.2.3).
6. Present cutting-edge power semiconductor technologies, power stage set-ups and driving concepts that bear advantages for the implementation in DBD pulse generators (Chapters 4.7 and 4.10) and provide design concepts for crucial circuit elements such as magnetic components (Chapter 4.13).

The scope of this work excludes the following aspects on purpose:

1. A precise, and in particular, quantitative benchmarking of power electronic topologies for the pulsed operation of DBDs. This is the consequence of the sheer wealth of variants of DBD operating points and inverter operation modes.
2. An all-embracing, in-depth description of all topologies already published. Also here, the variety is vast enough to fill a book solely on this topic. Therefore, only relevant key topologies representing the state-of-the-art are shortly presented.
3. The development and verification of control techniques for DBD pulse inverters. It is very clear to the author that this is an aspect of utmost importance which is crucial to achieve efficient and reliable operation. Even so, in the sense of a step by step development, the underlying principles of topologies and operation modes apriori appeared to be more important.

2 DBD LAMP CHARACTERISTICS AND MEASUREMENT TECHNIQUES

The discharge lamp is the central element of any UV/VUV generation system and as this it needs to be understood and characterised in order to maximise the efficiency and reliability of the system. Regarding excimer DBDs, considerable research has been conducted to investigate the plasma-physical background and to find multi-physical and electrical models describing DBD excimer lamp behaviour. This work will contribute to this research by discussing aspects of the electrical lamp structure and of the electrical parasitic elements appearing in every real-world structure.

2.1 PHYSICAL DBD LAMP SET-UP AND OPERATION PRINCIPLE

The principal dielectric barrier discharge (DBD) set-up is depicted in Figure 6A. In contrast to lamp and reactor set-ups where electrodes have galvanic access to the gas, DBD vessels are characterised by at least one dielectric barrier which lies in the current path between the electrodes and the gas.

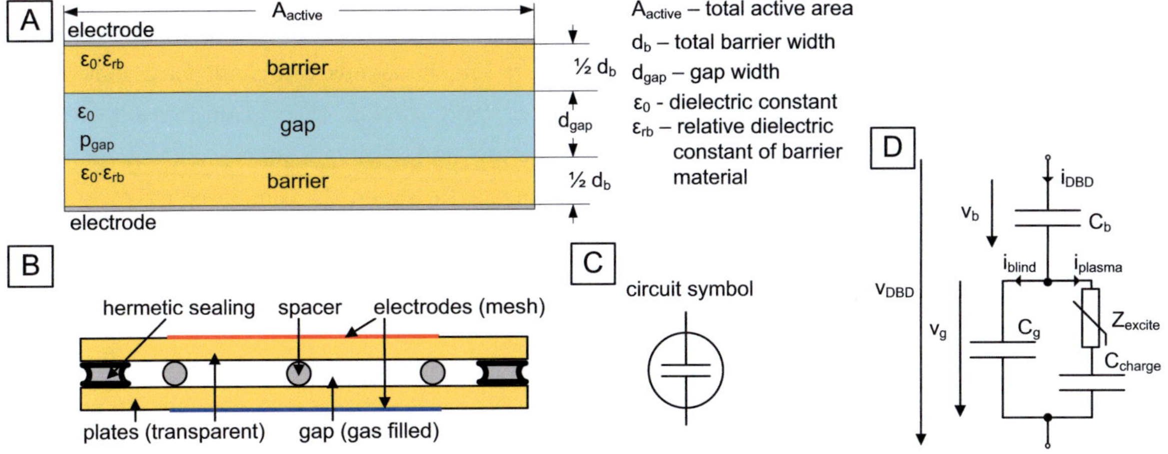

Figure 6A: Principal bilateral dielectric barrier discharge (DBE) lamp set-up. B: Physical set-up of a flat "plane-to-plane" DBD. C: Circuit symbol for DBD used in this work. D: Equivalent circuit including charge cloud capacitance C_{charge} (Trampert 2009; Paravia 2010).

The distance between the barriers filled by the gas is called gap width. The lamp vessel hermetically surrounds the gas filled volume. According to Figure 6B, the dielectrics of DBD lamps are mostly made of optically transmitting materials. Depending on the generated wavelengths, materials such as borosilicate glass (for VIS), quartz glass (for UV) or synthetic quartz glass (for VUV) are used. If optical transparency is not required, enamels, ceramics or even Teflon plates, plastic coatings or silicon rubber are suitable as dielectric. The dielectrics may also need to be chemically resistant and have a sufficiently high dielectric breakdown voltage.

In contrast to arc plasmas or arc torches, DBDs as well as corona discharges (Salvermoser 2003; Salvermoser 2004; Murnick 2009) represent self-sustained non local thermal equilibrium (non-LTE) electrical gas discharges (Wagner 2003). DBDs are consequently referred to as high-pressure glow discharges. The energy turnover of any discharge is limited by the dielectric barriers as no continuous current (DC) can pass through them. Instead, the transferred charge is limited and leads to an increasing voltage amplitude v_b across the barrier. Assuming the outer lamp voltage v_{DBD} is constant within the short time-domain of discharge, the reduced gap voltage v_g is defined as:

$$v_g(t) = v_{DBD}(t) - v_b(t) \tag{2.1}$$

During discharge the charges are transferred through the plasma and accumulate on the surfaces of the dielectric. Due to the fact that the barrier is a poor charge exchanger it also restricts the total charge that flows through the plasma. After available charges have been transferred, the discharge virtually dies out and with it the plasma state. Other than the charge, the (peak) current is not limited by the barriers, nevertheless sometimes stated faulty (Harry 2010). The dielectric barrier is also the reason for the poor power factor (PF) of DBD lamps which is typically in the range of 30 % (Sowa 2004). This low PF implies that a huge amount of apparent power needs to be transferred for lamp operation, while the real power that is actually consumed by the discharge is rather low. As a consequence, the efficiency of the pulse inverter that drives the DBD is restricted.

The term plasma is referred to as the fourth condition of matter and is an ionised gas that contains freely and randomly moving charge carriers. In case of low temperature plasmas and especially of non-LTE plasmas, the velocities of the lightweight electrons are much higher than that of the heavier ions. Due to the short-term transient excitation thermal equilibrium cannot evolve. In other words the electrical energy is mainly transferred to electrons that reach temperatures of ten thousands of kelvin while ions virtually remain at ambient temperature (Chabert 2011). Hence, even though the mean current density is comparatively low the short-term local electron temperature and electron density during ignition are sufficient to dissociate and ionise significant fractions of the gas. However, the remaining gas, also referred to as background gas, stays cold. Consequently, the electrical neutrality is restricted under non-LTE plasma conditions as charges build up by separation of slow (ions) and fast (electrons) charge carriers. The capacitor C_{charge} that is included in Figure 6D represents a charge cloud that is formed by this charge separation. Consequently, C_{charge} also acts as a level shifter: charged by the current drawn from a first "positive" ignition it internally reduces the external voltage level necessary for a second ignition induced by negative external voltage slope (Liu 2001).

DBDs are a common way to induce non-LTE plasma conditions. In this case, the plasma is also referred to as capacitively coupled plasma (CCP). Within this transient plasma so-called excimers (Stevens 1960) which represent excited unstable molecule complexes can be created. Excimers exist only for a very short time of ten to hundred nanoseconds (Haaks 1980). As soon as the excimers dissociate, UV or even VUV optical radiation is created that needs to be passed through the lamp vessel to be used for the processes and tasks described in Chapters 1.2.2 and 1.2.3.

The plasma behaviour and excimer generation is described in more detail in the related literature (Kogelschatz 2008; Trampert 2009; Paravia 2010) but will be shortly addressed in the following chapter.

Since the plasma is of non-equilibrium nature, DBDs bear a high impedance load characteristic prior to each ignition. Due to the low initial space charge density, each ignition requires a high electrical field strength within the gap. The ignition voltage is determined, beside others, by the partial gas pressure and the striking distance that is equal to the gap width d_{gap}. The maximum outer lamp voltage $v_{DBD,pk}$ is limited by technological reasons to approximately 5 kV. This limitation is mainly caused by the extensive costs of high-voltage cables, connectors and transformers. The reason for that is that the required voltage rating is not for DC but for high-frequency (HF), which puts enormous demands on the equipment.

In order to restrict the voltage necessary to ignite the gas volume, the DBD's gap width is rather narrow compared to other low-pressure lamps. Typical widths lay in the range of 0.1 mm to several millimetres (Kogelschatz 2000b). Alternatively, the gap may be locally diminished to create an ignition help (Schiene 2004).

The active area A_{active} of the lamp is determined by the size and shape of the electrodes. The gas volume that lies between the electrodes is called active volume. The electrodes may not necessarily enfold the whole DBD as indicated in Figure 6B. However, DBDs are mostly designed to maximise the active volume. This is done because of reasons of power density, gas and material cost reduction and gas purity.

The lamp vessel is filled with a pure rare gas, rare gas and halogens or multi-rare gas compositions (Guivan 2011). Main part of the filling is a gas or gas mixture that being in plasma state creates the excimers. The partial pressure of these gases is in the range of several mbar to tens of bar and has vast impact on the energy consumed by each ignition. The commonly used xenon gas is filled into the lamp with a partial Xe pressure p_{Xe} of some hundreds of mbar. Additional "buffer" gases may be added to increase the overall pressure in order to support the lamp's mechanical integrity (if partial gas pressure is below one bar) or to consume binding energy. However, for plane-to-plane DBDs according to Figure 6B, the major energy is spent into xenon excitation as long as the partial xenon content "exceeds some per cent" (Boeuf 1997). Hence, Penning mixtures play no role.

The capacitance of the "bare" DBD lamp according to Figure 6A is given by:

$$C_{DBD} = \frac{C_b \cdot C_g}{C_b + C_g} = \varepsilon_{rb}\varepsilon_0 \frac{A_{active}}{d_b + d_{gap} \cdot \varepsilon_{rb}} \tag{2.2}$$

The gap capacitance C_g is in most configurations much smaller than the barrier capacitance C_b. These capacitors create a capacitive voltage divider. As long as the gas is non-conductive, the ratio of lamp voltage to gap voltage is defined as:

$$\frac{v_{DBD}}{v_g} = \frac{C_g + C_b}{C_b} \tag{2.3}$$

Figure 6B shows the physical set-up of a flat DBD lamp with plane-to-plane electrode configuration. Spacers are placed between the glass sheets that form the dielectric barriers in order to prevent them from bending. The sides of the lamp need to be hermetically sealed, for example with a glass frit composition. Both the hermetic sealing of the gas volume and sustainable spacing of the barriers enclosing the gap and defining the gap width are critical components. Non-uniformities of the gap width cause preferred plasma starting points to occur and inefficient streamer discharges could develop. Flat excimer DBD UV-sources may be used in surface modification processes without (or reduced) inert gas flow or vacuum because, compared to tubular lamps, flat lamp geometries inherently allow for a smaller gap between lamp and substrate (Aiba 2008).

Comprising dielectrics in the electrical current path, DBDs can electrically be described as a capacitor with a certain highly non-linear and dynamic loss-component. For easy recognition of the lamp in the following schematics, the electrical symbol showed in Figure 6C will be used to represent the DBD lamp. This underlines that the DBD is here, above all, seen as black box with certain requirements and properties.

In order to tailor the spectral output of the lamp, besides using pure gas or gas mixtures generating the required output spectral distribution (Guivan 2011), highly efficient xenon gas can be combined with a phosphor or phosphor composition. The phosphor then transforms the short-wavelength radiation towards longer wavelengths and allows for tailoring of the output spectra to the requirements of the application processes. Alternatively, several lamps filled with gases or gas mixtures generating different output spectra may be combined in order to form the required cumulative spectra.

The lifetime of DBD lamps is mainly restricted by

1. phosphor degradation (especially in the case that phosphor is coated onto inner dielectric walls of the active area),
2. dielectric barrier material degradation including quartz glass degradation (Schreiber 2005) by VUV-excited development of non-bridging oxygen hole centres (NBOH) (Stockwald 1991) and VUV absorption by silica-bound hydroxyl groups (Morimoto 1999) and
3. gas purity degradation. A getter may be inserted into the gas volume to ensure gas purity over the whole lifetime (Hombach 2008).

It should be noted that the effect of the above mentioned lifetime limiters also significantly depends on the way the DBD lamp is operated. Inhomogeneous discharge mode and operation at very high lamp power enhance the influence of the lifetime limiters.

Despite of the flat plane-to-plane set-up of Figure 6B, other geometries for DBD lamps and reactors have been developed and applied in industry. Some variants are depicted in Figure 7. The lamp depicted in Figure 7B is more complex than the linear tubular DBD of Figure 7A. Additional to the actual lamp outer vessel, a centrally aligned second tube carries a centre electrode called inner electrode. Both electrodes and tubes are oriented coaxially, giving the name of this type of lamp. In this design the diameter of the inner electrode is minimal resulting in virtually no dead volume within the lamp. Because the whole gas volume is available for the discharge, the power density related to the lamp volume is outstanding. However, the maximum gap width restricts the maximum outer lamp diameter which also restricts the power per cm lamp length.

The inner surface area of the outer tube is by factor r_{outer}/r_{inner} larger than the outer surface area of the inner tube. The difference of the areas may affect the discharge behaviour which depends on the available surface charges and the secondary electron emission coefficient of the electrode material. Hence, this lamp design could require specialised pulse wave shapes for reliable operation.

If higher power is required the outer diameter needs to be enlarged to increase the volume available to the discharge. The inner tube diameter then needs to follow in order to restrict the gap width leading to a coaxial DBD design shown in Figure 7C.

The gas is enclosed by two tubes of dielectric material which are hermetically sealed at both ends. Depending on which side the optical radiation is used on, the respective electrode needs to have both a high optical transmission and high electric conductivity. Mostly, the optical radiation leaving the outer tube is used. Therefore, the inner electrode can be made of a conductive brush (Roth 2005), a conductive coating (Gold, Silver) occasionally with extra metal fabric support (Hombach 2010) or, favourably, of a coating with good reflection in the UV range. Referring to Figure 4, aluminium coatings have a reflection of approximately 69 % at 172 nm (Lorch 1987). The outer electrode can be made of metal fabric, metal mesh or metal coating in the form of stripes. Osram's Xeradex® (Radium) VUV lamp combines striped outer electrodes with a spiral electrode that is not covered with a dielectric. This lamp is an example for a commercially available lamp with only one dielectric barrier.

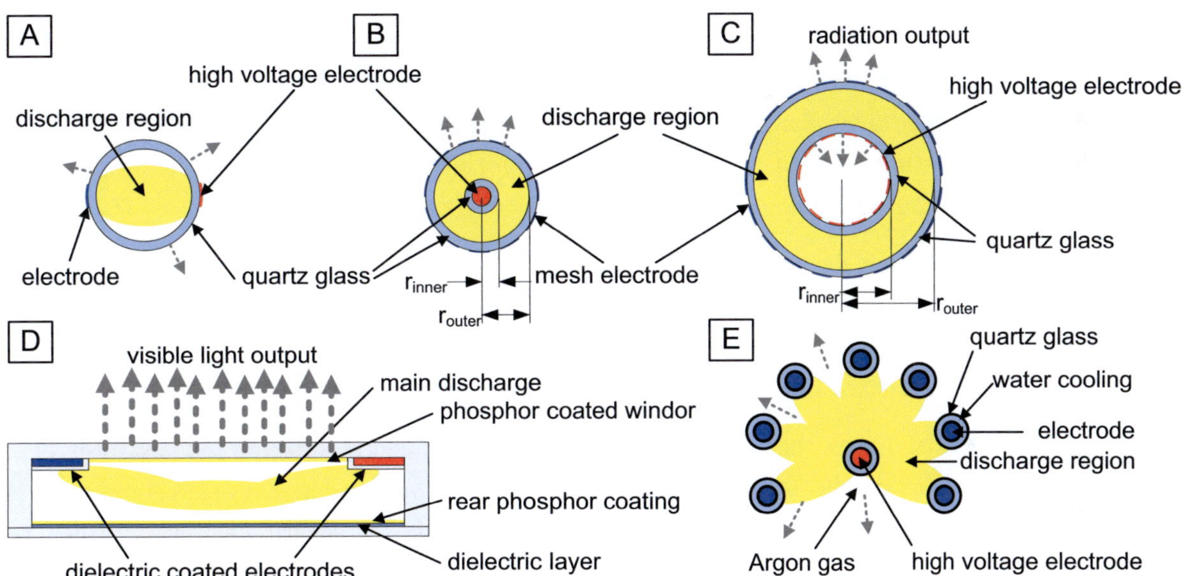

Figure 7: Excimer lamp set-ups. A: Tube type DBD-lamp with linear electrodes, similar to LINEX© (Osram 2002). B: Tube type DBD-lamp with coaxial electrode set-up. Inner electrode with minimum diameter (pen-type). C: Coaxial tube type DBD-lamp with non-minimum cross-sectional area of inner electrode. Interior can be filled with water for cooling or treatment (Jüstel 2002); inward or outward electrode may be used as UV / VUV reflector. D: Lamp setup comparable to PLANON©-lamp (Ilmer 2000) and design of Park (2007a) E: Open configuration where radiation does not need to pass windows (Köhler 1989; Sobottka 2008). Instead, work piece is in contact with operation gas.

Co-planar lamp configurations as shown in Figure 7D are preferably used for flat DBD lamps as the Planon (Ilmer 2000). Here, the dielectric-covered electrodes are placed alternately in one plane within the gas volume. This configuration is the base of PDPs and is also referred to as surface dielectric barrier discharge (SDBD) (Voeten 2011).

Figure 7E shows an open, windowless DBD configuration that is preferably used if the wavelength of the excimer radiation is below 172 nm (Kogelschatz 2003b). For example, the Ar_2 excimer system generates radiation with an intensity peak at 128 nm. As the costs of materials that have an optical transmission in the VUV superior to synthetic quartz glass is very high, open systems with continuous gas flushing are used

(Köhler 1989). Alternatively, the discharge gap can be distributed within a porous dielectric material which is claimed to achieve high process efficiency at low pressure loss (Eliasson 2000).

2.2 Radiation Generation Process

As introduced in Chapters 1.2.2 -1.2.3, DBDs may be operated as an efficient source of intense incoherent ultraviolet (UV) or even vacuum ultraviolet (VUV) optical radiation. These lamps are also referred to as excimer-DBDs or exciplex lamps. Excimers are excited two-atom complexes (dimer) while exciplexes are excited multi-atom complexes. Excimers exist for a very short period of time that lies in the range of some nanoseconds. They are created by three-body collisions that require special plasma conditions in dedicated gases or gas mixtures with determined partial pressure. Also, for the generation of sufficient amounts or predecessors, high electron energies are required. DBDs provide these conditions as they operate as high-pressure self-extinguishing discharge leading to the formation of a short-time non-LTE plasma. The dissociation of excimers leads to optical radiation with an energy equal to their binding energy. In the following paragraphs, the time-dependent behaviour of electrical parameters and of the plasma will be described for the Xe_2^*-excimer system.

Assume a two-side dielectric barrier discharge lamp set-up that is driven by a variable voltage source. Further assuming a comparatively long idle time in advance of the excitation, the initial space charge density is assumed to be infinitesimally small. Therefore, the electric field within the gas gap rises proportionally with enhanced applied outer voltage v_{DBD}.

In order to initiate a discharge, the v_{DBD} is raised above a certain value defined by the outer ignition voltage $v_{DBD,ign}$ (see Figure 6D). The discharge starts when the corresponding electrical field that is created inside the gap accelerates free electrons which then gain sufficient energy to sustain ionisation processes. Beside ionisation, electrons dislocate and excite species of the background gas.

Charge separation occurs due to the different mobility and consequently differing temperatures of the particles. While the comparatively fast electrons accelerate to the anode the ions move slowly towards the cathode. Therefore, a positive space charge is created which enhances the electric field even further. This space charge, or charge cloud is situated closely to the dielectric barriers and is also referred to as sheath.

The electrons that reached the anode create radial local fields that lead to a spreading of following electrons across the anode and a widening of the discharge channels. The electrons can solely be limitedly absorbed because of the dielectric property of the anode. For the same reason, the cathode cannot emit a sufficient amount of electrons to feed a spark. Therefore, trapped electrons are taken from the cathode surface by sliding surface discharges (Schorpp 1991).

It is important to note that in contrast to electrode-equipped discharges, the dielectric barriers of the DBD limit electron multiplication and therewith prevent thermalisation of the plasma. As long as the ignitions are sufficiently timely separated, the favourable non-LTE conditions are ensured.

The generated ionised matter relaxes following specific kinetic pathways as depicted in Figure 8. It is important to note that the energy flow is strictly directional. Electronic excitation is the only way to enhance the matter's energy level while the only available downward channel towards energy ground state is via optical radiation.

Ionisation that occurs due to the applied field strength is the basis of all kinetic processes that lead to excimer generation:

$$e^- + Xe \rightarrow 2e^- + Xe^+ \quad (2.4)$$

The resulting free electrons are the precondition for all following body collisions, whereas the high energetic ions recombine. With further excitation atoms may gain doubly excited status Xe^{**}. By collision with ground state atoms they form the important single excited metastable Xe^*. Alternatively, Xe^* is formed by direct electronic excitation according to Formula (2.5) or by radiant transition of Xe^{**} to Xe^* generating narrow infrared (NIR) light. The NIR radiation may be used to identify the moment of ignition and to investigate discharge homogeneity. Moreover, (Wammes 2013) claims to use long wavelength radiation to control the DBD illumination system. Referring to Figure 8, the NIR output cannot be taken as a measure of VUV output intensity since the radiation path is bypassed by heavy body collision and direct electronic excitation (Trampert 2009).

$$e^- + Xe \rightarrow e^- + Xe^* \quad (2.5)$$

Originating from the metastable Xe^* levels, excimers are formed by three-body collisions according to Formula (2.6). Within tens to hundreds of nanoseconds these homo-nuclear excited molecules relax to the energy ground state, emitting the first excimer continuum at 150 nm from the vibrationally excited level and the efficient second excimer continuum at 172 nm according to Formula (2.7).

$$Xe^* + 2Xe \rightarrow Xe_2^* + Xe \quad (2.6)$$

$$Xe_2^* \rightarrow 2Xe + hf \quad (2.7)$$

Experiments (Vollkommer 1998; Mildren 2001a) confirmed the previously claimed (Vollkommer 1994) high VUV efficiency of Xe excimer lamps. However, simulations (Beleznai 2008b) set the boundary of the 172 nm radiation efficiency to about 47 %.

The time-ranges in which the different interactions take place are also included in Figure 8. The time-dependence of the VUV radiation output has also been experimentally investigated (Liu 2003b). The spectrum of the optical VUV radiation is highly depending on pressure (50 - 680 mbar (Gellert 1991),1.3 - 53 mbar (Kindel 1996), 25 - 500 mbar (Paravia 2006)) but is not dependent on excitation wave-shape, input power or repetition rate (Liu 2003a; Sewraj 2009). At low pressure, the 147 nm resonance line dominates the spectrum whereas at pressures exceeding 100 mbar the 172 nm second continuum clearly dominates the total radiation output (Paravia 2010). This is due to the fact that up to 99 % of the 147 nm line is absorbed by ambient Xe ground-state atoms for which higher filling pressure is suggested in order to support the efficient second continuum (Lee 2008a). The electric potential distribution within the gap develops highly dynamically during ignition and discharge (Paravia 2010). Likewise, the temporal distribution of the VUV output of Xe excimer DBDs is pressure-dependent. Higher Xe pressure significantly shortens the VUV output pulse width and boosts the VUV output pulse slope (Carman 2004a). The main fraction of VUV output is delivered within 400 ns. Beside the required reactions leading to excimer radiation, several loss mechanisms adversely affect the plasma efficiency. According to Lorentz (1973) these losses are mainly caused by particle collision, IR photoionisation and energy absorption by ground-state molecules. Additionally, plasma efficiency is highly depending on the purity of the filling gas. If the main excimer-generating gas is contaminated by fractions of other gases, other less-efficient excimer systems may be excited. For instance, oxygen and water contents may lead to formation of XeO^*-excimers that create strong band radiation at around 535 nm (Park 2004) and a broad line at 307 nm (Stockwald 1991).

Unfortunately, the 535 nm radiation intensity is not monotonously rising with rising O_2 impurity which would be a unique mark for the grade of impurity. Instead, the maximum appears at around 1 ‰ O_2 content (Stockwald 1991). Impurities may also lead to inefficient optical radiation from radicals as nitrogen monoxide (NO, 236.33 nm), hydroxyl (OH, 308.3 nm) and nitrogen (N_2, 337.13 nm and 357.69 nm; N_2+ 391.44 nm) (Pipa 2007).

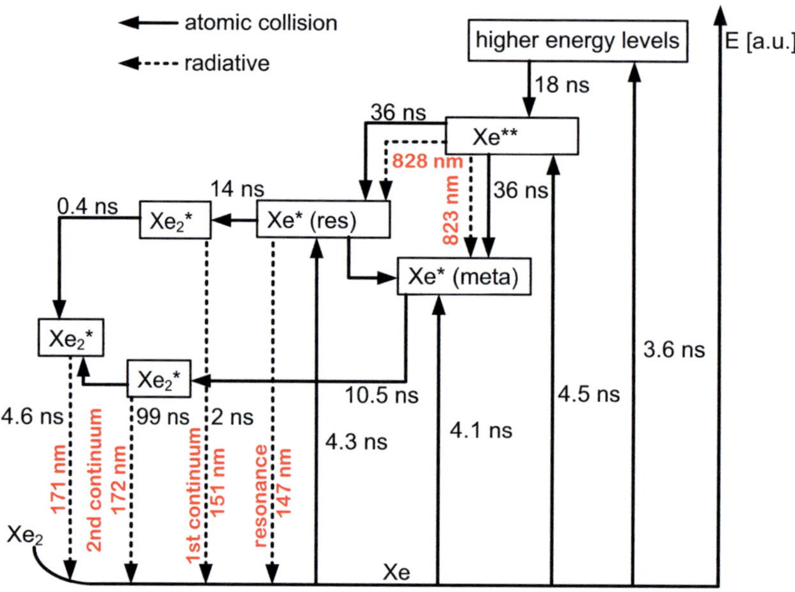

Figure 8: Simplified kinetic pathways for the formation and the decay of xenon excimers in DBDs based on Liu (2002) including calculated evolution over time within one excitation time-domain (1 µs) based on Carman (2004a). Typical transition times between the Xe energy states are labelled. Xe pressure was 1.5 bars and the pulse rate was 10 kHz. Background aspects taken from Jou (2011).

2.3 Discharge Modes

Assuming pulsed operation, the DBD's discharge intensity and its distribution and behaviour is determined mainly by the shape of the voltage waveform, the voltage amplitude, the impedance of the power supply (which can be modified e.g. by placing a capacitor C_p in parallel to the DBD), the pulse duration, the repetition frequency, the electrode configuration, the dielectric material and shape, and the filling gas composition. The discharge is classified into two main modes which are defined as follows.

2.3.1 Filamented Discharge

The filamented discharge, which is also referred to as patterned or non-diffuse discharge, is the dominant discharge mode as it is easier to be achieved compared to a homogeneous characteristic. It is likely that first investigations of DBDs showed filamented discharges, only. This is since this discharge mode occurs in most gases and gas mixtures at high (atmospheric) pressure and low frequency operation. Filamented discharges can be further sub-divided into dedicated groups (Schorpp 1991; Kling 1997) which are exemplarily depicted in Figure 9A. At atmospheric pressure, the filaments itself are bounded volumes of high current density (≈ 100 A/cm^2). These channels represent privileged locations for micro-discharges initiated by a subsequent ignition. The channel volume can be cylindrical reaching from (temporal) anode to cathode. After extinguishing, the channels leave conditions in the gas volume and the dielectric barrier surface that privilege them for discharge initialisation if voltage polarity is reversed. Consequently, at low-frequency alternating voltage and one ignition per half-wave, fixed discharge patterns are observed. The grade of filamentation can be characterised by means of spatially and time-resolved optical measurements as shown in Figure 9B. With that, filaments can be described qualitatively and the quantitative analysis can be interpreted as grade of homogenisation. Paravia (2010) could show that this grade of homogenisation is directly coupled to the plasma current density during ignition. Different discharge patterns and their spatiotemporal development were also investigated by Fan (2007). The patterns depicted in Figure 10 reach from single filaments to a stripe characteristic.

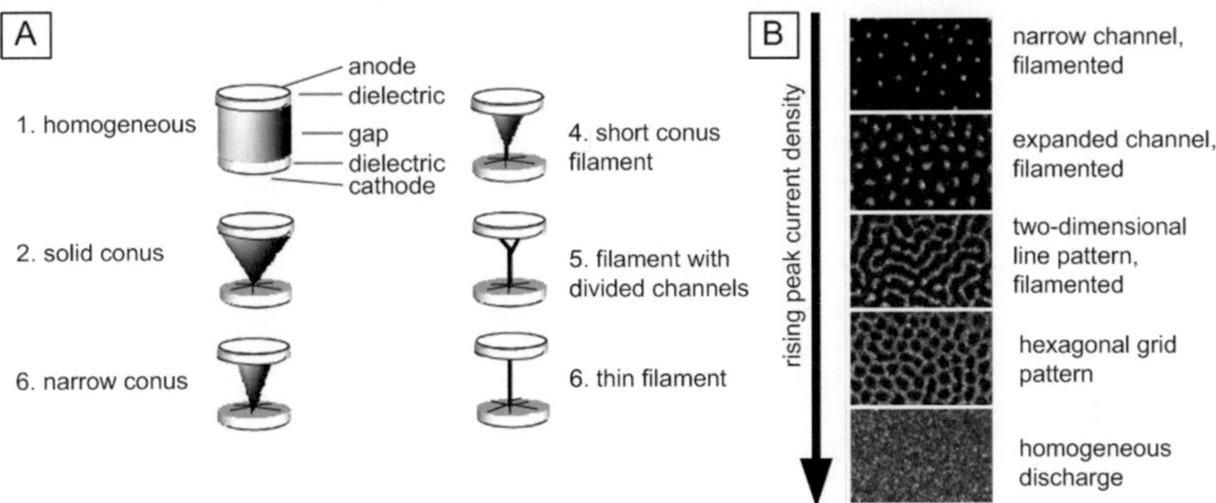

Figure 9: A: Selection of possible discharge characteristics based on (Kling 1997). B: Different discharge patterns depending on the peak current density. Graphic bases on (Paravia 2008b; Paravia 2010).

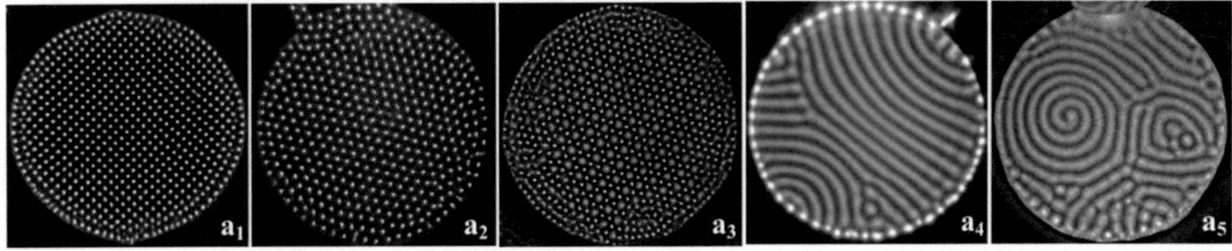

Figure 10: Different filamented discharge patterns obtained at 50-60 kHz square wave excitation; increasing grade of homogenisation from left to right (Fan 2007).

2.3.2 Homogeneous Discharge

Due to its proven higher efficiency, the homogeneous, diffuse or spatial uniform discharge mode is best suited for DBD excimer lamps. This is due to the more evenly spatially distributed electron energy and the homogeneous breakdown (Mildren 2000). Simulations showed that discharge homogeneity is essential to keep (peak) current densities low during ignition (Pflumm 2003). This is believed to be a precondition of high plasma efficiency (Mildren 2001c). A positive side effect of the homogeneous discharge is that because of the lower current density, also the ions are created and accelerated more uniformly across the discharge area. For phosphor-coated DBD lamps, this results in less phosphor material transport and phosphor wear-out is reduced which significantly enhances phosphor life-time.

Homogeneous discharges were reported by Borisov (1998) to occur in KrF lamps and obtained by Carman (2004a) in very small (A_{active} = 1.27 cm^2), narrow gap (d_{gap} = 1.6 mm) DBDs filled with pressures of up to 2.5 bar driven by unipolar short (200 ns) pulses. Works on filamented and homogeneous discharge modes were reviewed by Kogelschatz (2002). As of this writing, literature indicates different kinds of uniformity. The term homogeneity is actually not explicitly defined. In this work, it is distinguished between:

Firstly, discharges in DBD lamps with visually observable homogeneous appearance. This impression of homogeneity may be assisted by phosphors with large decaying time constants. Because of the time-integration occurring in the human eye, filamentary discharges can appear homogeneous if brightness is diffused by wandering filaments or by dynamic patterns that change from discharge to discharge. Using the visual assessment method, Schorpp (1991) identified spatially homogeneous discharges at pressures of up to 1 bar and distinguished between a large set of discharge modes.

Secondly, truly homogeneously distributed discharge from breakdown to extinguishing is solely ensured by optical measurements that are time and spatially resolved and not disturbed by phosphors. It is to mention that further distinction between a progressing discharge wave requiring tens of microseconds to embrace the whole active area A_{active} and an instant, spatially evenly distributed discharge is to be made. To make this distinction, multiple pictures with individual exposure time well below the total discharge time are necessary. Chapter 2.6.1 briefly describes an optical system for that purpose.

2.4 Electrical Modelling of DBD Lamps

This work focuses on efficient electrical driver concepts of excimer DBD lamps. In contrast to typical switching mode power converters, the load can neither be assumed to be of constant impedance nor to have linear resistive characteristics. Nevertheless, this inspired scientists to search for ways to model their behaviour in order to make it handy for converter optimisation. Besides the very complex, multi-physical modelling, the utilisation of pure electric components to mimic the behaviour of a DBD discharge under given operation conditions aroused international research interest. Some examples of those pure electrical models will be briefly introduced as in the following. It is equal to all models that the actual gap is represented by a capacitor C_g with fixed value as shown in Figure 11.

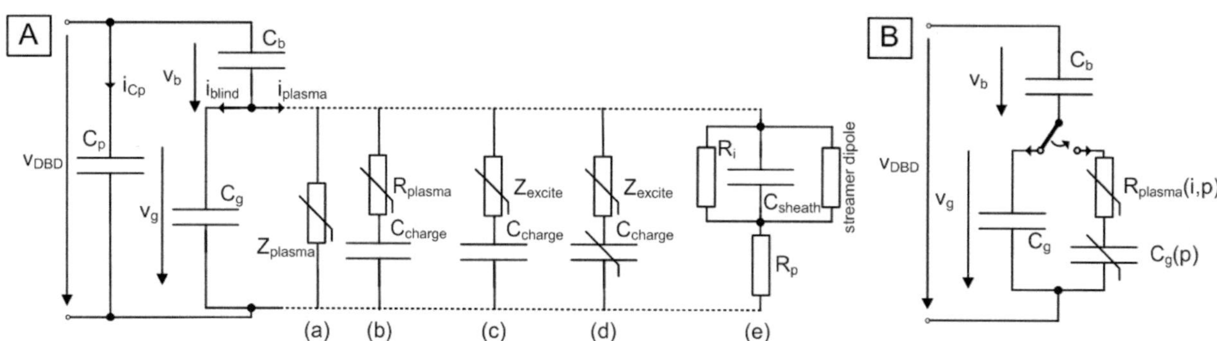

Figure 11: Pure electrical models of DBD lamps:
A(a): simplest model according to Diez (2007) and Pal (2011);
A(b): charge carrier transport model (Trampert 2009) and extended charge carrier transport model (Paravia 2010);
A(c): macroscopic lamp model according to Daub (2010);
A(d): model with efficiency indicator as published by Bhosle (2006)
B: model by Olivares (2007) distinguishing between different modes. C_g and R_{plasma} are power-dependent.

The value is defined by the vacuum capacitance of the gap. The simplest models use a resistor R_{plasma} to represent the consumption of real power by the ignited plasma (a). A charge carrier model (b) was introduced by Trampert (2009) and further modified by Paravia (2010). In this model, another capacitor

C_{charge} is placed in series to the dissipative element of the plasma in order to represent the space-charge cloud formed by charge carrier separation during ignition. It is also referred to as sheath capacitance. In his dissertation, Daub (2010) presented a further extended charge transport model (c) containing a variable capacitance C_{charge}. Depending on the plasma state this capacitance is supposed to dynamically change its value. A rather complex model had been presented by Bhosle (2006) also containing a sheath capacitance. Additionally, resistors were added representing ion drifting and recombining (R_i), the plasma bulk (R_p) and the streamer dipole that dynamically changes its value depending on the plasma state. Figure 11B shows an alternative model similar to Figure 11A(c). Here, the gap is either represented by a static capacitance or a resistance depending on discharge current and power and a series capacitance only depending on power.

It is common for those models to adapt their parameters to data acquired from unique DBDs operated in unique operation modes in order to match simulation results. However, physical and electro-chemical reactions inside the DBD are extremely complex and highly dependent on a variety of (not measureable) parameters. It is therefore believed that a transfer of simulation results to changed operation modes or different lamp structures is not feasible. Consequently, in contrast to earlier publications (Koudriavtsev 2000; Kudryavtsev 2004; Flores-Fuentes 2006; Piquet 2007; Daub 2010; El-Deib 2010b; Piquet 2010) within the scope of this work there was no attempt made to simulate the electrical behaviour of the excimer DBD.

2.5 Determination of Electrical Lamp Parameters

As is stated in Chapter 2.1, the electrical properties of the DBD lamp and the driving signal have significant influence on the discharge behaviour and characteristic. In this chapter, the determination of electrical lamp parameters by measurement and calculation, especially the so called "inner" lamp parameters will be described.

2.5.1 Measurement of DBD Voltages and Currents

DBD voltage v_{DBD} and DBD current i_{DBD} are referred to as outer values which can be directly measured. In Figure 12, the typical measurement set-up for determining electrical parameters of the DBD and the system input power level is depicted. The DBD voltage that can exceed several thousand volts is measured by a high-voltage, high-bandwidth probe. Up to 4 kV_{pk}, PMK's PHVS 1000-RO 400 MHz probes can be used. Their package is still compatible to probes of lower maximum voltage. This allows for favourable connection of probe tip and probe ground lead in order to prevent pig-tails that could deteriorate signal quality. In order to minimise error of multi-MHz high-voltage measurements, the probes need to be compensated with the small-signal square-wave generator of the oscilloscope and DC-calibrated by means of a high-voltage DC source. Additionally, compensation with a large signal square-wave generator (e.g. Quantum Composers' generators with 12 V_{pk} output) is recommended. This ensures that compensation is also valid for higher voltage levels.

Although currents within the inverter circuits were also measured by means of low-inductive shunt resistors and clamp-on current probes, the DBD current was always measured with high-bandwidth current transformers. Models from Pearson Inc. (2877) and Bergoz Instrumentation (CT-B0.25) were used. Although these current probes do not require calibration, it is important to de-skew the voltage and current measurement probes. In this work, the skew of voltage and current probes was eliminated with the help of the de-skew fixture Agilent 701936.

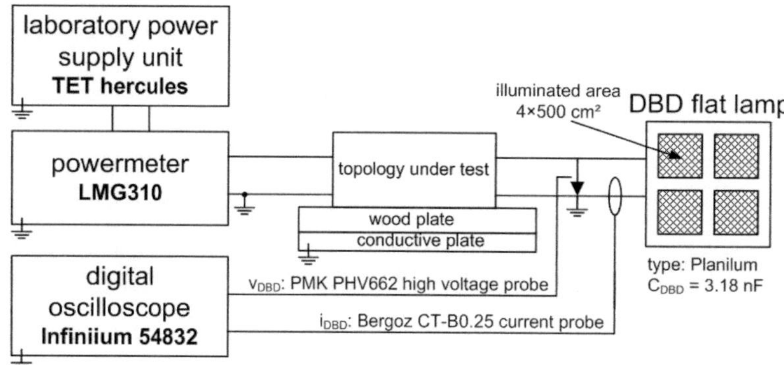

Figure 12: Set-up for electrical power measurements including the power source and a power meter for measurement of the input power, the topology under test driving the attached DBD lamp. A digital oscilloscope with attached voltage and current probes is used to measure DBD lamp power. Note that measurement of short pulses required careful de-skewing and calibration of the probes.

The used current transformers themselves do not insert significant inductance to the circuit as they virtually do not contain a magnetic core. However, because of the cross-sectional area of the CT, a certain current loop is created that generates inductance by itself. The standard loop that fits tight to the CT case produces 60 nH of additional inductance. Larger loops, as necessary for the connection of 4mm connectors in the high-voltage DBD lamp circuit generate inductances of rather 100 nH. If the current transformer should be used within the inverter, this significant introduced inductance needs to be considered. For both mentioned CTs, this inductance can be reduced down to 26 nH by embracing its package thoroughly with conductive foil used as low-inductive return path. Therewith, the magnetic field is partly cancelled out.

The voltage across the gap can be derived assuming the barrier capacitance to be constant. If further parasitic inductances are discarded, the gap voltage is given by the difference of total lamp voltage v_{DBD} and barrier voltage v_b:

As i_{DBD} has to pass C_b, v_b is clearly defined and the gap voltage can be calculated from the measured outer DBD voltage:

$$v_g(t) = v_{DBD}(t) - \frac{\int i_{DBD}(t)dt}{C_b} \tag{2.8}$$

Liu (2001) calculated the plasma current by treating total DBD voltage v_{DBD} and total DBD current i_{DBD} as time-dependent and C_b and C_g as time-independent variables. Differently, Paravia (2010) used a method rooting back to (Roth 2001) which requires an additional DBD lamp with identical properties except for the gas filling. The second lamp which is referred to as coupling DBD lamp is used to derive the total displacement current that also flows through the DBD lamp under investigation. By hardware subtraction of this displacement current from the actual lamp current, a time-dependent current signal proportional to the plasma current can be derived. Figure 13 shows the two possible DBD lamp set-ups from the electrical point of view. Adding a significant capacitance (C_P) in parallel to the DBD lamp dramatically changes the discharge energy support, enhances the total capacitance that has to be driven by the inverter and is therefore only suitable for investigations of very small laboratory DBD lamps. However, C_p can assist in terms of buffering v_{DBD} (Stockwald 1991) and supporting a homogeneous discharge. As will be shown, C_p favourably needs to be scaled in relation to the other DBD capacitances.

If information regarding the plasma current waveform of industrial-scale lamps is required, it is more suitable to use the following calculation method that bases on a publication of Liu (2001) and was used by Pal (2011).

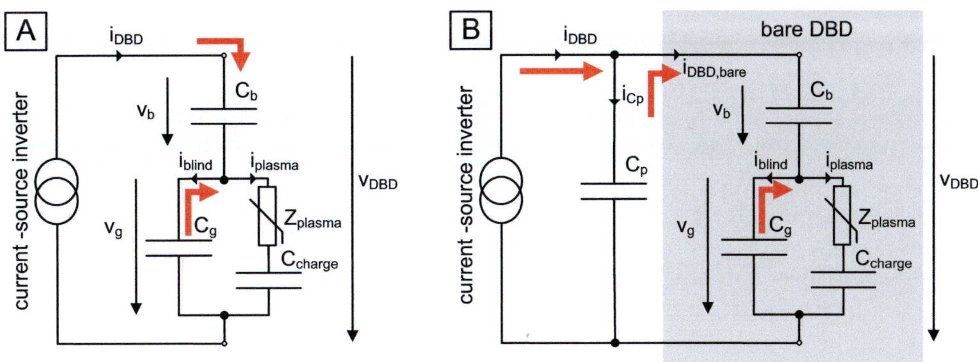

Figure 13: A: Simplified electrical equivalent circuit of a DBD lamp driven by a current-source inverter. B: Equivalent circuit of a DBD lamp with attached parallel capacitor C_P. Current that supports the discharge is marked in red.

In case of a bare DBD according to Figure 13A, the displacement current of the constant gap capacitance C_g is:

$$i_g(t) = C_g \frac{dv_g(t)}{dt} \tag{2.9}$$

The plasma current i_{plasma} is the difference of the total DBD current i_{DBD} and i_g:

$$i_{plasma}(t) = i_{DBD}(t) - i_g(t) \tag{2.10}$$

Substituting (2.9) in (2.10) the following equation is gained:

$$i_{plasma}(t) = i_{DBD}(t) - C_g \frac{v_g(t)}{dt} \tag{2.11}$$

In Formula (2.11), the gap capacitance is given by:

$$C_g = \frac{C_{DBD} \cdot C_b}{C_b - C_{DBD}} \tag{2.12}$$

Substituting $v_g(t)$ with (2.8) and C_g with (2.12), from Formula (2.11) we derive the plasma current from measureable values only:

$$i_{plasma}(t) = i_{DBD}(t) - \frac{C_{DBD} \cdot C_b}{C_b - C_{DBD}} \left[\frac{v_{DBD}(t)}{dt} - \frac{di_{DBD}(t)}{C_b} \right] \tag{2.13}$$

This Formula can be rewritten for C_g:

$$i_{plasma}(t) = i_{DBD}(t)\left[1+\frac{C_g}{C_b}\right] - \frac{v_{DBD}(t)}{dt}C_g = i_{DBD}(t)\left[\frac{C_g}{C_{DBD}}\right] - C_g\frac{v_{DBD}(t)}{dt} \tag{2.14}$$

The minus in front of the second term indicates that a falling DBD voltage (negative slope) supports the plasma current as charge is being drained from C_g.

Things change in case of a DBD with attached parallel capacitor C_p as shown in Figure 13B. In this case, the discharge is assisted by additional current that is provided by C_p. Therefore, Formula (2.14) expands to:

$$i_{plasma,Cp}(t) = \left[i_{DBD}(t) - C_p\frac{v_{DBD}(t)}{dt}\right]\left[1+\frac{C_g}{C_b}\right] - C_g\frac{v_{DBD}(t)}{dt} \tag{2.15}$$

In this case, the total DBD capacitance is enhanced to:

$$C_{DBD+P} = \frac{C_b \cdot C_g}{C_b + C_g} + C_p \tag{2.16}$$

Substituting (2.16) in (2.15) yields:

$$i_{plasma,Cp}(t) = i_{DBD}(t)\left[\frac{C_g}{C_{DBD+P} - C_p}\right] - \frac{v_{DBD}(t)}{dt}C_g - \frac{v_{DBD}(t)}{dt}C_g\left[\frac{C_p}{C_{DBD+P} - C_p}\right] \tag{2.17}$$

In contrast to (2.15), a third term describes the additional support provided by C_p. Rearranging yields:

$$i_{plasma,Cp}(t) = i_{DBD}(t)\left[\frac{C_g}{C_{DBD+P} - C_p}\right] - \frac{v_{DBD}(t)}{dt}C_g\left[\frac{C_{DBD+P}}{C_{DBD+P} - C_p}\right] \tag{2.18}$$

The effect of the additional capacitor becomes clear by comparing (2.14) and (2.18). While the (external) current source support stays constant, the (internal) voltage source support is enhanced by the voltage support enhancement factor (vsef) of:

$$vsef = \frac{C_{DBD+P}}{C_{DBD+P} - C_p} \tag{2.19}$$

Obviously, Formula (2.19) is nothing else than the ratio of total DBD capacitance to bare DBD capacitance.

2.5.2 DBD Electrical Power Measurement

The Lissajous figure method, reported by Manley (1943) and later modified by Falkenstein (1997) is still a common method to determine DBD power consumption (Kundu 2012). Although it had been shown that even for power measurements on Planon© DBD lamps LMG500 power meters are sufficient (Zimmer 2011), in this work state-of-the-art digital oscilloscope technology in combination with the probes mentioned in the previous chapter was used. With that, the real power and the apparent power of the DBD could be calculated automatically while keeping the raw signal data on storage for later use.

Measuring the DBD voltage and DBD current for one repetition period T_{rep} in a repetitive manner, the acquisition accuracy (quantisation error) could be drastically reduced and the DBD real power is given as:

$$P_{DBD} = \frac{1}{T_{rep}} \int_0^{T_{rep}} (v_{DBD} \cdot i_{DBD}) dt \tag{2.20}$$

Equally important, the apparent power accommodated by the DBD needs to be determined. Having acquired the data for DBD voltage and DBD current, the apparent power is given as:

$$S_{DBD} = \sqrt{\frac{1}{T_{rep}} \int_0^{T_{rep}} (v_{DBD})^2 dt} \cdot \sqrt{\frac{1}{T_{rep}} \int_0^{T_{rep}} (i_{DBD})^2 dt} \tag{2.21}$$

2.5.3 Influence and Measurement of DBD Lamp Parasitics

Despite of the key parameters of the lamp as total capacitance C_{DBD} and gap capacitance C_g, lumped parasitic electric elements have influence on the DBD's voltage and current wave-shapes and thus on the plasma behaviour. Since literature circumnavigated this aspect so far, the possible influences of parasitic elements as well as derived conclusions are found here. In that respect, Chapter 3 needs to be paid attention to. Based on Figure 13B, Figure 14A shows an extended electrical equivalent circuit of a DBD lamp including significant parasitic elements.

As can be seen in Figure 14A, the lamp is not only a capacitor with a highly dynamic and non-linear loss component Z_{plasma}, but a whole LCR (multi-) resonant circuit. This is due to the fact that the structure of the electrodes and contact cables not only involves electrical resistances but also inductive elements. The LC-resonant behaviour of bare DBD lamps had been found in rather early publications (Müller 1996) but haven't been embraced in later publications, although they dealt with high-frequency operation of DBDs where parasitics loom large.

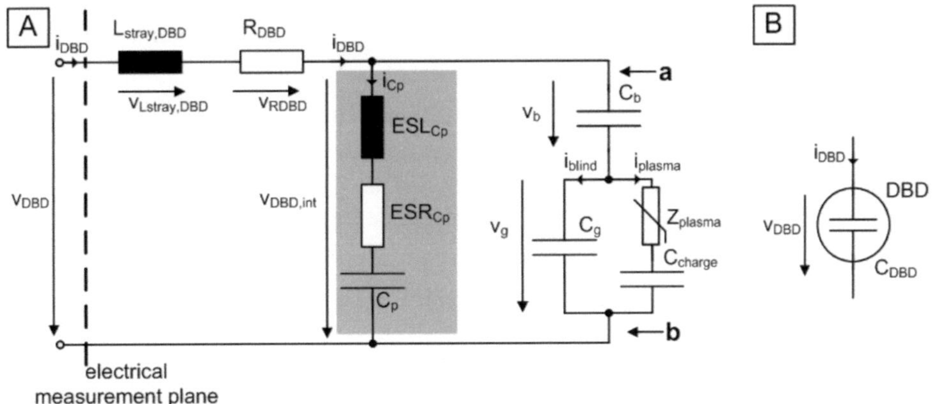

Figure 14: A: Electrical equivalent circuit of a DBD lamp including electrical lamp parasitics. B: Electrical DBD lamp schematic symbol with key parameters.

Electrical measurements can practically only be performed from the left of the defined measurement plane (in Figure 14A) while the voltage and current waveforms obtained at the mesh points a and b are inaccessible and determined by an internal pulse forming network (PFN). Beside the capacitances included in C_{DBD} this PFN consists of the lamp's stray inductance $L_{stray,DBD}$ and wiring resistance R_{DBD}.

Either by a certain DBD lamp structure or intentionally, an additional parallel capacitor C_p may be added additionally to the bare DBD set-up. The purpose of this capacitor is to provide additional energy to the discharge which is discussed in Chapter 2.7.2. If C_p is included, the lamp represents a two-stage RLC low-pass filter. In other words, operation of the lamp is only possible below the respective corner frequency. In case of a bare lamp for which C_{DBD} is given in formula (2.2), this frequency is given as:

$$f_{res,DBD} = \frac{1}{2\pi\sqrt{C_{DBD} \cdot L_{stray,DBD}}} \quad \text{with } C_{DBD} = \frac{C_b \cdot C_g}{C_b + C_g} \tag{2.22}$$

In most configurations, C_g is much smaller than the barrier capacitance C_b, meaning it dominates the actual value of C_{DBD}. If a parallel capacitor C_p is attached, this frequency reduces to:

$$f_{res,DBD+p} = \frac{1}{2\pi\sqrt{(C_{DBD} + C_p) \cdot L_{stray,DBD}}} \tag{2.23}$$

Coming back to a bare DBD, any discharge that occurs within the DBD's gap leads to a dramatic reduction of impedance Z_{plasma}. In case of a discharge that virtually shortens the gap, for the time duration of this discharge, the DBD capacitance can rise to the maximum value of:

$$C_{DBD} = \frac{C_b \cdot (C_g + C_{charge})}{C_b + C_g + C_{charge}} \tag{2.24}$$

Assuming C_{charge} reaches an infinite value, the DBD would in this case have a total capacitance purely determined by the barrier capacitance C_b:

$$C_{DBD} = C_b \tag{2.25}$$

Combining Formulae (2.22) to (2.25), one can obtain the highest possible frequencies that, depending on the configuration, could pass through the parasitic elements up to the DBD capacitances enclosed between mesh points a and b depicted in Figure 14A. Depending on the topology and driving scheme that is used to operate the DBD, the merely negative impact of the parasitic DBD inductance differs. This is addressed in Chapters 6 and 7.

Beside the cable inductance and the inductance in the C_p-path, the cable resistance is enhanced by current displacement effects such as skin effect and proximity effect. Both lead to a frequency-dependent reduction of the effective cross-sectional conductor area that can be referred to as AC resistance R_{AC} which is discussed in Chapter 4.13.

The impedance measurement of the commercial Saint-Gobain flat DBD lamp proved that beside the main capacitive part, the connecting cables and the internal circuit structure contribute to a significant inductive part. In applications with resonant overshoot where an inductor is permanently series connected to the lamp anyway, this solely contributes to the series inductance. The parasitic inductance becomes an issue in topologies with inductor pre-charge or in cases where the ignition should be additionally supported by externally connected extra capacitors.

In order to quantify parasitic elements of DBD optical radiation sources they can be characterised by means of impedance spectroscopy. The frequency-resolved impedance measurements of DBD lamps with connection cables were conducted using an Agilent 4395A network analyser with impedance test-set according to Chapter 4.8.

The results presented in Figure 15 indicate that the internal electrode structure and the power supply cords add a significant inductance to the intrinsic capacitance of the lamps. For series resonant topologies (refer to Chapter 4.7), this additional inductance only enhances the value of the total series inductance and thereby reduces f_{pulse} and enhances the quality factor of the pulse forming resonant circuit. Even though the presented USP-topology (see page 170) follows the same mechanism, it holds a severe drawback: the parasitic series inductance of the DBD lamp restricts the maximum voltage slope while disturbing the pulse wave shape by an additional parasitic oscillation. As a result, the maximum achievable pulse frequency is not higher than the self-resonance frequency of the DBD lamp itself which is in this case 2.7 MHz. The parasitic DBD inductance also has to be considered as major source of magnetic field radiation.

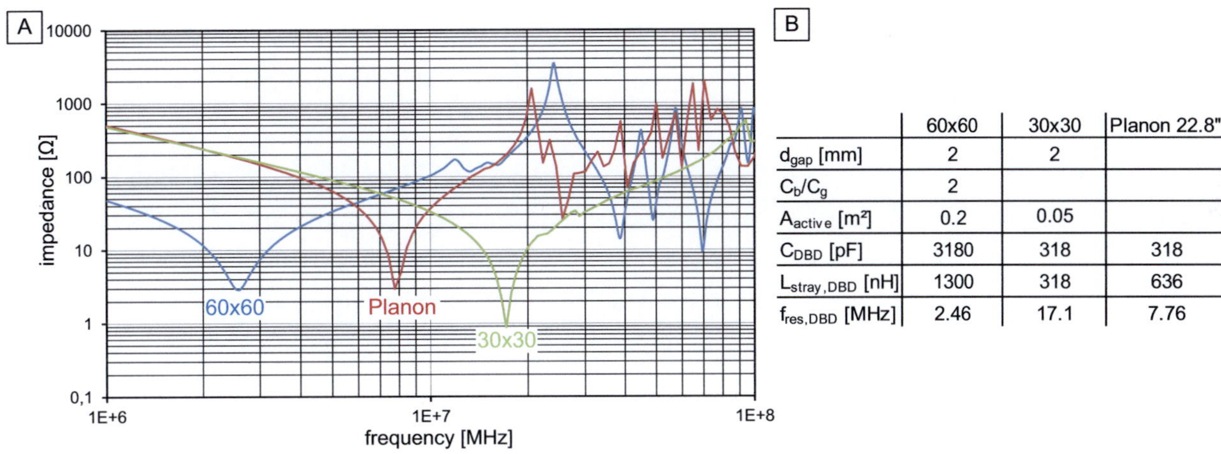

	60x60	30x30	Planon 22.8"
d_{gap} [mm]	2	2	
C_b/C_g	2		
A_{active} [m²]	0.2	0.05	
C_{DBD} [pF]	3180	318	318
$L_{stray,DBD}$ [nH]	1300	318	636
$f_{res,DBD}$ [MHz]	2.46	17.1	7.76

Figure 15: A: Frequency dependent impedance of used flat Planilum® DBD lamps "60x60", "30x30" and Planon® lamp. B: Table of respective electrical key properties.

The intrinsic wiring resistance, which in case of the measured 60x60 DBD lamp has a high value of 3 Ω, reduces the quality factor of the overall resonant circuit and adds ohmic loss. At typical values of lamp current of 3 A_{rms}, significant power loss of 27 W would be generated.

The rather bad characteristics of the used lamp mainly result from the application requirement that large luminous areas alternate with transparent areas. This introduces an electrode finger structure that increases resistance and inductance of the device. Coaxial type geometries, in comparison, have significant advantages in that respect. The internal voltage $v_{DBD,int}$ across C_{DBD} is reduced by the influence of R_{DBD} and $L_{stray,DBD}$ to:

$$v_{DBD,int}(t) = v_{DBD}(t) - L_{stray,DBD} \frac{di_{DBD}(t)}{dt} - R_{DBD} \cdot i_{DBD} \qquad (2.26)$$

The DBD's stray inductance can be reduced by shortening the connecting cables to the possible minimum and by using high-voltage twin cables or coaxial cables. The latter would also dramatically reduce EMI and would have very low inductance. However, due to the restricted skin depth at elevated frequencies, especially the centre conductor would need to have a comparatively large diameter in order to be able to handle HF ohmic losses.

2.6 Determination of Optical DBD Lamp Parameters

Although this work focusses on the inverter that operates the DBD lamp, optical measurements were necessary in order to characterise the discharge behaviour and the efficiency of the DBD lamp. The discharge behaviour was investigated by measuring the NIR radiation, while the effective fraction of the radiation was (relatively) measured in the VIS and VUV wavelength range.

2.6.1 NIR Discharge Characterisation

When it comes to experiments, for instance regarding the pulse shape that is applied to the DBD, a very important property to be determined is the homogeneity of the discharge. To prove homogeneity, high speed ICCD imaging of the discharge's NIR emissions can be used (Mildren 2001c; Liu 2003a; Shao 2011). In this work, a DiCAM-PRO camera system with attached RG780 filter pane triggered by a gate signal used to control the respective pulse generator acquired time and area-resolved NIR pictures. The minimum exposure time of this system is 10 ns and up to 250 subsequent exposures were combined to optimise picture contrast. The high-speed camera system was already used to identify homogeneity in earlier works (Trampert 2009; Paravia 2010).

NIR radiation of the lamp was also used to locate times at which ignitions occur (Shuhai 2003). For that purpose, the NIR radiation of the DBD was detected with a photo-multiplier tube (PMT) and an attached RG780 filter. The supply voltage of the PMT was set to a value between 500 and 700 V in order to assure a sufficient signal to noise ratio. Additionally, the whole PMT including the optical window was shielded in order to minimise the impact of electromagnetic interference generated by the pulsed DBD.

2.6.2 Measurement of Luminous Flux

The light output of commercial tubular UV lamps is standardly measured by means of UV irradiance meters that are geometrically arranged to the lamp in a way described by Lawal (2008). With this set-up, he claims a maximum measurement error of 5 %.

Instead of spectrally resolved VUV measurement set-ups using a nitrogen-flushed monochromator (Lambrecht 1998) or an evacuated VUV-photo-goniometer (Trampert 2007) in this work the phosphor-converted lamp light output has been measured by a luminance meter. The plane geometry of the phosphor-coated DBD lamps facilitated this measurement.

The luminous flux of commercial DBD lamps was determined by measuring the luminance with a luminance camera. The respective luminous flux was then calculated by integration over the active area A_{active} of the lamp.

2.7 Influence of DBD Capacitances

At the instant of time of ignition, the gas gap inside the DBD lamp changes rapidly from high-ohmic gas state to low-ohmic plasma state. The drastic enhancement of conductivity leads to a steeply rising current though the established plasma. As the achievable peak current and the current slope have an influence on lamp efficiency and homogeneity, this chapter investigates **the available energy sources** that can provide this current.

It is important to distinguish between plasma current support and plasma energy support. While the current source inverter that drives the DBD provides a relatively continuous current, the internal capacitances of the DBD are voltage sources and therefore provide an instantaneous peak current but the voltage amplitude falls as soon as charge is delivered.

According to Figure 16 the plasma is supported by the total parallel capacitance $C_{support}$ and also depends on the difference between plasma ignition voltage $V_{g,ign}$ and plasma extinguishing voltage $V_{g,ext}$. From the viewpoint of inverter efficiency, it is beneficial to enhance the power factor of the lamp PF_{DBD}. As energy is stored in the capacitances and only a part of that energy is actually accessible to the plasma, PF_{DBD} never reaches 100 %. However, by choosing favourable ratios of the DBD capacitances, the ratio of reactive energy to energy available to the discharge can be optimised.

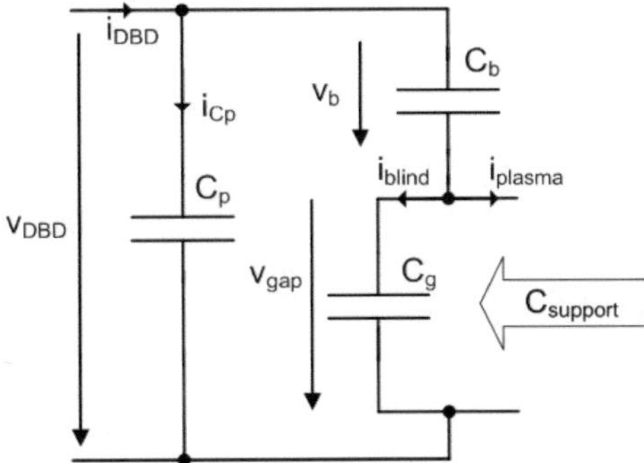

Figure 16: Illustration of the discharge support by the DBD capacitances. $C_{support}$ is supposed to be the equivalent capacitance accessible by the discharge.

The total electrical energy stored in the DBD capacitances is defined as:

$$E_{DBD} = \frac{1}{2} V_{DBD,pk}{}^{2} C_{DBD} \tag{2.27}$$

The DBD energy E_{DBD} is directly related to the values of C_b and C_g. Since these capacitors create a capacitive voltage divider, the ratio of lamp voltage to gap voltage if the gas is not ionised is:

$$\frac{V_{DBD,ign}}{V_{g,ign}} = \frac{C_g + C_b}{C_b} \tag{2.28}$$

2.7.1 Energy Ratio and Discharge Support of the Bare DBD

If a bare DBD lamp is operated with a current-source type pulse inverter, energy is contributed by the current that is fed-in by the inverter and by a charge transfer from C_g to the plasma. The influence of the externally impressed current is very complex and requires multi-physics modelling as the time-dependent plasma impedance Z_{plasma} has superior impact. Here, it will be assumed that for the short time of DBD ignition both v_g and i_{DBD} are constant. In this case, all external current i_{DBD} also passes through the plasma and transfers the energy:

$$E_{g,ext} = i_{DBD} \cdot v_g \cdot t_{discharge} \tag{2.29}$$

Regarding the capacitive contribution to plasma energy support, in case of a bare DBD lamp, only C_g acts as a voltage source at the moment of ignition. Thus, the energy available to the plasma is defined as:

$$E_{g,ign} = \frac{1}{2} V_{g,ign}^{\;2} C_g \tag{2.30}$$

Formula (2.30) can be rewritten to define the electrical energy density $\varpi_{g,ign}$ of the gap capacitance:

$$\varpi_{g,ign} = \frac{V_{g,ign}^{\;2} C_g}{2 \cdot d_{gap} \cdot A_{active}} = \varepsilon_0 \frac{V_{g,ign}^{\;2}}{2 \cdot d_{gap}^{\;2}} \tag{2.31}$$

$\varpi_{g,ign}$ heavily depends on the ratio of gap ignition voltage $V_{g,ign}$ and gap width d_{gap}. $V_{g,ign}$ rises almost proportional with rising d_{gap} but also with increasing xenon pressure. Therefore, $\varpi_{g,ign}$ is merely pressure dependent. Combining (2.27) and (2.30) with (2.2) yields the ratio EE of energy that is directly accessible by the discharge $E_{g,ign}$ and electrical energy stored in the DBD shortly before ignition E_{DBD}:

$$\frac{E_{g,ign}}{E_{DBD}} = EE = \frac{V_{g,ign}^{\;2} \cdot C_g}{V_{DBD,ign}^{\;2} \dfrac{C_b \cdot C_g}{C_g + C_b}} \tag{2.32}$$

This formula is valid if the gap ignition voltage is much higher than the gap extinguishing voltage ($V_{g,ign}$ »$V_{g,ext}$) leading to $V_{g,ext} \rightarrow 0$.By using the relation given by the capacitive voltage divider (2.28) the following simple expression is gained:

$$\frac{E_{g,ign}}{E_{DBD}} = EE = \frac{C_b}{C_g + C_b} \tag{2.33}$$

The relation is plotted in Figure 17. The two asymptotes at $E_{g,ign}/E_{DBD}$ = EE = 1 and EE = 0 are connected via inflection point C_b/C_g = 1 and $E_{g,ign}/E_{DBD}$ = ½. Since this formula cannot be solved for C_b/C_g, the graph data was gained by numerical simulation. For favourable configuration, the ratio C_b/C_g should be at least in the order of ten. A large ratio is advantageous as it reduces the energy the inverter has to deliver to the DBD lamp during time of plasma discharge. Although C_b enhances C_{DBD}, the ratio of v_{DBD} to v_g is favourably reduced by the same degree. However, in reality, the plasma extinguishes at a defined voltage that is much higher than zero volt. The maximum available energy is therefore reduced to:

$$E_{g,ign} = \frac{1}{2} C_g \left(V_{g,ign}^{\;2} - V_{g,ext}^{\;2} \right) \tag{2.34}$$

For defined extinguishing voltage $V_{g,ext}$, Formula (2.32) therefore gains complexity:

$$\frac{E_{g,ign}}{E_{DBD}} = \frac{\left(V_{g,ign}^{\;2} - V_{g,ext}^{\;2} \right) \cdot C_g}{V_{DBD,ign}^{\;2} \dfrac{C_b \cdot C_g}{C_g + C_b}} \tag{2.35}$$

Combining (2.35) with (2.28) yields the voltage-depending energy ratio EE:

$$EE = \left(1 - \left(\frac{V_{g,ext}}{V_{g,ign}}\right)^2\right) \frac{C_b}{C_g + C_b} \tag{2.36}$$

The first term in Formula (2.36) can be defined to be the characteristic factor CF:

$$CF = \left(1 - \left(\frac{V_{g,ext}}{V_{g,ign}}\right)^2\right) \tag{2.37}$$

CF is the distinction between (2.36) and (2.33) and has a value range of 0 – 1. According to Paravia (2010), the extinguishing voltage is approximately 60 % of the gap ignition voltage. Hence, the value of CF is in the range of 0.64.

Another approach comes to similar statements as (2.33). Inserting (2.2) and (2.27) in (2.28), one derives the dependence of the total energy stored in a bare DBD in relation to C_g, C_b and the gap voltage v_g:

$$E_{DBD,rel} := \frac{2 \cdot E_{DBD}}{v_g^{\;2}} = \frac{C_g (C_b + C_g)}{C_b} \tag{2.38}$$

The gap capacitance depends on the active area and the gap distance while the gap voltage depends on the gap distance and the xenon pressure. If these parameters are assumed to be fixed, based on (2.38) the relationship between ratio C_b/C_g and the relative energy stored in the DBD $E_{DBD,rel}$ can be calculated. Beside EE according to Formula (2.33), $E_{DBD,rel}$ is displayed in Figure 17.

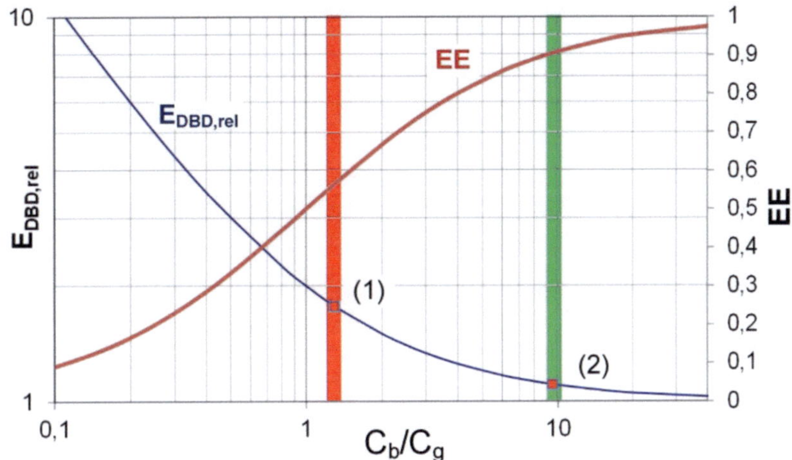

Figure 17: Dependence of relative DBD energy $E_{DBD,rel}$ (logarithmic, left y-axis) and EE ratio (linear, right y-axis) on ratio C_b/C_g. Marked are lamps presented in literature: 1: (Sugimura 2006) 2: (El-Deib 2010a).

The curve approaches the asymptote at $E_{DBD,rel} = 1$ which is the optimum since stored energy per gap ignition voltage is irreducible – or in other words, C_{DBD} is per definition greater than 0. As the figure shows, further enhancement of C_b/C_g exceeding the ratio of the lamp used by El-Deib (2010a) does not lead to significant improvement.

To conclude, it is favourable to ensure a large ratio of C_b to C_g with the rationales:

- v_{DBD} in relation to v_g needs to be minimised according to Formula (2.28). This reduces the maximum voltage required by the DBD lamp. Consequently, the efficiency of the pulse inverter rises.
- The relative energy (Formula (2.38)) is favourably reduced while the energy ratio EE (Formula (2.36)) of energy available to the plasma to total energy stored in the DBD is advantageously higher than 90 % which occurs if C_b/C_g exceeds 10.

2.7.2 Energy Ratio and Discharge Support of DBD with Parallel Capacitor

If the DBD lamp is equipped with an external parallel capacitor C_p, as of now three sources provide energy to the plasma. Beside the energy is contributed by the current that is fed-in by the inverter, C_g and C_p transfer charge to the plasma. If it is also assumed here that both v_g and i_{DBD} are constant, the current now splits between C_p and the bare DBD:

$$i_{DBD} = i_g + C_p \cdot \frac{dv_{DBD}}{dt} = i_g \left[1 + \frac{C_b \cdot C_p}{C_g^2} \right] + C_p \frac{dv_g}{dt} \qquad (2.39)$$

Since dv_g/dt is assumed to be zero, compared to a bare DBD, here the externally fed-in energy $E_{g,ext}$ is reduced to:

$$E_{g,ext} = \frac{C_b}{C_b + C_p} i_{DBD} \cdot v_g \cdot t_{discharge} \qquad (2.40)$$

Regarding the capacitive plasma support, in case of an attached parallel capacitor C_p, the equivalent capacitance that is in parallel to the gap is enlarged to:

$$C_{support} = \frac{C_p \cdot C_b}{C_p + C_b} + C_g \qquad (2.41)$$

C_g directly supports the plasma, whereas because of the series connection of C_p and C_b, their support depends mainly on the lower of these two capacitances. Therefore, C_p and C_b should have values at least in the same order of magnitude to be efficient in supporting the plasma.

In case the plasma could access the full energy down to $V_{g,ext} = 0$ V, the energy ratio becomes:

$$\frac{E_{g,ign}}{E_{DBD}} = \frac{C_g + \dfrac{C_p \cdot C_b}{C_p + C_b}}{\dfrac{(C_g + C_b) \cdot C_g}{C_b} + \dfrac{(C_g + C_b)^2 \cdot C_p}{C_b^2}} = \frac{C_b^2}{(C_b + C_g) \cdot (C_p + C_b)} \qquad (2.42)$$

The plot of the energy ratio relating to the important capacitance ratios C_p/C_{DBD} and C_b/C_g in Figure 18 reveals the nature of this relation. It visualises the coherence of the energy ratio with the relative values of the DBD capacitances. One key question is for which relative values of the capacitances the energy ratio is high, e.g. exceeds 90 %. If the extinguishing voltage is higher than zero volts, the characteristic factor given in Formula (2.37) needs to be multiplied:

$$\frac{E_g}{E_{DBD}} = EE = \frac{C_b^{\ 2}}{(C_b + C_g)\cdot(C_p + C_b)}\left(1 - \frac{V_{g,ext}^{\ \ 2}}{V_{g,ign}^{\ \ 2}}\right) \tag{2.43}$$

EE is called energy ratio. In order to evaluate the required ratio of capacitances in order to gain a high energy ratio, this term is solved for C_p:

$$C_p = -\frac{C_b\left(\frac{EE}{CF}\cdot C_b + \frac{EE}{CF}\cdot C_g - C_b\right)}{\frac{EE}{CF}(C_b + C_g)} = \frac{C_b^{\ 2}}{\frac{EE}{CF}(C_b + C_g)} - C_b \tag{2.44}$$

In this formula, the term EE/CF defines the energy ratio in relation to the characteristic gap voltage factor defined in (2.37). The value of CF is in the range of 64 % (Paravia 2010). Because Formula (2.44) is valid for positive values of C_p only, EE/CF must not exceed one: Hence, Formula (2.46) manifests that the available energy is fundamentally restricted by the gate voltage drop.

$$\frac{EE}{CF} \overset{!}{<} \frac{C_b}{C_b + C_g} \tag{2.45}$$

$$\frac{E_g}{E_{DBD}} \overset{!}{\leq} 1 - \frac{V_{g,ext}^{\ \ 2}}{V_{g,ign}^{\ \ 2}} \tag{2.46}$$

By dividing (2.44) by (2.2) the important relation according to Formula (2.47) is gained.

$$\frac{C_p}{C_{DBD,bare}} = -\frac{\frac{EE}{CF}C_b + \frac{EE}{CF}C_g - C_b}{\frac{EE}{CF}\cdot C_g} = \frac{C_b}{C_g}\left(\frac{1}{\frac{EE}{CF}} - 1\right) - 1 \tag{2.47}$$

Formula (2.47) is plotted in Figure 18 as array of curves for different values of EE/CF. The individual curves meet asymptotes that are defined by EE/CF:

$$\frac{C_b}{C_g} = \frac{\frac{EE}{CF}}{1 - \frac{EE}{CF}} \tag{2.48}$$

In other words, Formula (2.48) defines the minimum C_b/C_g ratio to meet a certain EE/CF ratio. As soon as C_p reaches the order of magnitude of C_{DBD}, the barrier-to-gap capacitance ratio C_b/C_g needs to grow proportionally with the rise of C_p`s value. C_p is favourably of the same value as C_{DBD}, higher values are not reasonable. Another interpretation of Figure 18 is that for a given C_p/C_{DBD} ratio, the ratio C_b/C_g needs to be increased in order to meet the intended EE/CF margin. C_b/C_g rises naturally with wider gap distances. However, along with enhanced active volume, total energy support needs to rise.

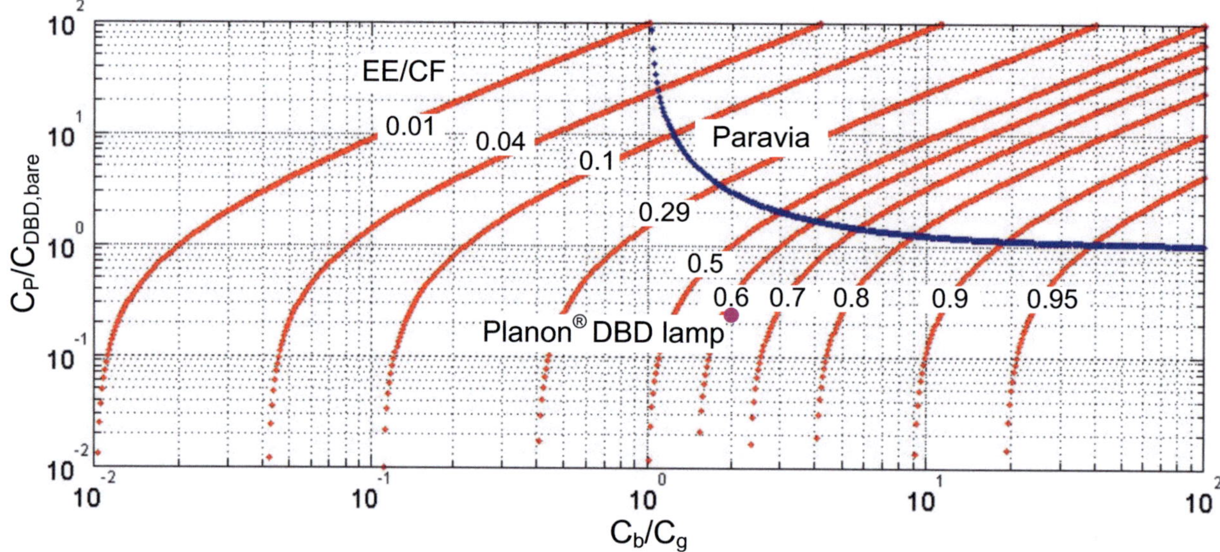

Figure 18: Array of curves C_p/C_{DBD} depending on C_b/C_g and ratio EE/CF according to Formula (2.47) and curve visualising Formula (2.50) according to (Paravia 2010). Based on its properties, the Planon® DBD lamp is marked in the chart.

In conclusion it can be stated that if C_p is at least one order of magnitude larger than C_g, then C_b is favourably one order of magnitude higher than C_p. This ratio would be found at the intersection of the blue curve with the right y axis in Figure 18. The blue curve visualises Formula (2.50) which is based on an earlier investigation of Paravia (2010). There, the discharge support had been determined by means of electrical charge transfer from a parallel capacitor C_P treated as ideal voltage source. This resulted in the rather fixed relation:

$$C_p = \frac{C_g}{1 - \frac{C_g}{C_b}} = \frac{C_b \cdot C_g}{C_b - C_g} \tag{2.49}$$

Paravia (2010) concluded that the value of C_b needs to exceed the value of C_p and that C_p must have at least the value of C_g to be effective. Furthermore, it had been advised to maximise C_b in order to allow for a lower C_p value. Modifying (2.49), the relation C_p/C_{DBD} becomes:

$$\frac{C_p}{C_{DBD,bare}} = \frac{C_b + C_g}{C_b - C_g} \tag{2.50}$$

Formula (2.50) has an asymptote at $C_b/C_g = 1$, approaches 1 for high C_b/C_g and is defined for $C_b > C_g$, only.

To conclude, it can be stated, that from the standpoint of an optimum ratio of energy accessible by the plasma to total energy stored in the DBD with attached parallel capacitor, the ratio C_b/C_g needs to be very high if C_p exceeds a value of 10% of $C_{DBD,bare}$. In the end, it appears that it is more favourable to achieve a large barrier capacitance (thin sheets of glass or even thin film dielectric coating) than to connect a parallel capacitor. This is even more important since the external current support is lowered by C_p (Formula (2.40)).

2.8 ENHANCEMENT OF DBD LAMP POWER DENSITY

Regarding DBD radiation sources, their power density is defined as the power of optical emission divided by the volume enclosed by the lamp. The power density of cylindrical lamps with radius r_{lamp} according to Figure 7A and B can also be expressed as irradiance times $r_{lamp}/2$. This chapter deals with methods that bear a possible enhancement of power density.

The enhancement of continuous power density can be broken down to two objectives: discharge efficiency and comprehensive use of the available volume.

The first objective can be achieved by improving the electrical operation mode (also see Chapter 3.4). In the past, power density and efficiency were significantly enhanced by using pulsed operation instead of continuous low-frequency sinusoidal excitation (Mildren 2001a). Kawanaka (2001) operated a 1.5 kW_{pk} Krypton DBD flash lamp which reveals that, also for DBDs, power density is a question of duration of power input. The same power level is maintained by a coaxial KrCl DBD excimer lamp irradiating up to 60 mW/cm^2 in continuous operation (Tarasenko 2009). This lamp relied on a combined water and air cooling concept. A further step could be the burst mode excitation introduced by Beleznai (2008c). As Figure 19 depicts, the burst mode appears to be the paramount operation mode as for a given efficiency level it allows for a significant enhancement of power density. The reason for that is that multiple efficient ignitions occur during a burst while plasma thermalisation which would reduce the discharge efficiency is prevented by the short burst duration. Please see Figure 29A on page 62 for details regarding the burst wave-shape.

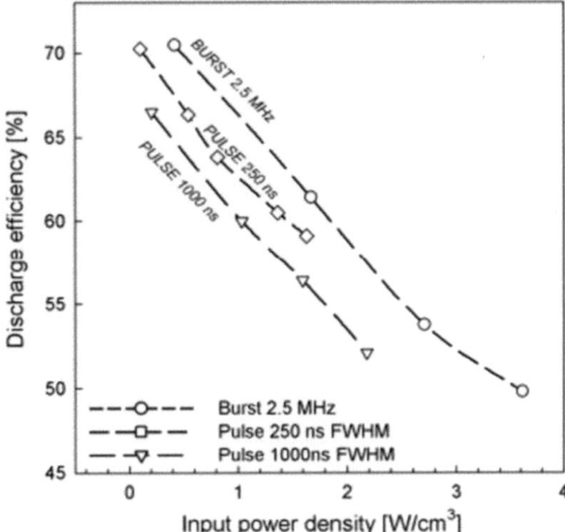

Figure 19: Comparison of different DBD operation modes by Beleznai (2009). Note that the highest efficiency per achieved power density is obtained with the burst operation mode. The burst wave-shape is depicted in Figure 29 on page 62.

The DBD set-up also plays a major role regarding discharge efficiency. The structural design has to minimise optical losses and maximise plasma efficiency. Coaxial designs, for instance, benefit from large gap widths compared to the diameter of the inner tube. Efficiency can rise from below 30 % at d_{gap} = 1 mm gap width to more than 60 % at d_{gap} = 8 mm (Paravia 2009a). Rising pressure and gap width both dramatically increase DBD power density and up to a pressure of approximately 800 mbar also efficiency. However, in practical lamp designs, ignition voltage exceeds four kilovolts at a pressure-gap product of approximately $p_{Xe} \cdot d_{gap}$ = 1 bar·mm. Voltages above this margin would disproportionately boost system costs.

The second objective is to maximise the discharge volume to total available volume ratio. In order to meet this demand, a patent application has been submitted claiming DBD structures as depicted in Figure 20 to enhance power density. The invention focuses on coaxial lamp designs and bears the advantage of utilising volume which would otherwise be useless for excimer generation. The background is that due to the limitations reported in the previous paragraph, the gap width cannot randomly be extended. Therefore, if the outer diameter of the lamp, which is defined by the application, by far exceeds twice the intended gap width, a significant volume is left unused for the discharge.

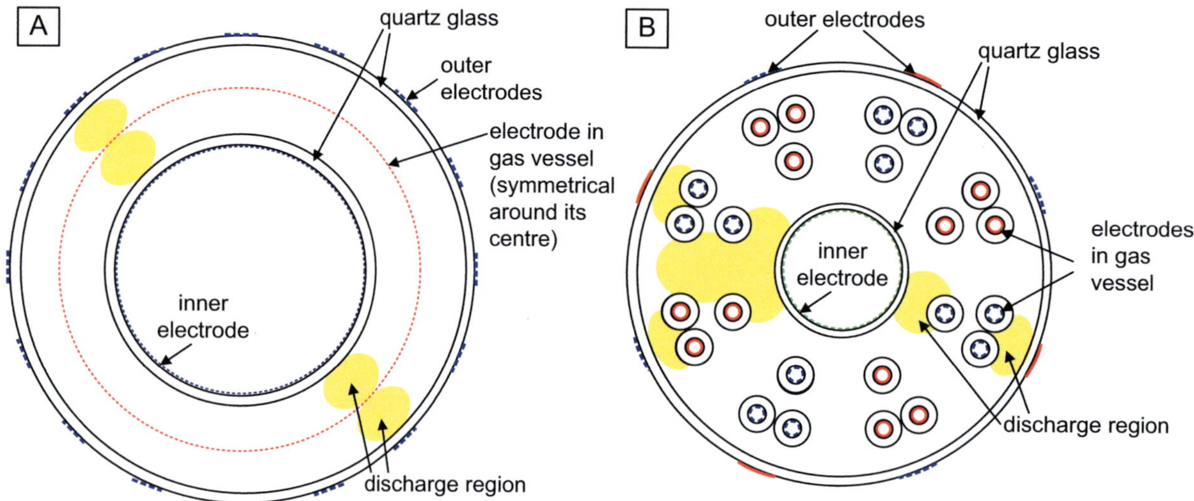

Figure 20: Ways to improve DBD power density according to pending patent (Meisser 2012c). A: Inventive DBD structure where the total gas volume is separated into two (an inner and an outer) discharge regions by a metal mesh electrode (red). If only outer optical radiation is required, the inner electrode favourably is made of a thin layer of Al. B: Inventive DBD structure where the total gas volume is sub-divided into multiple discharge areas with axial and radial orientation. Due to the inter-digital structure of the outer electrode also surface discharges can be excited.

The inventive structure bases on the novel principle of de-coupled discharge regions that are either situated as layers in a coaxial manner as depicted in Figure 20A or with axial and radial orientation as illustrated in Figure 20B. In case of Figure 20A, the discharge volume is enlarged by adding a third electrode. This electrode stretches another axisymmetric discharge volume inside the otherwise unused volume. If radiation from the outer lamp surface is intended to be used, the outer and the middle electrode need to be optically transparent. Beneficially, the inner and outer electrodes are connected to ground which acts as shield reducing EMR. The alternative inventive set-up shown in Figure 20B contains internal finger electrodes stretching discharge zones not only in radial but also in axial direction. This can be combined with outer electrodes arranged in an inter-digital structure. In all cases, the generated radiation should be impeded as less as possible by the electrodes, which sets limits to the number and structure of the electrodes.

Compared to another approach of Reich (2005) where hexagonal electrode structures solely within the lamp vessel are claimed, the approach presented here also includes external electrodes attached to the lamp walls. It is believed that hexagonal electrode structures are not suited to achieve a high optical transmittance.

3 DBD UNDER PULSED EXCITATION

The early investigations on DBDs (Honda 1955; Volkova 1984) as well as many contemporary industrial installations operate DBD lamps with low or medium frequency continuous waves (CW). The wave-shape can be sinusoidal (Koudriavtsev 2003) or of square wave-shape.

Although these operation principles have their respective advantages, pulsed operation of DBDs was found to be superior in terms of discharge homogeneity and lamp efficiency (Vollkommer 1994; Mildren 2000). As discussed so far, the homogeneous operation of a DBD as described in Chapter 2.3 relies on many factors. Beside gas purity and electrode uniformity, the applied waveform is of high impact. Therefore, this chapter summarises the state-of-the-art of pulses for DBD lamp operation in order to identify key waveform parameters that support discharge homogeneity and efficiency.

3.1 GENERAL PULSE WAVEFORM PROPERTIES

As far as this work is concerned, the term pulse relates to a voltage pulse. This pulse is defined as having one or multiple peaks of high voltage amplitude with the same or a different polarity and amplitude. Pulses are separated by an idle time with a comparatively constant voltage level.

Regarding this voltage level, it needs to be distinguished between interim zero voltage (IZV) and interim non-zero voltage (INZV). Achieving IZV or a low voltage level during idle time is a way to minimise the RMS voltage measureable at the lamp's cables. This value becomes important if the DBD lamp is intended for the consumer market. In this case, higher RMS voltage could entail stricter regulations regarding EMC and safety, for example. If low voltage during idle time needs to be achieved, the lamp capacitance cannot intentionally be used to store energy during that time, which is done in square-wave operation of DBDs (Kyrberg 2007a). Instead, electrical energy needs to be removed from C_{DBD}. This may favourably be done by means of an energy recovery that transfers energy back to the power supply to be used for the initiation of the next pulse.

Coming back to the pulse shape and its properties, Figure 21A and B show the two major types of DBD pulse wave-shapes: the unipolar pulse (Figure 21A) and the bipolar distributed pulse (Figure 21B). Both wave-shapes are characterised by the following key parameters:

1. pulse repetition time period T_{rep}, $T_{rep} = 1/f_{rep}$, with f_{rep} being the pulse repetition frequency
2. idle time t_{idle}, the time between the pulses during which almost no current flows through the DBD, e.g. no DBD voltage change occurs
3. pulse time t_{pulse} , $t_{pulse} = T_{rep} - t_{idle}$, $f_{pulse} = 1/(2{\cdot}t_{pulse})$
4. rise time t_{rise} when v_{DBD} has positive slope that defines the positive current amplitude of DBD current i_{DBD}

5. ceiling time t_{ceil} which follows t_{rise} and marks the time between the positive and the negative DBD voltage slope
6. fall time t_{fall} when v_{DBD} has negative slope that defines the negative current amplitude of DBD current i_{DBD}
7. intermediate time t_{int} that is defined as time between positive and negative voltage plateau in case of a bipolar pulse.

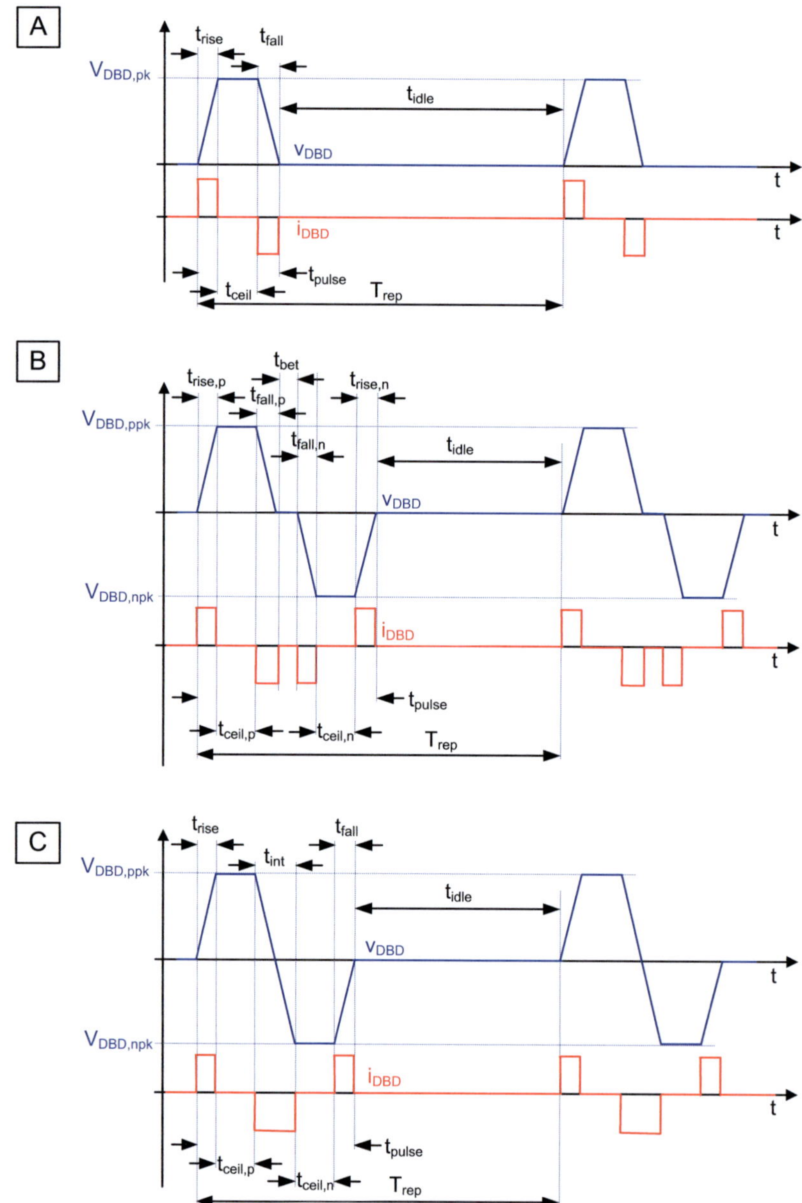

Figure 21: Principal pulse wave-shapes with constant slopes and marked key properties, A: unipolar pulse, B: bipolar distributed pulse and C: bipolar pulse with immediately following negative half-wave. Note that depending on the driving circuit v_{DBD} can exhibit an offset voltage.

The unipolar pulse is referred to as being generated by simpler circuitry and to support homogeneity especially of unilateral DBD lamps. Unipolar pulses are typically generated by a resonant flyback topology that is reviewed in Chapter 6.1.

The variety of bipolar distributed pulse shapes exceeds that of unipolar pulses by far which is due to the extended parameters as:

1. rise time $t_{rise,p}$ which describes the positive voltage slope of the positive pulse
2. fall time $t_{fall,p}$ which describes the negative voltage slope of the positive pulse
3. rise time $t_{rise,n}$ which describes the positive voltage slope of the negative pulse
4. fall time $t_{fall,n}$ which describes the negative voltage slope of the negative pulse
5. ceiling time $t_{ceil,p}$ which follows t_{rise} and marks the time between the positive and the negative DBD voltage slope of the positive pulse
6. ceiling time $t_{ceil,n}$ which follows t_{rise} and marks the time between the positive and the negative DBD voltage slope of the negative pulse
7. time between positive and negative pulse t_{bet}
8. pulse time t_{pulse} , $t_{pulse} = T_{rep} - t_{idle}$, $f_{pulse} = 1/(t_{pulse})$

Figure 21C depicts a special bipolar pulse wave-shape in which the negative pulse immediately follows after the positive pulse. The time between the voltage plateaus is defined by t_{int}. So far, this coherent bipolar pulse is assumed to combine most of the aspects discussed in Chapter 3.4 which are favourable for the discharge properties. Note that the pulse frequency of bipolar pulses is defined as $f_{pulse} = 1/(t_{pulse})$ while the frequency of unipolar pulses is defined as $f_{pulse} = 1/(2 \cdot t_{pulse})$. This refers to the fact that the voltage wave-shape of real-world pulses is sine-derived, e.g. a bipolar pulse equals a full sinusoidal period.

The square-edged waveforms of Figure 21 are not practical as power electronic converters are not able to generate them with reasonable efficiency. Both rectangular current waveform and strictly linear voltage wave-shapes could only be generated by circuits in which power semiconductor switches are operated in linear region of their transfer function. This would lead to excessive ohmic losses.

Therefore, for efficient pulse generation, resonant circuits with power semiconductors that act as on-off switches are used. In this case, losses only occur due to the non-ideal properties of the power semiconductors but ideal efficiency is 100 %. The principles of resonant pulse generating circuits are discussed in Chapter 4.

3.2 State of the Art: Operation with Unipolar Pulse

In this and the following chapter, state-of-the-art pulse wave-shapes as they appeared in scientific publications or patents will be presented. No emphasis will be put on the way they were generated nor the efficiency of the respective pulse generation.

Operated with a thyratron-based peaking-circuit, the square-wave voltage waveform presented in (Riese 1991) allows for very high voltage slopes of 1300 kV/µs but shows extensive ringing as visible in Figure 22A. A resonant flyback-derived topology had been used to create the unipolar pulse shape depicted in Figure 22B. A DBD voltage slope of 7 kV is reported.

Unipolar square-wave pulses with duty-ratios down to 20 % have been described by Schwarz-Kiene (2001). The peak voltage of 2.5 kV is reached within a rise time of 50 ns resulting in a slope of 50 kV/µs.

Lower duty cycles were used to operate Xe excimer DBD lamps with square-wave pulses having peak voltages of up to 15 kV and rise times down to 20 ns (Liu 2001). Steep voltage slopes of 400 kV/ns were generated. Beside the primary discharge a secondary discharge which is claimed to consume energy from space charges created by the primary discharge had been obtained.

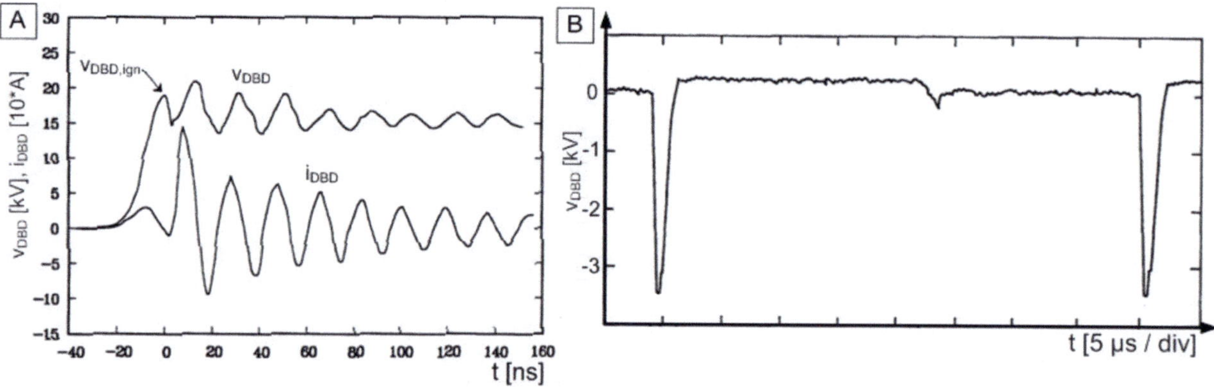

Figure 22: A: Square-wave unipolar pulse with steep slope and heavy ringing presented by Riese (1991). B: Sinusoidal half-wave pulse generated by a resonant flyback-derived topology (Huber 1996).

Later, double discharges were observed in Xe excimer DBDs that were driven by unipolar pulses shown in Figure 23A with steeply rising and falling voltage (Liu 2003a). In the same year, unipolar pulses with sinusoidal wave-shape were used to investigate the discharge properties of Xe DBD discharges at different repetition frequencies, pulse frequencies and buffer gas pressures (Jinno 2003). Unipolar pulses shown in Figure 23B where used.

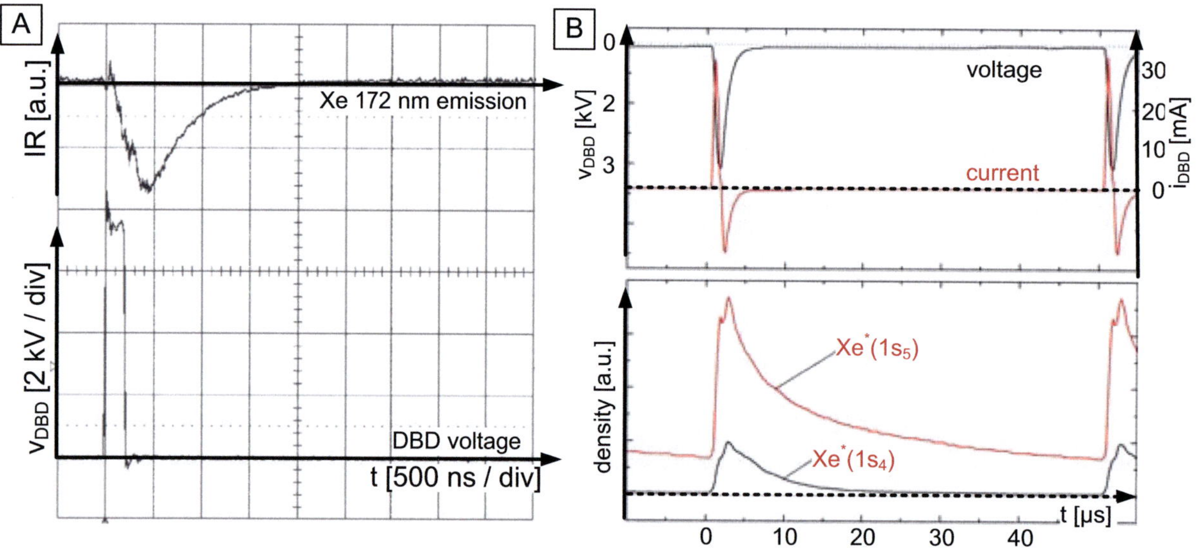

Figure 23: A: Square-wave unipolar pulse leading to double discharges (Liu 2003a). B: Sinusoidal half-wave pulse used by Jinno (2003) to investigate luminance enhancement by variation of pulse duration.

A unipolar square-wave pulse is presented by Shao (2010). Voltage slopes of up to 800kV/µs with $27kV_{pk}$ voltage are claimed. Carman (2003) presented a pulse wave-shape starting with a 48 kV/µs slope migrating into heavy ringing as shown in Figure 24. By using experimental data obtained with this wave-shape and simulations, Carman gained insight into the plasma-physical discharge behaviour. Figure 24A shows the first 300 ns of the pulse where the first intense ignition is followed by a second ignition with only small power consumption.

A unipolar pulse without subsequent oscillation and only 400 ns duration was presented by Beleznai (2008a) and is depicted in Figure 25A. The pulse consists of a steep slope with maximum dv/dt at its beginning, a short voltage plateau and a subsequent negative slope, again with maximum dv/dt at its beginning, that terminates the pulse.

A very special unipolar waveform is depicted in Figure 25B. This two-level sinusoidal pulse includes a variable voltage plateau and is intended to be used for the drive of PDPs. Although the multiple voltage levels restrict the overall voltage slope, it is claimed that its generation is very efficient because low-voltage switches can be used (Chung-Wook 2003). The individual voltage slopes that link the different voltage plateaus are the step answers of resonant circuits.

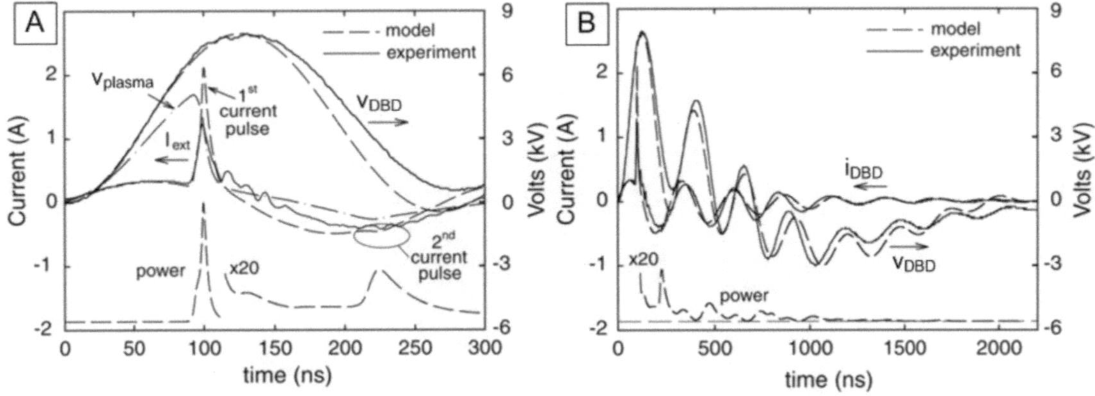

Figure 24: Unipolar pulse with sharp rising edge and subsequent ringing according to Carman (2003). A: Initial voltage slope of v_{DBD} provoking first ignition indicated by current pulse and intense power dissipation. Its decay results in a weak second ignition. B: Full excitation period where decaying oscillation is visible.

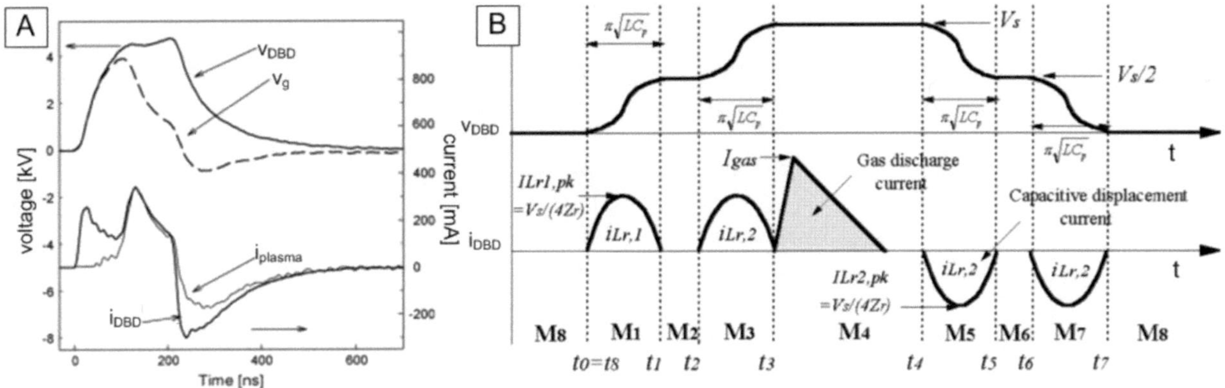

Figure 25: A: Unipolar short pulse with maximum displacement current at the beginning of the particular slope (Beleznai 2008a) B: Two-level sinusoidal pulse according to (Chung-Wook 2003).

Finally, a very interesting approach is disclosed in a patent of Park (2008). The topologies presented are claimed to generate various wave-shapes as depicted in Figure 26. The pulses can be divided into different sections which are the positive slope, the plateau and the negative slope. The presented topologies allow for a start and a termination of the pulses by different wave-shapes. The pulse shape depicted in Figure 26A starts with a maximum slope that decays until reaching the plateau voltage level. Then, pulse termination is started with a zero slope and ends with a maximum slope. Except for the plateau placed in between the slopes, this pulse shape is found in resonant flyback DBD drivers which are discussed in Chapter 6.1. The second pulse shape, shown in Figure 26B, starts with a low voltage slope which monotonously increases until the peak voltage is reached. This is favourable for the lamp since i_{DBD} has highest level around the time of ignition. This is supposed to support a powerful first ignition (refer to Chapter 3.4). The same is true for the pulse termination. Finally, the pulse shape of Figure 26C combines a sinusoidal slope leading to a voltage rise following a negative cosine with a pulse termination having the maximum slope at the beginning. Regarding generally possible pulse shapes it is referred to the previous chapter.

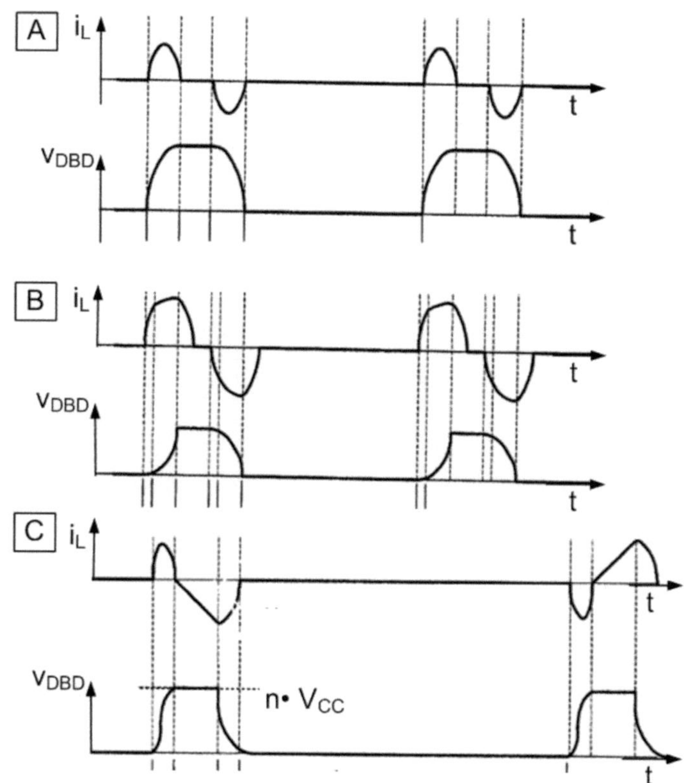

Figure 26: Unipolar pulse wave-shapes disclosed in a patent of Park (2008). Note that i_L is inductor current and not DBD lamp current. Picture processed for clarification.

3.3 State of the Art: Operation with Bipolar Pulse

It is a rather new concept to drive DBDs with bipolar pulses. Development started with distributed bipolar pulses (Chapter 3.1) consisting of positive and negative voltage half-waves separated from each other by a significant time delay. Figure 27A shows a pulse presented by Lomaev (2002) belonging to that category. Tastekin (2011a) presented a DBD drive based on a Marx generator. This allowed for transformer-less generation of distributed bipolar 10 kV pulses with 100 kV/µs slopes as shown in Figure 27B. This massive voltage slope was only limited by parasitic inductances of the circuit and a current-limiting resistor. The same author also used 10 kV matrix switches to drive DBDs with a full-bridge configuration. By avoiding the recharge times required by the Marx generator, this topology allowed to initiate the negative pulse directly after the positive forming a bipolar pulse (Tastekin 2011b).

Distributed bipolar pulses were also presented by Schwarz-Kiene (1998). The full-bridge excited transformer-equipped topology allowed for independent tuning of rising and falling edge slew rates up to 100 kV/µs. Another distributed bipolar pulse wave-shape is shown in Figure 28. Somekawa (2005) utilises the parasitics of a transformer being used to amplify the output of a low-voltage pulse square-wave generator to design decaying sine-waves. The first two "ringing parts" are supposed to lead to ignitions. The special wave-shape is further characterised by alternating polarities of the respective first and second sinusoidal half-wave. A similar pulse with low pulse frequency of 100 kHz and same pulse polarity is used by (Yukimura 2007) for NO_X removal.

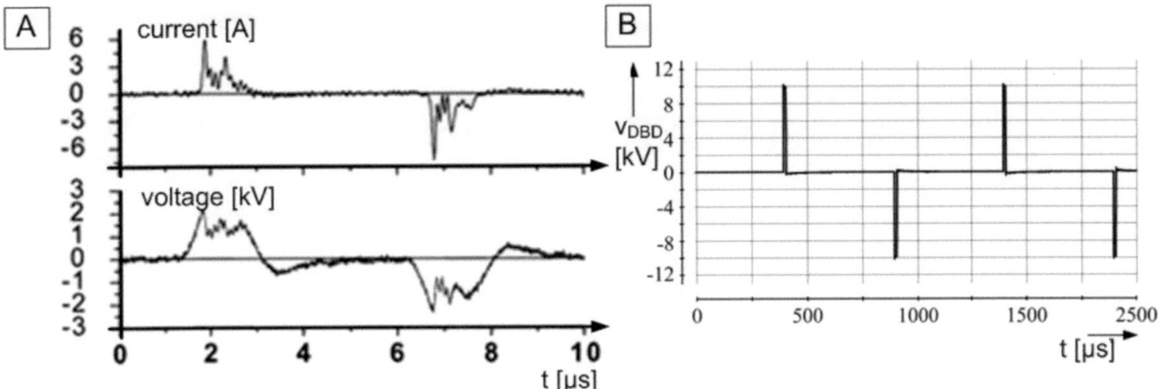

Figure 27: A: Distributed bipolar voltage pulse (Lomaev 2002) similar to (Panoisis 2008). B: Distributed square-wave pulses of a Marx generator according to Tastekin (2011a). Pictures modified for clarity.

Despite of unipolar variants, Park (2008) presents in his patent also a distributed bipolar pulse wave-shape which is depicted in Figure 28B. Slew rates both of the rising and the falling edge of the DBD voltage are maximum at their respective end which can be assumed to support efficient discharge behaviour while the subsequent zero-current time-range (voltage slope equal to zero) avoids glow-phase losses (Paravia 2008a).

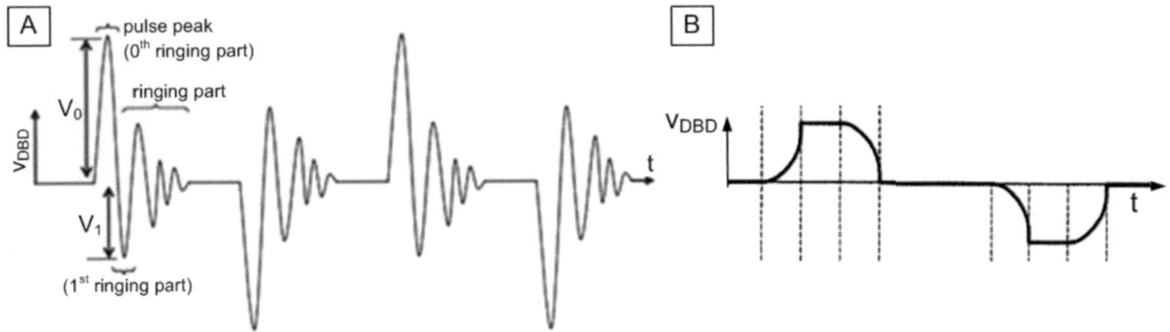

Figure 28: A: alternating bipolar pulse and determination of positive (V0) and negative (V1) voltage peak; chart basing on (Somekawa 2005), B: distributed bipolar pulse according to Park (2008). Pictures modified for clarity.

A very different strategy is pursued by (Beleznai 2009). The burst wave being depicted in Figure 29A is generated by a resonant push-pull converter and takes advantage of the fact that space charges travel with restricted speed. The objective here is to prevent the space charge region to hit the barriers by rapidly changing the polarity of the applied electrical voltage. For instance, at a gap distance of 8.5 mm and a xenon pressure of p_{Xe} = 0.1-0.3 bar, the space charges need roughly 100 ns to reach the barrier (Beleznai 2008c). Consequently, collision with the barriers is not totally prevented using a burst frequency of 2.5 MHz. Also, because of multiple discharges occurring randomly during the burst wave excitation, an accumulated movement of the space charge region cannot be prevented. Hence, the burst is terminated after a determined amount of wave periods. Compared to pulsed excitation, higher DBD power density would theoretically be possible with this excitation mode. Beleznai (2009) proved that compared to operation with unipolar 200 ns pulses, the 2.5 MHz burst allows for a more than 60 % increase in power density while keeping lamp efficiency constant. Large gaps are required to keep the necessary burst frequency low due to the high speed of the space charge region movement. However, this calls for high operation voltage that does not seem to be commercially practicable.

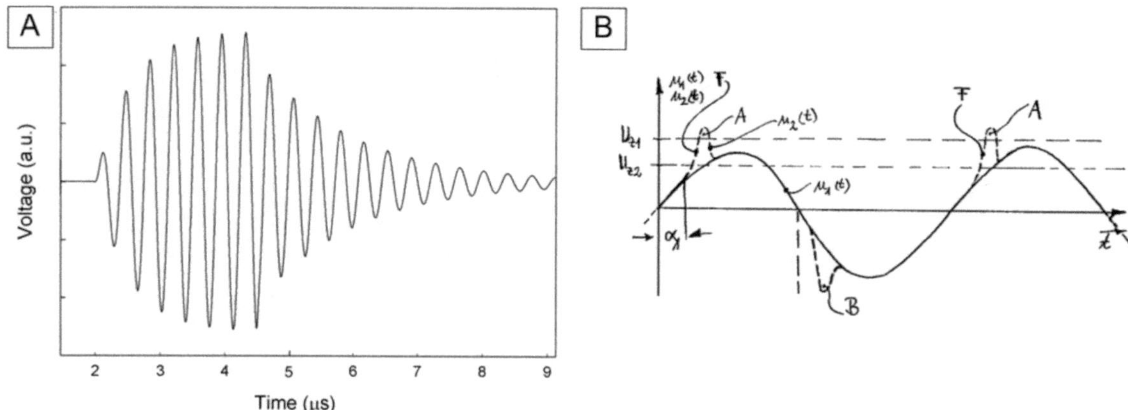

Figure 29: A: 2.5 MHz burst wave suggested by Beleznai (2009) allowed for a significant increase in efficiency. B: Low frequency sinusoidal continuous wave superimposed by voltage peaks (Neff 1997).

More than 10 years earlier, Neff (1997) suggested to tailor time-wise and local discharge distribution by superimposing voltage peaks to continuous wave voltage excitation of DBDs according to Figure 29B. The published patent application mentions both unipolar and bipolar voltage pulses. This technique is claimed to favourably operate the DBD with low losses. This approach is partly in line with the idea to monotonously increase the voltage slope until lamp ignition in order to achieve maximum current at this instant of time.

One of the first mentions of bipolar pulses intended to support a back-ignition occurs in an Osram patent (Wammes 1999a). The voltage wave-shape depicted in Figure 30A with corresponding current in Figure 30B also consists of a low-frequency carrier and a high-frequency pulse. The superposition of both forms the bipolar wave-shape with high voltage slopes during times of ignition and subsequent reduced slopes. The next chapter further will discuss certain aspects mentioned here which will lead to a suggestion of an optimum wave-shape.

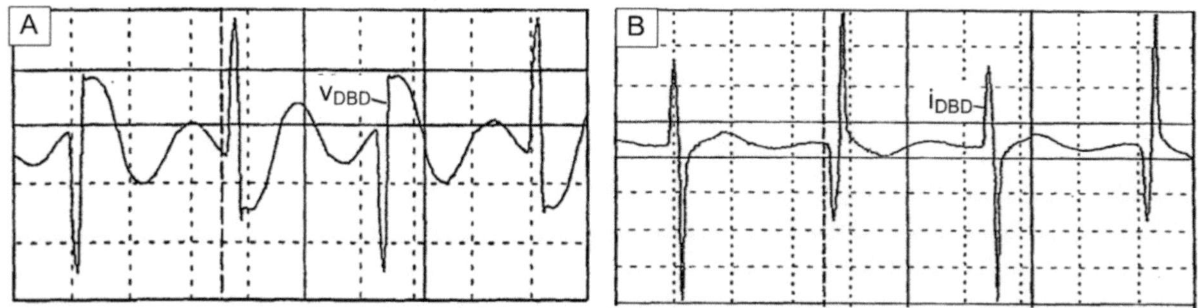

Figure 30: Time-dependent graphs of voltage (A) and current (B) of a wave-shape proposed by Wammes (1999a). The wave-shape is intended to support a back-ignition and is generated by an inverter according to the forward converter principle. Pictures are processed for clarity.

3.4 Proposed Optimal Pulse Wave Shape

While worldwide research activities led to the insight that pulsed operation supports DBD efficiency (Vollkommer 1998) and homogeneity, the question regarding favourable pulse wave-shapes remains. The last chapter presented some common wave-shapes and Chapter 3.1 introduced the key waveform properties. Based on available literature, the likely properties of a pulse wave-shape that is tailored to the requirements of DBD drive will be explored in this chapter. With this end in mind, key pulse parameters are discussed as follows.

3.4.1 Pulse Frequency

Short pulses support the more efficient, shorter and more intense discharge occurring at high Xe pressures. The higher voltage slope provided by high-frequency pulses enhances the necessary ignition field strength (also external ignition voltage) and lowers the fraction of low-energy electrons (Roth 2001). Gas heating effects are reduced and higher three-body reaction rates are achieved (Neiger 1992). This enhances the average electron energy which is said to increase excitation and radiation efficiency. Shorter pulse widths were found to support higher luminance of DBD lamps (Jinno 2003).

Another aspect of applying high-frequency pulses is the shortening of the glow phase that appears shortly after the main discharge (Paravia 2008a). The glow phase, indicated by a low-amplitude tail current (Roth 2001) was found to be comparatively lossy. Contrary to that approach, other work groups tried to extend the glow phase by impressing current through the established plasma (Piquet 2012).

Additionally, high pulse frequency also reduces the time between primary and secondary ignition which not only reduces glow phase loss but also maximises utilisation of the energy stored in the charge cloud left from the primary ignition. Importantly, also the resonant current level rises with shorter pulses. Depending on the moments of ignition, the externally fed-in resonant current may assist C_g and C_p in supporting the discharge (see Chapter 2.7).

3.4.2 Repetition Rate

In order to obtain high plasma efficiency, an idle time with constant voltage level between the dynamic voltage changes that lead to ignitions is required (Vollkommer 1998). This is met by pulsed operation instead of continuous sine wave excitation. A long idle time after the discharge excitation is advantageous for discharge homogeneity as residual charges have time to recombine and to distribute uniformly (Beleznai 2008c; Paravia 2010). On the contrary, high repetition frequency and voltage changes prior to the pulse that leads to DBD ignition enhance gas conductivity by pre-ionisation and prevention of gas solidification. Low-efficient glow discharges could be the consequence of the resulting (partial) loss of the non-LTE plasma property (Paravia 2009b). However, restricted gas ionisation prior to ignition may also favourably reduce the necessary ignition voltage, which could bear an efficiency gain for the inverter.

3.4.3 Support of Secondary Discharge (Back Ignition)

Like the pulsed operation of DBDs (Vollkommer 1994), also the idea of favourable DBD lamp operation with pulses that support a back-ignition roots back to early Osram patents (Wammes 1999b). As soon as pulsed excitation of excimer DBDs was reported to enhance performance (Mildren 2001b), it was found that the so-called secondary discharge which is initiated by opposite voltage polarity compared to the primary discharge (Kogelschatz 2000a) favourably consumes energy provided from space charges left from the primary discharge (Liu 2001; Liu 2003a; Beleznai 2008b). The space charges are also referred to as counter-polarisation (Wammes 1999b). Although beneficial in PDPs (Meunier 1995) space charges, also referred to as memory charges, were suspected to account for filamentary discharge progress. Research unveiled that the secondary discharge can be supported by an immediate subsequent negative voltage forming a bipolar wave-shape (Shuhai 2003). Although secondary discharges were initially mistaken as being of parasitic nature (Carman 2004a), double discharges induced by unipolar short square pulses were further investigated. Since the secondary discharge expels charge carriers from the dielectric surfaces conditions virtually free of space-charges prior to the next discharge are gained. This was found to support the spatial evenly distributed discharge development (Somekawa 2005; Paravia 2007; Paravia 2010). DBDs operated with bipolar pulses were reported to show higher efficiency (Lomaev 2002). Using repetitive voltage pulse trains, it was found that the recovery behaviour of the extinguished discharge channels has dominating influence on the distribution of following discharges (Kogelschatz 2008).

3.4.4 Voltage Wave-Shape

Already in 1991, the voltage slope was found to affect the electron energy (Gellert 1991). Later, steep voltage slopes were applied in order to generate excited Xe atoms. During a subsequent phase of low voltage slope the excimers efficiently emit optical radiation. This operation mode had been claimed to be favourable and had been patented (Yokokawa 1998). In scientific investigations DBDs operated with unipolar pulses showed higher VUV output if peak voltage had been increased above outer ignition voltage. However, Carman (2004a) reported that lamp efficiency dropped by 25 %. The same author made aware that the electric field applied to the DBD should be removed soon after occurred ignition in order to reduce the ion heating loss process (Carman 2004b). Furthermore, DBD voltage rise after ignition was later claimed to maintain a lossy glow discharge (Paravia 2010).

Discharge homogeneity was related to voltage slope and consequent current density (Paravia 2006). The existence of a threshold current density $i_{plasma,tresh}$ as essential condition for homogeneous discharge was claimed by Paravia (2008b). However, inductive energy feed-in (current-source) after ignition is estimated to be less advantageous than capacitive energy provision (voltage-source). This is explained with a likely higher glow-phase current being pushed by the inductance. In contrast, voltage source feed-in as provided by a capacitance in parallel to the DBD, referred to as C_p, is suggested (Paravia 2010). Additionally to C_g, C_p feeds the plasma with a restricted amount of energy. In this case, less externally impressed (inductively fed-in) current after occurrence of ignition is necessary. From the point of view of the system efficiency, also the efficiency of the inverter plays an important role. Beside others, its efficiency is directly affected by the apparent power that has to be transferred to the DBD. Tailoring the pulse wave-shape by removing any voltage change that is not required for efficient discharges, the lamp's power factor can be significantly improved (Meisser 2010c; Meisser 2011a).

3.4.5 Conclusion

Combining the discussed aspects it is concluded:

1. The peak current during ignition and the distance between first and second (back) ignition plays a major role regarding the discharge behaviour. These parameters are directly affected by the pulse frequency.
2. Favourable high pulse frequency induces the need for fast polarity change. However, this causes high current to flow through the DBD which would inherently lead to increased glow losses.

3. As long as plasma conditions are established, current can flow. After ignition and efficient excimer generation, this current maintains the inefficient glow phase. Zero crossing and concomitant polarity change of the current is not extinguishing the plasma. Zero plasma current needs to be maintained over comparatively long time, favourably during the whole idle time.
4. Oscillations during the idle time should be prevented or at least minimised. With this, zero or at least low current subsequent to the second ignition is achieved. The glow discharge would be avoided.
5. The current at ignition time should be high which goes in hand with required high dv_{DBD}/dt slope.
6. A back-ignition should be initiated shortly after the first ignition. In order to achieve this, C_{DBD} needs to be discharged implying current flow. However, it is assumed to be more favourable to reduce the time between first and second ignition than to enlarge this time for the sake of reduced current flow. The back-ignition should also be supported by large short-term current (see point 5).
7. The current after the second ignition should instantaneously fall to zero or at least small values. This in turn implies that DBD voltage reaches zero volt level only slowly.

3.4.6 Introduction of an Optimum Wave-Shape

The proposed wave-shape builds upon a patent of Park (2008). In this comprehensive publication, a wave-shape as depicted in Figure 31A is suggested which comprises voltage rise and fall times having highest slope at their respective termination. The slope peaks fall together with the current as long as the wave-shape is generated by resonance. Distinct from continuous resonance operation, the resonant circuit is artificially divided at the respective moments t_{co}. This scheme is called current cut-off (Meisser 2010c) and will be presented in Chapter 5.2.3. In contrast to the even polarity of the pulses shown in Figure 31A, the suggested wave-shape depicted in Figure 31B combines the current cut-off feature with the properties of a bipolar pulse. That is the pulse consists of a first positive and a second negative voltage half-wave. According to the current cut-off scheme, the current waveform consists of two quarter-waves and a subsequent half-wave.

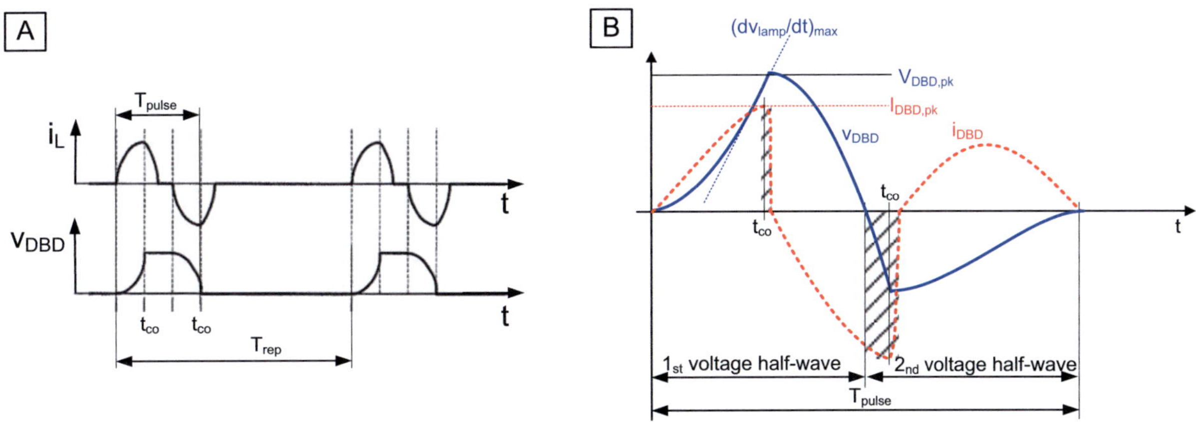

Figure 31: A: Wave-shape suggested in a patent of Park (2008); B: Novel bipolar pulse wave-shape derived from Xe excimer DBD requirements. The times t_{co} indicate the times when current is cut off.

Two ignitions are supported – one at the positive and one at the negative maximum voltage amplitude. The scheme permits maximum current at the moments of the respective ignitions. After the respective ignition, the absolute current value rises monotonously starting from zero amplitude.It should be noted that since the presented pulse wave-shape likely requires more switch transitions than a pure sinusoidal wave-shape, its application at high pulse frequency could contradict the main purpose to achieve a higher system efficiency. This becomes definitely relevant at pulse frequencies exceeding one MHz.

The wave-shape can be generated by tailoring the switching scheme of either the novel SPSP topology presented in Chapter 5.2.3 or the novel high-voltage full-bridge topology presented in Chapter 5.2.4. It is suggested to start the DBD lamp by bursts of such pulses in order to achieve lower initial ignition voltage. After initial ignition, it should be continued with pulsed excitation in order to ensure the intended high efficiency.

Albeit the theoretic advantages of this pulse wave-shape, first experiments (see page 140) did not demonstrate a higher DBD lamp efficiency. Even worse, due to the multiple current cut-offs required, the inverter efficiency is also comparatively low. Though, not surprisingly efficiency significantly rises with enhanced pulse frequency as experimentally investigated in Chapter 5.2.3 on page 138.

4 Resonant Circuit Fundamentals, Analysis and Components

To facilitate the understanding of the operation of the novel topologies presented in Chapters 5 - 7, the following chapters give insights into resonant circuit behaviour and the relevant limitation factors of resonant topologies. Here, the emphasis is initially put on series resonant circuits.

Starting with Chapter 4.2, crucial circuit components as well as critical properties of high-frequency inverter systems are discussed. In parallel to the topology investigations carried out for this dissertation, the development of reliable gate drives and efficient magnetics went on. A novelty in this field is the broad use of impedance spectroscopy in order to identify and quantify device properties vital for the system. Impedance spectroscopy provided insights into parasitic properties of the power stage (Chapter 4.7), power semiconductors (Chapter 4.10) and power magnetics (Chapter 4.13). SiC power semiconductor switches became broadly available in 2010 and since then have been implemented in the high-voltage topologies presented in the previous chapters. A characterisation rig, presented in Chapter 4.11 was designed and used to gain information regarding the real-world performance of the SiC devices. Differently to datasheet measurements, the devices were mounted on a thermoelectric cooler/ heater in order to continuously measure losses at different heat-sink temperatures.

4.1 Overview

As the DBD lamp is mainly a capacitive load, the generation of pulses requires both high currents to charge the lamp capacitance and also techniques to recover the energy stored in the lamp capacitance in order to terminate the pulse. Charging and discharging currents need to be limited in order to meet the maximum currents of the electrical and electronic circuit components. It can be shown that also from the efficiency point of view inductive current limitation is more preferable than resistive current (Chapter 4.3). This is mainly because inductive current limitation permits:

- low-loss current limitation
- resonant operation enabling ZVS
- low-loss recovery of energy stored in DBD capacitance
- filtering of high frequencies causing EMI.

Due to these benefits, this work solely deals with resonant topologies for pulse generation. Figure 32 shows a block diagram of the whole electronic system that drives the lamp. The system is divided into the three parts: front-end, high-frequency inverter and resonance circuit. The front-end rectifies and smooths the AC line voltage and ensures a high power factor by means of power factor correction. The high-frequency circuit is supplied by the front-end with DC-voltage and converts this into a multi-level output waveform. In many topologies, the power stage consists of two power semiconductor switches connected in half-bridge configuration, a DC-link capacitor network to buffer the voltage supply and driving circuits attached to the power semiconductor switches. Due to the switching action of the power semiconductors, the power stage generates highest frequencies which are filtered by the attached resonance circuit. The resonance circuit which is mainly made of reactive elements such as inductors (chokes) and capacitors is responsible for the pulse shaping and is able to store a significant amount of energy.

It is important to note that the pulse generation topologies presented in this work excite the respective resonance circuit dynamically leading to transient reactions. Therefore, the mathematic tools differ from continuous wave (CW) maths (Grebennikov 2007).

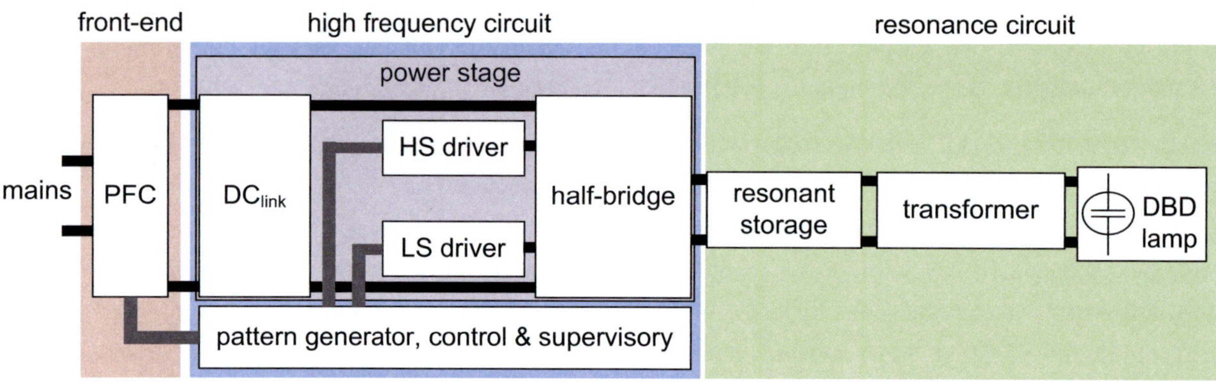

Figure 32: Building blocks of the electronic control gear with connected DBD to be driven. Here, the variant of a transformer-equipped resonant half-bridge topology as further discussed in Chapter 5.1 is depicted.

The typical basic electrical resonant circuit consists of at least two energy storage elements storing different energy types. These energy storing elements are capacitors that store energy in terms of an electrical field and inductors that store energy in terms of a magnetic field. Between these elements, a continuous energy exchange occurs as long as the circuit is electrically closed. In contrast to resistors which **dissipate** power, ideal capacitors and inductors are purely reactive and therefore solely **exchange** energy. Figure 33A shows a series LCR-resonant circuit that is excited by an AC sine voltage source. Inductor L and capacitor C also act as a second-order low-pass filter. Due to resistor $R_{L,s}$, the circuit is damped meaning that it has not only a reactive but also a dissipative nature. If the resonant circuit is excited by power semiconductors as shown in Figure 33B, a classical series resonant inverter (SRI) is formed. This converter is of class D, which means that the included power semiconductors are used as binary switches. Within this switching action lies the reason for the high efficiency of switching power converters. In contrast to class A, B or AB operation where transistors are used as adjustable resistors in the linear region of their transfer function, in resonant converters transistors act purely as switches, meaning they are either in fully on or in fully off-state which dramatically enhances the converter efficiency. However, real world semiconductor devices are not ideal. For operation at higher frequencies (> 50 kHz) and elevated voltages (> 100 V), MOSFET switches are preferably used.

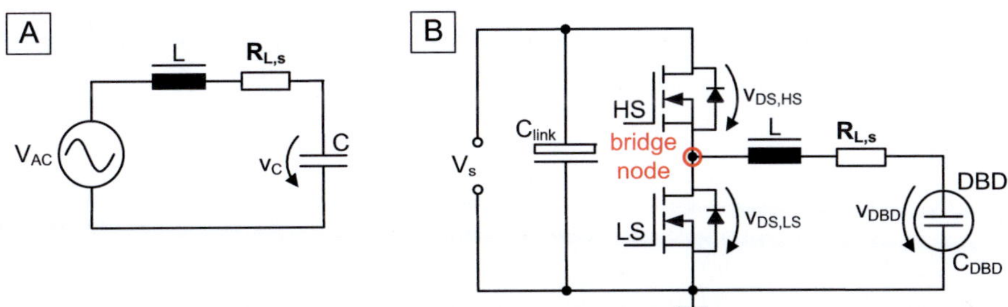

Figure 33: Series resonant circuits. A: Continuously excited RLC-circuit. B: Half-bridge excited series resonant circuit with DBD lamp both as capacitive reactive part and load.

MOSFETs consist of an intrinsic body diode which conducts if the MOSFET is reverse-biased. Favourably, MOSFETs are arranged in a half-bridge configuration as shown in Figure 33B, where the supply voltage V_S sets an upper limit to the individual drain-source voltage V_{DS}. Chapter 4.7 deals with the characterisation and implementation of power semiconductors.

Another major benefit of using resonant circuits is that the DBD lamp itself can already be used as capacitive storage element. In Figure 33B, the resonant circuit is connected to two semiconductor power switches that form a half-bridge which is the core element of the circuit. The upper switch that is connected to the positive supply rail is named high-side switch (HS), while the lower switch that is connected to ground is named low-side switch (LS). The drain of LS and the source of HS are connected to the same point as the resonant circuit. This point is called bridge node. The half-bridge acts as voltage source to drive the resonant circuit consisting of L, $R_{L,s}$ and the DBD capacitance. If switched on, the switches can conduct current in both directions, while switched off, current flow still occurs from source to drain through the intrinsic body diode. In case of the basic SRI, the switches are alternately turned on and off or are both in off-state. By this switching action, a square-wave voltage is generated and applied to the resonant circuit. In this circuit, energy can be transferred from the voltage source to the inductor (magnetic storage) and the capacitor, namely the DBD (capacitive storage). Beside these two locations of intended energy storage, additional storage elements are either placed intentionally in the circuit or have parasitic nature. An example for intermediate energy storage in an extra capacitance before transmission to the DBD is described in the patent of Park (2008).

As in the lamp capacitance, energy storage in the inductor is necessary during any resonant rising or falling edge. However, in some topologies, the inductor is used as storage element in idle time. This feature is named energy freezing.

From the viewpoint of the DBD lamp, the resonant circuit behaves like a current source in which L limits the current. In fact, nearly all resonant converters for DBD operation are of current source type in that respect. However, this fact is still discussed in literature (El-Deib 2010c; Diez 2012; Florez 2012). Depending on the ratio of control frequency f_c to resonance frequency of the circuit f_0, three different operation modes are possible. Regarding the switch transition from HS to LS, the three modes of operation are described as:

$f_c < f_0$ - capacitive operation:

The current is negative during switching action as the moment of switching lags behind the moment of zero crossing of the current. No soft switching can be achieved at turn-on. Therefore, high losses occur during switching in both switches. The energy stored in the LS output capacitance C_{OSS} is dissipated. The reverse recovery of the LS body diode can lead to excessive current spikes through HS. As a result, high losses can occur in both HS and LS. Even more unfavourable, it can lead to switch destruction as the parasitic BJT inherent in the MOSFET structure can show "second breakdown" failure. The switching speed is reduced due to the miller-effect occurring in both switches.

$f_c = f_0$ - in phase operation:

Current is close to zero during switching and leads to advantageous zero-current (ZCS) switch transition. Related losses are reduced to dissipation of energy stored in the LS output capacitance C_{OSS}. Diode reverse recovery does not affect switching. Despite of the miller-effect, the switching losses are low since current is close to zero.

$f_c > f_0$ - inductive operation:

Current is still positive during turn-off switching action leading to natural current commutation from HS to the LS body diode. Since drain-source voltage of LS $v_{DS,LS}$ is then close to zero, zero-voltage switching (ZVS) is achieved which drastically reduces losses. By the flowing current limited by the resonance inductance, output capacitance of LS is naturally discharged and output capacitance of HS is charged lossless. No reverse recovery of the body diodes occurs. LS is turned on without being slowed down by the miller-plateau since $v_{DS,LS}$ is close to zero during switching action. While the turn-off transition of the HS is caused by its gate drive signal, turn-on of LS is initiated by its forward conducting body diode.

This comparison makes clear that for obtaining high efficiency of switching converters, either in-phase or inductive switching modes should be met as often as possible during normal operation. This is especially true for the resonant topologies discussed in this work which are based on the temporal excitation of resonant circuits. In contrast to the aforementioned CW operation of the class-D converter, the time-dependent values of currents and voltages in temporarily excited resonant circuits are transient and can only be calculated in steady-state. Initiation and termination of the resonance actions are performed by the semiconductor switches. In this case, waveforms can solely be calculated by splitting the whole time-domain into time ranges in which the respective equivalent circuit remains the same. As the equivalent circuit changes in case of any switching action of diodes or transistors, this switching defines the boundary between the different time-domains for which calculation is possible. For further reading, it is directed to the extensive investigations of Kazimierczuk (2008).

4.2 Characteristic Frequency Domain Parameters

Resonant circuits can be described in the frequency and in the time-domain. Previously to the next chapter providing maths that characterise the voltage and current waveforms in the different time-domains, this chapter provides frequency-domain formulae that permit the determination of circuit restrictions and performance.

Coming back to Figure 33B, due to the dissipative nature of $R_{L,s}$, the natural angular resonance frequency ω_0

$$\omega_0 = 2\pi \cdot f_0 = \sqrt{\frac{1}{L \cdot C_{DBD}}} \tag{4.1}$$

is reduced to the angular frequency of the damped circuit ω_d:

$$\omega_d = \sqrt{\frac{1}{L \cdot C_{DBD}} - \frac{{R_{L,s}}^2}{4L^2}} \tag{4.2}$$

The eigenfrequency f_d is given by:

$$f_d = \frac{1}{2\pi}\sqrt{\frac{1}{C_{DBD}\cdot L} - \frac{R_{L,s}^{2}}{4L^2}} = \frac{1}{2\pi}\sqrt{\omega_0^{2} - \frac{\omega_0^{2}}{4Q_S^{2}}} = \frac{\omega_d}{2\cdot\pi} \tag{4.3}$$

The quality factor Q_S which is found in this definition of f_d also determines the ratio of reactive power Q_b to real power P at resonance:

$$Q_S = \frac{|Q_{b,\omega_o}|}{P_{\omega_o}} = \frac{\sqrt{S_{\omega_o}^{2} - P_{\omega_o}^{2}}}{P_{\omega_o}} \tag{4.4}$$

Q_S is related to the circuit's attenuation α as follows:

$$\alpha = \frac{R_{L,s}}{2L} = \frac{\omega_0}{2Q_S} \tag{4.5}$$

The eigenfrequency f_d is inversely proportional to the resonance period T_d:

$$f_d = \frac{\omega_d}{2\pi} = 1/T_d \tag{4.6}$$

The circuit's quality factor Q_S defines the ratio of the two frequencies to:

$$\frac{\omega_d}{\omega_0} = \sqrt{1 - \frac{1}{4Q_S^{2}}} \tag{4.7}$$

Furthermore, Q_S is a measure of the amplitude of the voltage overshoot $V_{C,pk}$ in case the resonance circuit shown in Figure 33A would be continuously excited with its resonance frequency given in (4.2). The overshoot factor in this continuous case is defined to:

$$\frac{V_{C,pk}}{V_s} = Q_S = \frac{Z_0}{R_{L,s}} \quad \text{with} \quad Z_0 = \sqrt{\frac{L}{C_{DBD}}} = \frac{V_{pk}}{I_{pk}} \tag{4.8}$$

Z equals the impedance of the circuit at the resonance frequency and also defines the ratio of the peak voltage V_{pk} (of L **or** C_{DBD}) to peak current I_{pk} (of L **or** C_{DBD}).

Q_S highly depends on the equivalent series resistance $R_{L,s}$ (copper loss, on-resistance of semiconductor switches) and also on the square-root of the ratio L/C_{DBD}, which is called the characteristic impedance Z_0 of the resonant circuit. Q_S needs to be greater than ½ in order to permit resonance. To meet this condition, the value of $R_{L,s}$ must be smaller than twice the value of Z_0:

$$R_{L,s} \overset{!}{<} R_{bound} = 2\sqrt{\frac{L}{C_{DBD}}} \tag{4.9}$$

Assuming Z_0 to be constant, the relative frequency shift ω_d/ω_0 is only depending on the resistor value as shown in Figure 34A. Thus, there is a considerable frequency shift if $R_{L,s}$ approaches half the value of R_{bound}.

In case of a critically damped circuit (Q = 0.5), f_d approaches zero Hz, whereas for Q > 35.5, f_d is closer than 99.99 % to f_0. The eigenfrequency is equal to the natural frequency if $Q \rightarrow \infty$. The frequency shift has an impact on the procedure for component identification as described in Chapter 4.8. The series resonance circuit of Figure 34B with a series load resistor can be changed to a configuration where the load resistor is placed in parallel to the resonance capacitor as depicted in Figure 34C. The quality factor of both circuits is given by:

$$Q = \frac{1}{R_{L,s}}\sqrt{\frac{L}{C}} = R_{L,p}\sqrt{\frac{C}{L}} \tag{4.10}$$

At resonance frequency, the relationship of series to parallel resistance is then:

$$R_{L,p} = \frac{1}{R_{L,s}}\frac{L}{C} = \frac{1}{R_{L,s}}Z_0^{\,2} \tag{4.11}$$

The time-dependent total energy of the undamped circuit is the sum of the energy partitions stored in the inductor and the capacitor:

$$E_{tot}(t) = \frac{1}{2}\left(C \cdot v_C(t)^2 + L \cdot i_L(t)^2\right) \tag{4.12}$$

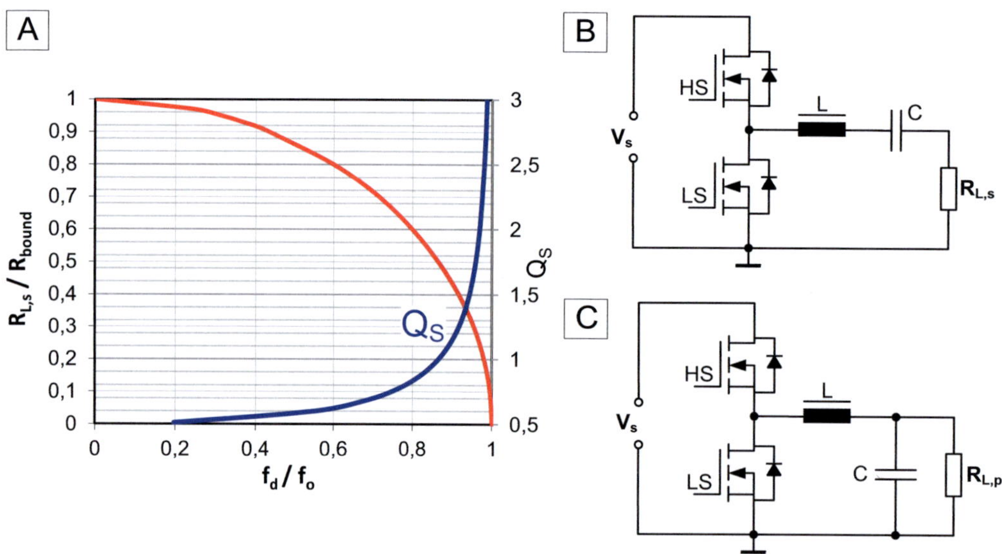

Figure 34: A: Dependence of resonant frequency shift due to change of dissipative element R. R_{bound} marks the boundary case to aperiodic behaviour. B: Basic half-bridge series resonant inverter with series connected load. C: Load connected in parallel to resonant capacitor.

4.3 Time-Domain Parameters

In contrast to continuously operating resonant converters, the topologies presented in this work are used to generate pulses – which means they operate **discontinuously**. Therefore, instead of steady state calculations, the transient behaviour and especially the step response of the circuits are of superior interest. The presented formulae are valid as long as the DBD (here represented as pure capacitor) is not ignited and thus consumes no energy. Furthermore, the term voltage step implies a change between two different levels of voltage within a comparatively short time, e.g. high voltage slope.

Figure 35 shows the relative voltage and current waveforms resulting from the step. The peak of the capacitor voltage slope occurs at the point in time of maximum current taken from Formula (4.19). Note that in the resonant topologies presented in Chapters 4.7 - 7, the equivalent circuit is changed by semiconductor switching action latest at the moment $t/T_d = ½$, meaning that only one half-period is shaped by this equivalent circuit. The transient answer of the circuit to a voltage step applied can be calculated using the Laplace transform. The voltage across the capacitor v_C is then given by:

$$v_C(t) = V_s\left[1-\left[\cos(\omega_d \cdot t)+\frac{\alpha}{\omega_d}\sin(\omega_d \cdot t)\right]e^{-\alpha t}\right] \tag{4.13}$$

The waveform is described as an oscillation with decaying amplitude and centre at step level V_S. The decay is described by the e-function. The sinusoidal term adds a time-dependent frequency shift. The current wave-shape is sinusoidal with zero-crossings at times 0, π/ω_0, $2\cdot\pi/\omega_0$ ... .

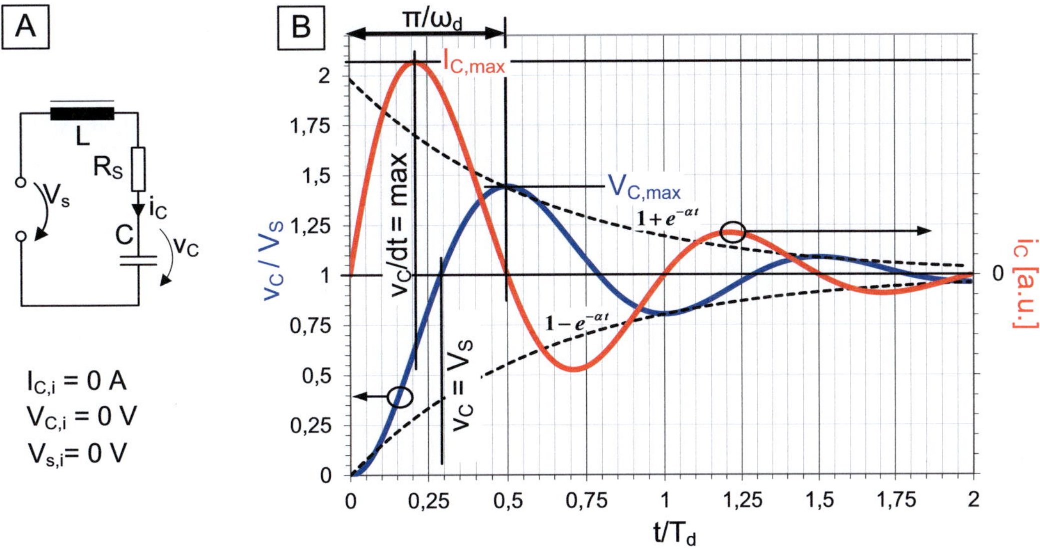

Figure 35: A: Simplified resonant circuit of Figure 34B and initial conditions. B: Time-dependent progress of DBD voltage and voltage slope which is proportional to DBD current. Values are standardised with regard to supply voltage V_S and time period T_S.

An interesting point is which overshoot factor could theoretically be achieved if the series resonant circuit was excited by a voltage step with height V_S. From Formula (4.13) the amplitude of v_C at time $t = \pi/\omega_d$ can be calculated yielding the first peak of the capacitor voltage:

$$V_{C,pk} = V_s\left(1+e^{-\alpha\frac{\pi}{\omega_d}}\right) \tag{4.14}$$

By assuming $\omega_0 \approx \omega_d$ (high Q_S) and inserting (4.5) in (4.14), we obtain the dependence of overshoot ratio osr from the circuit's quality factor:

$$osr = \frac{V_{C,pk}}{V_S} = \left[1+e^{-\frac{\pi}{2\cdot Q_S}}\right] \tag{4.15}$$

This relation is plotted in Figure 36. As can be seen, the formula shows restricted accuracy in the region close to $Q_S = ½$, which is the critically damped case where the overshoot factor is close to 1. The reason for that is the taken approximation. A value of $Q_S = 10$ is defined as minimum quality factor of the circuit in order to get a voltage gain of 1.85.

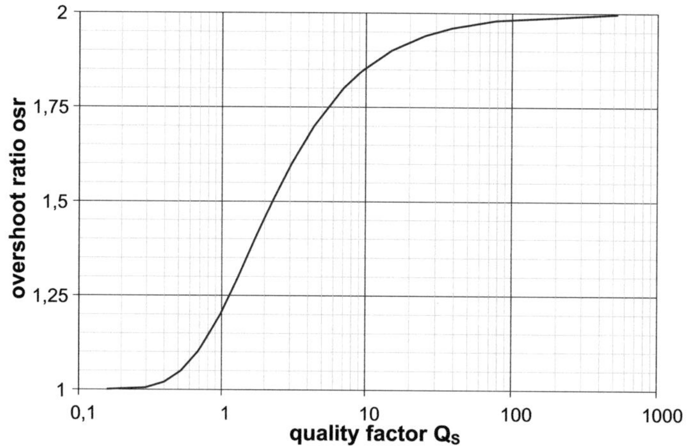

Figure 36: Overshoot-factor osr dependent on the quality factor of the series resonant circuit Q_S. A minimum Q_S value of 10 is suggested in order to achieve a gain of 1.85.

The slope of the capacitor voltage v_C follows a mere sinusoidal function which is quite similar to the current wave-shape defined by Formula (4.16):

$$v_C(t)' = \frac{V_s \cdot e^{-\alpha t}}{\omega_d}\left[\left(\alpha^2 + \omega_d^{\ 2}\right) \cdot \sin(\omega_d \cdot t) - \alpha \cdot \omega_d\right] \tag{4.16}$$

Referring to Figure 35B, the voltage v_C starts as a negative cosine while the current has an almost 90-degree phase shift:

$$i_C(t) = \frac{V_s \cdot \sin(\omega_d \cdot t) \cdot e^{-\alpha t}}{L \cdot \omega_d} \tag{4.17}$$

The current slope is determined by a more complex function:

$$i_C{}'(t) = -\frac{V_s}{L \cdot \omega_d}\left[\frac{\omega_d \cos(\omega_d t) + \alpha \sin(\omega_d t)}{\omega_d^{\ 2} + \alpha^2}\right] e^{-\alpha t} \tag{4.18}$$

The moment of the first current half-wave's maximum is found by setting (4.18) equal to zero:

$$i_C{}'(t) = 0 \quad \rightarrow \quad t_{i=max} = \frac{\arctan\left(\frac{\omega_d}{\alpha}\right)}{\omega_d} \tag{4.19}$$

The maximum current is then given by:

$$i_{C,max} = \frac{V_s \cdot \sin(\arctan\left(\frac{\omega_d}{\alpha}\right) \cdot e^{\frac{-\alpha}{\omega_d}\arctan\left(\frac{\omega_d}{\alpha}\right)}}{L \cdot \omega_d} \tag{4.20}$$

An important aspect is that for real world designs driving DBDs, resonant (LCR) capacitor charge is superior to resistive (RC) capacitor charge. Compared to RC networks, RLC networks may offer higher maximum slope and a better slope distribution. Resonant charge provides the maximum slope in the middle of the charge time which is more favourable than the initial maximum slope of RC charge (see Chapter 3.4). Schwarz-Kiene (2001) compared resistive charge with charge by a constant current source. Here, resistive charge is compared with resonant charge according to Figure 37A and D, respectively. Figure 37B and C compare the rise times of the different concepts for a real-world parameter set. A capacitor of C = 3 nF and a fixed resistor of 0.2 and 2 Ω is assumed for both cases. Figure 37C exemplarily compares the wave-shapes for Q = 1 and R = 0.2 Ω. It can be seen that for this low quality factor resonant charge is much faster than the resistive charge. Figure 37B shows the ratio of the circuits` rise times depending of the resonant circuit's quality factor. For the case R = 0.2 Ω the resonant circuit allows for shorter rise times as long as Q is below 1.6. The graph was plotted with simulated data.

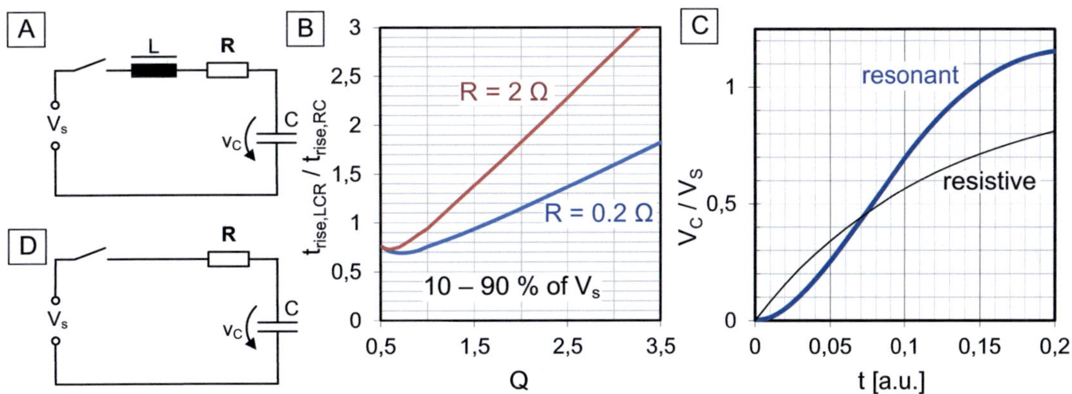

Figure 37: A: Circuit for resonant charge of a capacitor. B: Ratio $t_{rise,RCL}$ to $t_{rise,RC}$ in dependence on Q for fixed values R = 0.2 Ω and 2 Ω, C = 3 nF with rise times defined as time required to rise V_C from 10 % to 90% of V_S level. C: Exemplary wave-forms for resistive and resonant capacitor charge for Q = 1 and R = 0.2 Ω. D: Circuit for resistive charge of a capacitor.

For practical purposes, Q will be chosen to by far exceed these low margins in order to obtain a significant voltage overshoot according to Figure 36. Another aspect to compare is the efficiency of the charging process. Charging a capacitor through a resistor as shown in Figure 37D dissipates - independent from the actual time constant - losses equal to the energy stored in the capacitor:

$$P_{R,RC} = f_d \frac{C}{2} V_S^2 \tag{4.21}$$

Taking the conservative estimation of Formula (5.4) for the maximum current, its RMS value is:

$$i_{RMS} = V_S \sqrt{\frac{C}{2 \cdot L}} \tag{4.22}$$

Therewith the power dissipated in the series resistor R is determined. Formula (4.23) is modified by Expression (4.5) to:

$$P_{R,LCR} = R \frac{C}{2 \cdot L} V_S^2 = f_0 \frac{\pi \cdot C}{Q} V_S^2 \tag{4.23}$$

The ratio of Formula (4.21) and Formula (4.23) yields the simple expression:

$$\frac{P_{R,RC}}{P_{R,LCR}} = \frac{Q}{2\pi} \tag{4.24}$$

Consequently, the losses generated in R of the LCR resonant circuit are smaller compared to that generated in R of the RC circuit as long as $Q > 2\pi$. In practical inverter circuits presented in this thesis, Q exceeds 2π in any case.

4.4 Possible Rising and Falling Edge Waveform Types

To set up a resonant circuit, at least two storage elements of different energy types are necessary. Electric resonant circuits consist of at least one capacitor, which for DBD drive applications is favourably the DBD capacitance C_{DBD}, and one inductor L. The two elements can be connected to each other to form either a series or a parallel resonant circuit and may be dynamically connected and disconnected by semiconductor switches. Both elements essentially form the wave-shape that is applied to the DBD lamp. The basic voltage and current waveforms that can be obtained for charging and discharging the inductor are systematised in Figure 38. The DBD lamp capacitance requires a current source, namely the inductor, to be charged and discharged efficiently. Therefore, this action always bases on the resonance phenomena. In Figure 38A, the capacitor is charged resonantly by a full sinusoidal current half-wave while the voltage waveform appears to be a negative cosine.

The maximum voltage slope occurs at the time of maximum current in the centre of the time range. This waveform is typical for series-resonant pulse generators. In contrast, Figure 38C shows a pre-charging of L by a voltage source and subsequent charging of capacitor C with the energy stored in L. Parallel pulse generator topologies base on that principle. Of course, capacitor C can also be charged resonantly and then L can be discharged by connecting it to a voltage source as shown in Figure 38D.

In contrast to the capacitor, the inductor can be directly charged and discharged by a voltage source with comparatively high efficiency. This is depicted in Figure 38B. Between inductor charge and discharge the energy stored in L could be used to shape a pulse that may consist of elements of Figure 38B or C. A typical example is the resonant flyback which belongs to the parallel topologies.

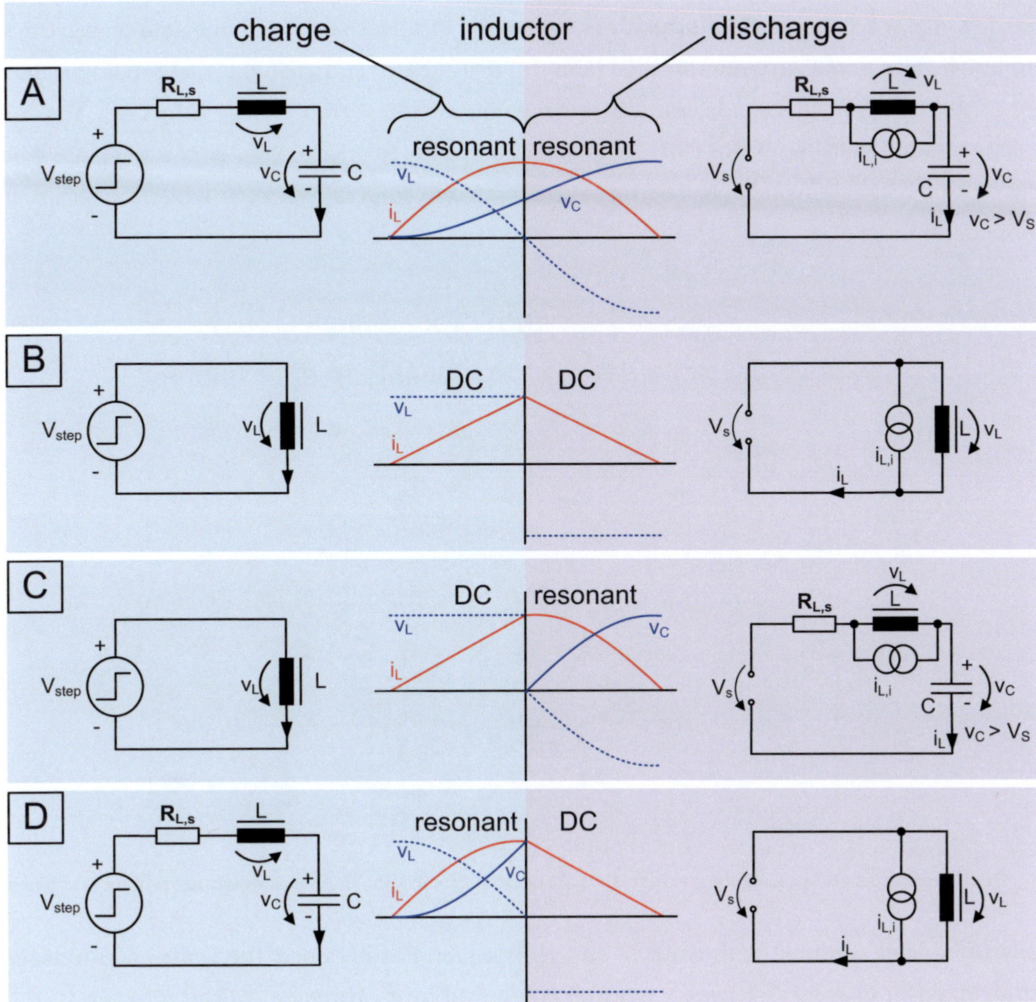

Figure 38: Overview of principal connection schemes of inductor L with corresponding wave-shapes for charging (left) and discharging (right). Power semiconductor switches may be used to combine the presented schemes in order to generate and shape a pulse. A: resonant inductor charge and discharge, B: DC inductor charge and discharge, C: DC charge and resonant discharge of L, D: resonant charge and DC discharge of L.

The above mentioned waveforms are of importance to both parallel and serial converters. Hence, they are analysed in more detail within the following chapters. It is important to note that the presented waveforms base on simplified circuits. Depending on the parasitic elements introduced by real-world components, the main wave-shape may be disturbed by high-frequency ringing or low-frequency bounce. The arising challenges were met by investigations presented in Chapters 5 - 7.

4.5 Comparison of Series and Parallel Inverter Fundamentals

In the context of inverters for the pulsed operation of DBDs, a novel classification scheme is presented in this work which divides related topologies into two major types: the series type converters and the parallel type converters. A third group combines the aspects of both. This hybrid is called mixed mode converter. Figure 39 gives an overview and compares the different converter types. It should be noted that all presented converters principally have current-source characteristic. The DBD, in no circumstance, is to be connected directly to a voltage source as represented by the supply rail or a DC-link capacitor.

The classification bases on the way the energy is fed into the DBD, which is the manner of the energy transfer. In series topologies, a resonant charge of L and C_{DBD} leads to a sinusoidal current waveform and a negative cosine v_{DBD} wave-shape. The inductor is connected in series to the voltage source, e.g. the switched supply and limits the current through the DBD lamp capacitance. In contrast, parallel topologies base on the pre-charge of the inductor and a subsequent freewheeling of the stored energy into C_{DBD}. The inductor is statically connected in parallel to the DBD lamp capacitance. In this case, v_{DBD} is of sinusoidal wave-shape and the current of cosine wave-shape. The initial DBD current $i_{DBD,i}$ is non-zero while the series converter starts a zero current.

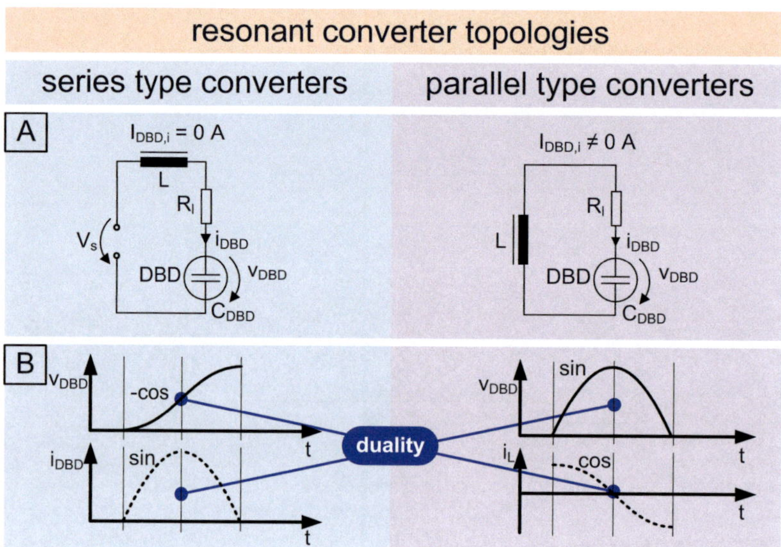

Figure 39: Duality of series and parallel type converters. A: equivalent circuits, B: characteristic dominant wave-shapes.

Mixed mode topologies combine both ways of energy transfer. For instance, the Universal Sinusoidal Pulse topology presented in Chapter 7.1 uses a parallel configuration to transfer energy into the DBD but then switches to a series configuration for energy recovery.

Classic topologies as well as novel approaches that had been developed within this work are assigned to these topology types.

A key question is how efficiently a pulse could theoretically be generated. In order to get a general answer, Figure 40 compares the three major possibilities to raise the capacitor voltage up to a certain peak value within a determined time. The options of Figure 40A and B, resonant charge and charge by a pre-charged inductor refer to the fundamental topology families presented in Figure 39. The option depicted in Figure 40C named current cut-off is related to Figure 40A but implies the difference that the current is cut off around its peak value where also voltage slope is maximum. This mode is the basis of the already suggested optimum pulse shape (Chapter 3.4.6) and had been experimentally investigated. The results are presented in Chapters 5.1.5 and 5.2.3. The RMS-value of inductor current i_L is assumed to be the main loss-related parameter. In all cases A-C, i_L is equally determined according to Formula (4.25) which is due to its sinusoidal wave-shape.

$$I_{L,rms} = I_{L,pk} / \sqrt{2} \qquad (4.25)$$

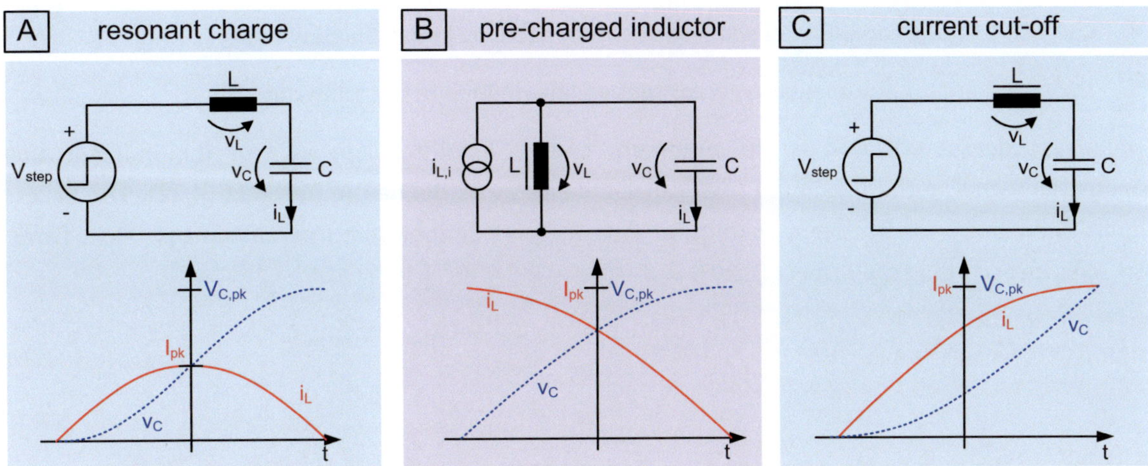

Figure 40: Comparison of different inductor current waveforms regarding RMS-current value. C, L and $V_{C,pk}$ are constants, time range is equal for all cases.

For reasons of simplification, the following calculations assume a non-damped LC-resonant circuit. For the pure resonant voltage slope of case A, the peak current is defined as

$$I_{L,pk}(t) = \frac{V_s\sqrt{C}}{\sqrt{L}} = \frac{V_s}{Z_0} = \frac{V_{C,pk}}{2 \cdot Z_0} \tag{4.26}$$

However, the calculations yield the peak current to be two times higher for cases B and C. In case B, the energy stored in inductor L needs to be equal to that stored in the capacitor at the end of the time range and therefore the double value is calculated:

$$I_{L,pk}(t) = \frac{V_{C,pk}}{Z_0} \tag{4.27}$$

In case C, V_S is equal to $V_{C,pk}$. Since $I_{L,pk}$ is also defined as ratio of V_S/Z_0, the result is identical to the result of case B (4.27). Additional losses occur during the pre-charging of the inductor in advance of the pulse in case B and the discharging of the inductor after pulse termination in case C.

In summary, it is found that the resonant principle of Figure 40A outperforms inductor pre-charge and current cut-off in terms of lowest conducting loss. This finding needs to be put into relation to the requirements of the DBD lamp as presented in Chapter 3.4.

4.6 Highest Possible Resonance Frequency

Taking the example of the series resonant pulse converter according to the SP-topology, which will be presented in Chapter 5.1, this chapter investigates its fundamental restrictions. These restrictions base on parasitic elements of the electric and electronic components involved and fundamental coherencies of the resonant technique. The highest frequency of generated pulses is limited by:

1. self-resonance of the DBD lamp (refer to Chapter 2.5.2),
2. parasitic stray inductance of an implemented transformer (refer to Chapter 4.13.2),
3. parasitic inductance of the power stage (refer to Chapter 4.13),

4. current-carrying capability of semiconductors, reactive elements and wiring,
5. the necessary minimum quality factor that will be defined in the following.

An important design criterion is the minimum energy handling capability of the inductor that is approximately equal to the maximum DBD capacitor energy as defined in Formula (4.28). The necessary minimum inductance for a certain **maximum current** and defined lamp parameters is taken from the energy balance according to Formula (4.29).

$$E_{L,pk} \approx \frac{1}{2} C_{DBD} \cdot V_{DBD,pk}^{2} \quad (4.28) \qquad L_{min} = \frac{C_{DBD} \cdot V_{DBD,pk}^{2}}{I_{DBD,pk}^{2}} \quad (4.29)$$

$I_{DBD,pk}$ is the absolute maximum negative value of the DBD current which flows through the DBD and L. It is referred to Figure 72 in Chapter 5.1.2 where the maximum negative current occurs in time range 2. $V_{DBD,pk}$ is the maximum voltage at the DBD that would occur without ignition. Formulae (4.28) and (4.29) are conservative estimations because damping by the occurring discharges within the DBD lamp and by other losses in the circuit is disregarded. Combining Formulae (4.29) and (4.1) the maximum resonance frequency of a non-damped circuit for a given DBD lamp and a maximum current is calculated:

$$f_{res,max,I} = \frac{I_{DBD,pk}}{2\pi \cdot C_{DBD} \cdot V_{DBD,pk}} \quad (4.30)$$

By rewriting Formula (4.30) the peak DBD current is then found:

$$I_{DBD,pk} = f_{res,max,I} \cdot 2\pi \cdot C_{DBD} \cdot V_{DBD,pk} \quad (4.31)$$

Combining Formulae (4.29) and (4.3) one obtains the maximum eigenfrequency of the damped resonant circuit which farther depends on its Q-factor :

$$f_{res,max,I} = \frac{I_{DBD,pk}}{2\pi \cdot C_{DBD} \cdot V_{DBD,pk}} \cdot \sqrt{1 - \frac{1}{4Q_S^{2}}} \quad (4.32)$$

The circuit shows resonant behaviour as long as Q_S exceeds the value of ½ which also marks the boundary case, where no overshoot exists. Inserting (4.8) in (4.32) yields a definition of the maximum resonance frequency that is independent from the peak current. In formula (4.33), Q_S defines an upper boundary for the resonance frequency as a minimum **quality factor** needs to be maintained in order to achieve the required voltage overshoot.

$$f_{res,max,Q} = \frac{1}{2 \cdot \pi \cdot Q_S \cdot R_S \cdot C_{DBD}} \quad (4.33)$$

The relation of Q_S and the achievable voltage overshoot was calculated in (4.15) and visualised in Figure 36. There it was found that a Q_S of 20 is advantageous. Formulae (4.32) and (4.33) imply that for a given lamp the maximum resonance frequency can be enhanced by increasing the current handling capability of involved power semiconductors, magnetics and wiring and by reducing (ohmic) losses in the same components. Equally important, the maximum operating frequency is restricted by frequency-dependent loss mechanisms such as skin effect and proximity effect that occur mainly in the lamp (Chapter 2.5.2) and the magnetics (Chapter 4.13) of the converter. In addition, bridge circuit configurations intrinsically set limits to the maximum operable frequency. Indeed, the parasitic effects of the high-side switches (HS) and especially their gate drives are the reason why bridge-less topologies such as class E are suggested for operating frequencies above 3 MHz (Sokal 2001). Unfortunately, class E inverters are not capable of pulse generation as the included DC-choke prevents any short time energy backflow into the bulk capacitor. Consequently, other topologies with solely one ground-referenced power semiconductor switch should be taken into consideration. For instance, the parallel topologies of Chapter 6 could be candidates for operation at multiples of 10 MHz.

4.7 High-Frequency Power Stage Design

The power stage is the heart of the pulse inverter. It is the section where energy is modulated with the system's highest frequencies and also the majority of power loss occurs. According to the block diagram depicted in Figure 32, the power semiconductor switches, their respective gate drives and the DC-link capacitor C_{link} belong to the power stage. For reasons of simplicity, this chapter will focus on power stages with half-bridge configuration. Other topologies may have different demands regarding the critical traces. This will be mentioned in the respective topology discussion (Chapters 5 - 7).

The critical traces of the half-bridge inverter are those which enclose the power commutation loop (Figure 41A) and the respective gate drive loops (Figure 41B). The node where the drain of LS is connected to the source of HS shows highest dv/dt within the circuit. In direction of the connected DBD, the output inductor L limits di/dt and with connected DBD also dv/dt.

The objective here is to minimise the marked areas of Figure 41A and B in order to both minimise parasitic inductance and EMR. As visible from Figure 41C, the parasitic bridge inductance (ESL) limits the usable frequency range towards higher frequencies while the value of the DC-link capacitor limits the frequency range towards lower frequencies. Figure 41D visualises the impact of the corner frequencies f_{low} and f_{high} regarding the wave-shapes that can be generated by the power stage. While f_{low} sets limits to the minimum repetition frequency, f_{high} restricts the maximum slope of the generated waveform.

The value of C_{link} is mainly restricted by space limitations. For instance, to achieve an impedance of 0.1 Ω, a capacitor value of 80 μF is sufficient for f_{low} = 20 kHz. This requirement is easy to meet since common repetition frequencies will exceed 20 kHz.

The situation is different regarding the upper frequency boundary. An impedance of 0.1 Ω is exceeded by a parasitic inductance of 100 nH for frequencies higher than 150 kHz (equivalent to a rise time of 3.13 μs). This example discloses that achieving minimum circuit inductance is mandatory.

The objective can be met by optimising the PCB layout as well as the package style of the power semiconductors. The challenge is to meet this demand with the requirement of the power semiconductors for good heat dissipation. For this thesis, several power stage set-ups were tested and regarding parasitic inductance, a remarkable advance was made.

Different novel topologies and operation modes for the intended high-frequency pulsed operation of DBDs were investigated. Topology development went hand in hand with the improvement of power stage concepts in order to minimise parasitic elements, especially the above mentioned parasitic inductance. As follows, different power stage designs will be described and qualitatively rated regarding their properties.

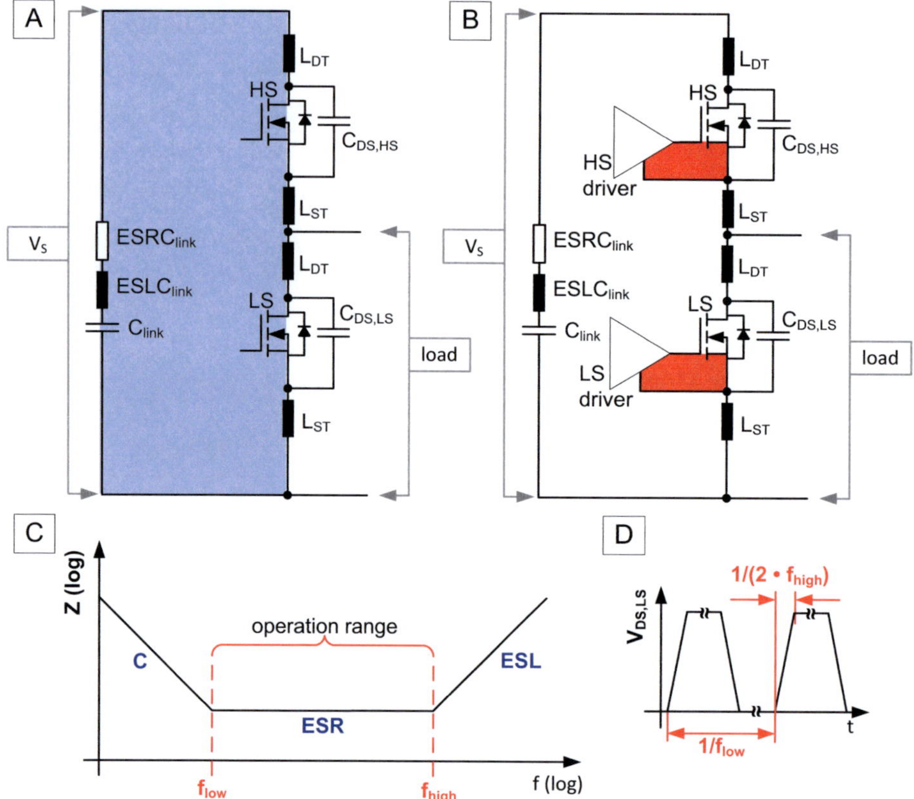

Figure 41: Half-bridge power stage – critical traces and component parameters. A: Power commutation loop (blue area). B: Gate drive commutation loops (red areas). C: Impedance spectra of a D_{Clink} capacitor. D: Time-dependent $v_{DS,LS}$ due to repetitive half-bridge operation. Marked are characteristic frequencies.

Figure 42B depicts a power stage set-up with very high configuration flexibility. The power semiconductor switches and diodes are attached to an active heat-sink. The V-shape attachment of the power semiconductors permits short traces between the devices. Holding clamps dedicated for TO247 and TO220 packages press the power semiconductors onto the heat-sink. The length of the heat-sink can be tailored to accommodate the required number of power devices. The power PCB accommodates only the power traces and can therefore be kept small.

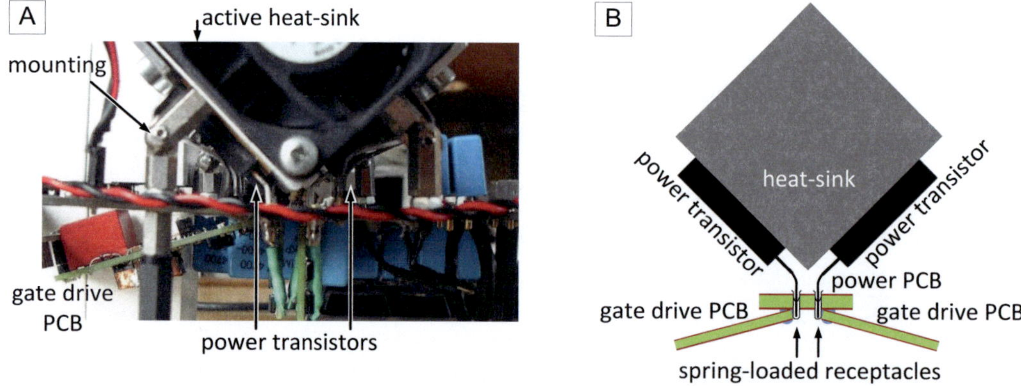

Figure 42: Flexible power stage set-up. Power semiconductors are mounted on two orthogonal sides of an extruded aluminium profile heat-sink which is actively cooled.

The gate drivers are directly attached to the spring-loaded receptacles which make an easily detachable connection to the power semiconductor leads. The chosen receptacles provide an almost zero inductance contact with sufficient current carrying capability and facilitate power semiconductor replacement. Typpical parasitic bridge inductance values are in the order of 100 nH. Figure 42A shows a photograph of the power stage from side view. On the left side a gate driver according to Figure 53 can be seen. In variations, this power stage set-up was used for all topologies presented in this work. However, in order to enhance the performance of topologies containing a half-bridge (e.g. SP- and SPSP-topology) further connection schemes were evaluated.

Figure 43B depicts a second approach which builds upon Si RF power transistors. These devices are claimed to be operable up to more than 30 MHz and are housed in HF-tailored flat packages with ribbon contact leads. Gate drivers are available in similar packages which allows for an in-line set-up. The cooling plates of the devices are pressed onto the joint heat-sink while the contact leads are directly soldered onto the PCB which carries the power traces as well as the gate drive circuitry. In Chapter 5.1.7, experimental results obtained with this set-up are presented. Surprisingly, the circuit tended to ringing and exhibited a low efficiency. Further investigations revealed that the transistor chips seem to have a very bad thermal contact to the cooling pad which is underlined by measurements presented in Figure 43C.

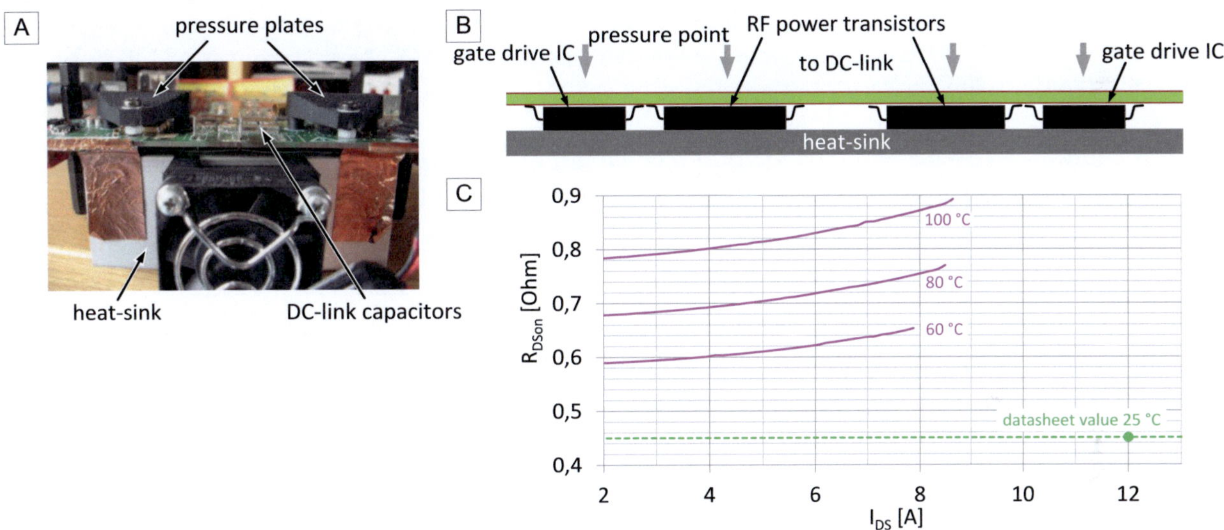

Figure 43: Power stage with two DE475-102N21A RF power MOSFETs and DEIC 420 driver. A: Power stage side view. Respective MOSFET and driver are pressed onto the heat-sink by insulating press-plates. B: Sketch of principle set-up from side-view. MOSFETs and drivers are directly soldered onto the PCB. Their exposed cooling pads are pressed onto the heat-sink. C: Static on-resistance measurement results obtained at different heat-sink temperatures showing the bad thermal path inside the MOSFET package. See for comparison the datasheet value obtained by pulsed measurement.

The on-resistance is highly influenced by the heat-sink temperature. It is likely that even within the short times of pulsed operation the chip exhibited excessive temperature which explains the visible rise of v_{DS} during on-time in Figure 85. Another major disadvantage is related to the availability of power semiconductors in RF packages. In fact, only a small number of Si types with up to 24 A rated current and 1 kV rated voltage is available. Due to these drawbacks, this concept has not been followed up.

A third alternative is depicted in Figure 44. Here, again TO247 devices were used for power semiconductor switches whereas diodes in SMD packages were implemented. Both device types were directly mounted on the power PCB – the anti-parallel diodes on top, the power switches on bottom. Each power semiconductor switch was attached with a discrete heat-sink which left space to connect gate drive PCBs as sketched in

Figure 44B. The power PCB further contains low-ESL ceramic capacitors and a film capacitor as DC-link capacitor. This design proved reliability and performance. The total parasitic bridge inductance was measured to be 18 nH. Flexibility is given as half-bridge PCBs can be combined to compose a full-bridge, for instance. The design is compatible to the standard TO247 package which enables for flexible installation of Si and SiC power switches.

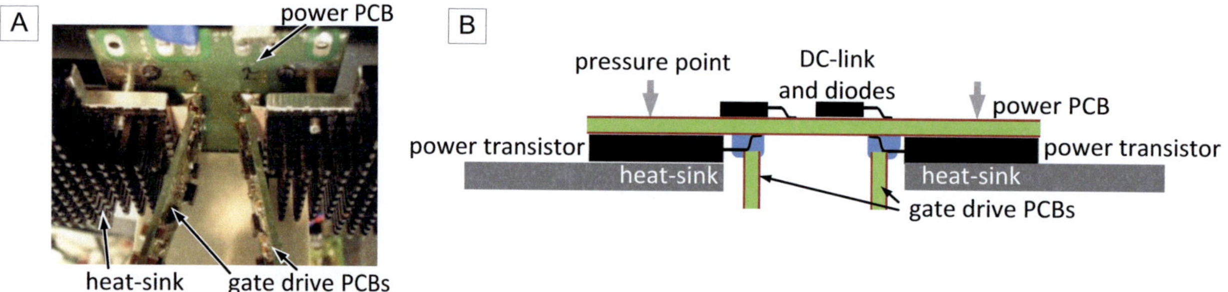

Figure 44: Power stage with power semiconductors directly soldered on the two-side assembled power PCB. Gate driver PCBs are mounted orthogonal to the power switches and the power PCB. This 3D-set-up ensures shortest trace lengths for all relevant signals. A: Real-world set-up with mounted passive heat-sinks. B: Sketch of principal set-up where TO247 transistor packages are mounted plane on the PCB. Typical inductances: 2 nH to ceramic capacitors, 20 nH to film capacitors.

The last innovation step was set towards DCB module technology. Since SiC power semiconductor switch dies became commercially available within this work, an attempt was made to implement them into an optimised Aluminium-Nitride DCB module design. The module design and manufacture was conducted at the PEATER lab of the University of Warwick during a three-month research stay. The result is visible in Figure 45. In Figure 45A, two different module designs are shown. On the left, a half-bridge consisting of two SiC power MOSFETs and respective anti-parallel diodes is shown. The design on the right side additionally comprises three ceramic DC-link capacitors which are connected to the bridge circuit by a low-inductive copper bridge. As Figure 45 suggests, the connection between DCB module and the PCB where the DC-link capacitors and gate drive circuits are mounted is made through a press-contact, only. The gold surface finish both of the DCB and the PCB traces ensures a reliable low-ohmic contact. In contrast to state-of-the-art modules which are equipped with connection pins, receptacles or even screw terminals adding inductance due to their considerable geometric dimensions, the presented modules prevented this additional inductance. As a result, the module of Figure 45A (left) exhibits a bridge inductance of 3.2 nH. This low inductance is also achieved by the extensive number of bond wires used which mimics low-inductive ribbon bond. The multiple bond wires also minimise bond wire resistance. The modules also show very good static performance being characterised by the set-up shown in Figure 45C. In order to measure the R_{DSon} of the switches, the modules were mounted onto the characterisation rig described in Chapter 4.11. The obtained results for the module of Figure 45A (left) are depicted in Figure 45D. Both HS and LS exhibit an R_{DSon} below 110 mΩ at 20 A and 100 °C heat-sink temperature. Due to limitations of time, the modules could not be implemented in pulse inverter designs, so far. However, future publications will report about this application and will provide further characterisation results.

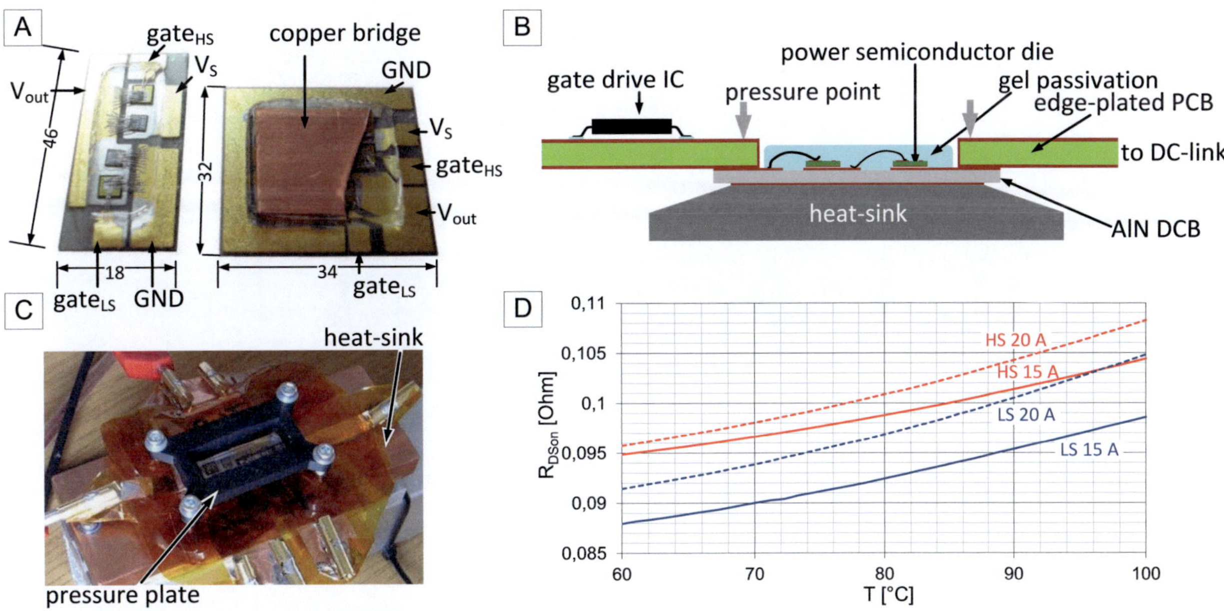

Figure 45: Power stages comprising of DCB half-bridge modules. The modules shown in A were designed and manufactured by the author at Warwick University using SiC MOSFET and diode chips from CREE Inc.. Right picture shows module design with included ceramic DC-link capacitors and copper bypass. The left module design achieves a bridge inductance of 3.2 nH. The respective MOSFET's source is connected by 10-14 500 µm bond wires. Top-side passivated by Elpeguard® DSL1705FLZ. Dimensions are given in mm. B: Sketch of principal connection of DCB modules with FR4 PCB. The edge-plated PCB is directly pressed onto the DCB. Contact is made without any connectors, just by the gold-finished contact surfaces. C: Module (left in A) in static-on characterisation rig. D: Static-on characterisation results. HS and LS have different thermal downlink. LS shows slightly better performance. At 20 A continuous current, the R_{DSon} stays below 110 mΩ at the elevated heat-sink temperature of 100 °C.

4.8 Characterisation by Means of Spectral Impedance Measurement

On the way towards high-frequency power electronics, the AC performance of the circuit, especially of the power stage, has to be known and improved. Support can come from impedance measurements taken from the real-world circuit. By that, parasitics can be identified and managed. Although recent research focuses on time-domain reflectometry (TDR) (Hashino 2010) for circuit characterisation, in this work an impedance analyser was used to measure the resonance spectra of the investigated circuits. In order to characterise the transient response and to improve the performance of the resonant pulse generator topologies presented in Chapters 4.7 - 7 as well as of the gate drivers described in Chapter 4.12, an impedance measurement environment being depicted in Figure 46A had been set up. The set-up was also used to measure the parasitic electrical DBD lamp properties as presented in Chapter 2.5.3.

The measurement set-up consists of an impedance analyser Agilent 4395A with attached impedance test kit 43961A. Before performing impedance measurements within a power electronic circuit, the impedance analyser must be calibrated up to the impedance test kit. It is also necessary to perform the compensation procedure with all cables, fixtures and probes connected. This ensures that the measurement setup is error corrected right up to the measuring plane. Example data of an impedance spectrum of a damped multi-resonant circuit is shown in Figure 46B. From the spectrum, inductances, capacitances and AC resistances can be derived. The example circuit consists of two resistive and three reactive circuit components yielding one series resonance and one parallel resonance. Inductive and capacitive reactances are represented as graphs with positive/ negative slope. Their point of intersection marks the characteristic impedance (Z_S, Z_P) and the natural resonance frequency of the respective resonant circuit.

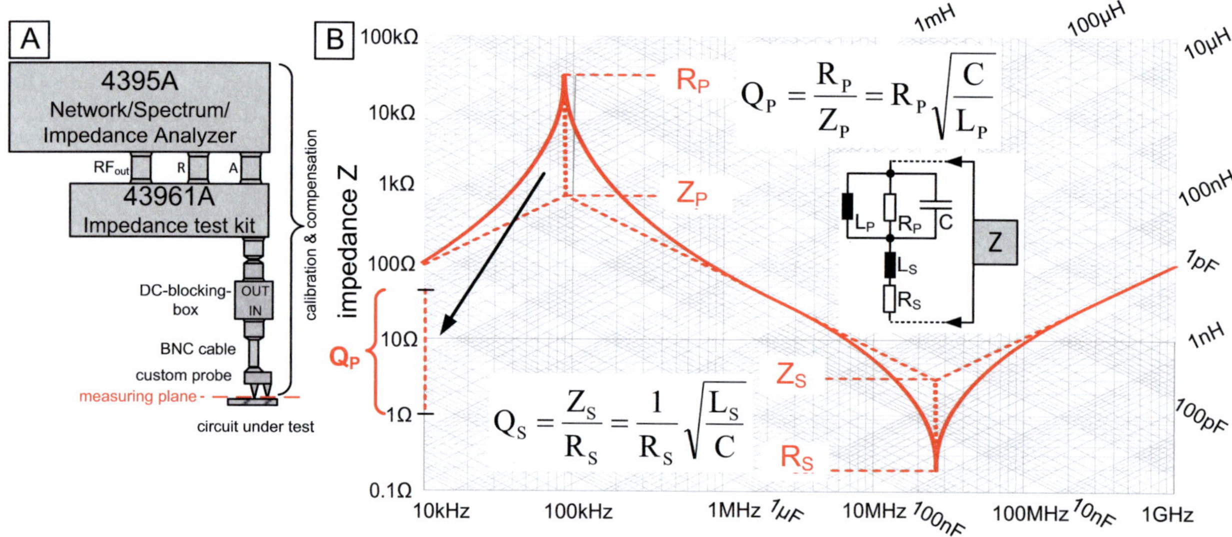

Figure 46: A: measurement setup including Agilent 4395A impedance analyser, impedance test kit, DC-blocker and cable. B: Example impedance spectrum of a multi-resonant circuit.

The minimum (and measured) impedance at resonance frequency of the damped series resonant circuit is determined by the AC resistance (R_S) of the circuit at this frequency. The AC resistance is higher than the DC resistance as the skin effect (see Formula (4.37)) reduces the effective cross-sectional area that conducts current. The quality factors Q_P and Q_S can be geometrically gained by measuring the difference of R and Z_0 at resonance and shifting the straight line to the one-Ohm point of the vertical axis. In order to ensure a low error, Q must be greater than 3.6. In that case, the discrepancy between the resonance frequency of the damped circuit and the natural resonance frequency is below 1 % (see Formula (4.7)). This is the case in Figure 46B. Hence, intersections can also be found by marking the maxima/ minima of the measured impedance. The use of small signal (low voltage) impedance measurements is generally restricted as:

- resistance of conductors changes with frequency due to skin and proximity effects
- reactance and inductance of inductors changes with the electric current they conduct
- reactance especially of ceramic capacitors drifts significantly with applied voltage

However, as long as magnetics are not operated close to saturation and capacitance shift due to operation voltage is known, impedance spectra give a helpful insight to AC circuit behaviour.

Beside of impedance measurements at zero DC offset voltage which are suiting to characterise magnetic components, especially two applications required a DC-offset. These are low voltage measurements of the gate drive outputs at high and low level as presented in Chapter 4.12.3 (Meisser 2012a) and characterisation of DC-link capacitor arrays (Meisser 2012b). Measurement results related to DC-link capacitor characterisation were presented in Chapter 4.8.

Impedance measurements under DC bias require protection of the impedance analyser from the high DC voltage that is applied to the circuit under test.

This protection is provided by a custom DC-blocker being placed between the impedance test kit and the custom probe that connects to the circuit under test. According to Figure 47B, the DC-blocker consists of fast RF diodes that clamp the voltage of the analyser port to 1 V and capacitors C_{block} to conduct the AC signal injected by the analyser. This means, the capacitors level-shift the input voltage and thereby remove the DC content. Broadly available are wideband bias-tees with maximum voltage ratings below 100 V. Many of them use MLCCs made of ferroelectric ceramics that show high temperature and voltage dependence. Unfortunately, the specific drift data is not provided in the datasheets. Following the work of (Tiggelman 2007), for the DC-blocker depicted in Figure 47A, low-drift NP0 ceramic high-voltage capacitors were implemented. Standard FR4 substrate was used. The bottom layer forms a continuous ground plane while the top layer also contains the signal path.

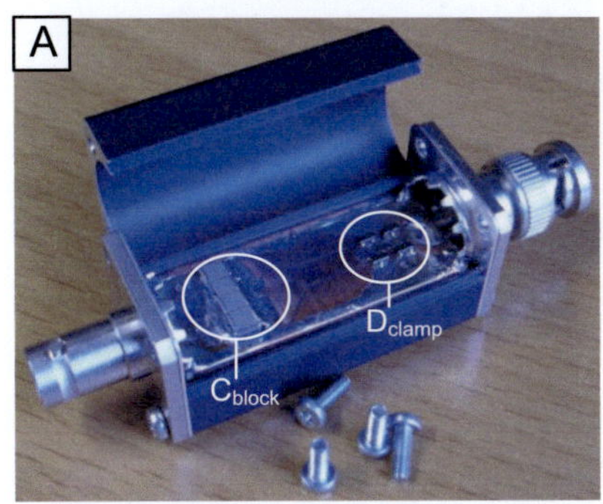

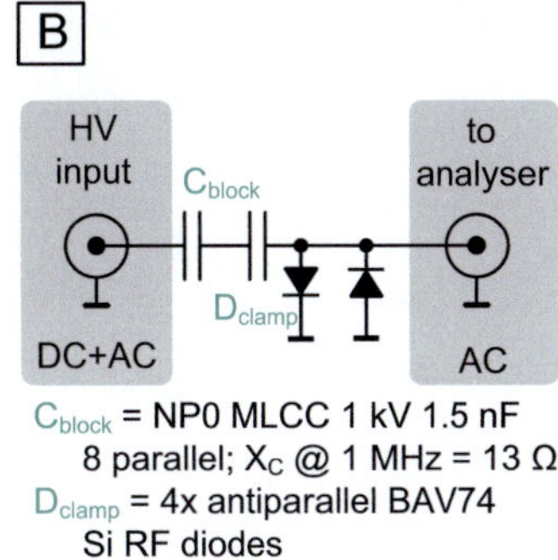

Figure 47: Constructed DC-blocking box. A: Picture of assembled HF-grade box. B: Schematic and components. Note that in order to ensure linearity over the whole input voltage range low-drift NP0 ceramic capacitors were implemented.

Top and bottom grounds are connected across the full length by copper foil and soldered to the BNC-connector shields. The peak output voltage of the analyser's generator must set to be low enough to prevent protection diodes to be operated in conducting mode. Source power of 1.5 dBm was found to meet that requirement. The presented DC-blocker can also be used to measure the voltage drift of the MOSFET output capacitance C_{oss}. A high value high-voltage resistor is used as DC feed-in. For the intended broadband usability and low DC current levels, the resistor suits more than an inductor as the latter could lead to resonance problems. The resistor restricts maximum current in case of unintended short circuit and also decouples the power supply from the circuit by its high impedance. The DC-blocker was used in particular for measurements presented in Chapters 4.12 and 5.1.7.

A comparison of open and short measurements at the probe tips with analyser source power P_S of 0 dBm and 15 dBm after calibration/ compensation is given in Figure 48. With low source power, the open impedance is above 100 kΩ up to 40 MHz but the short impedance is measured to be below 30 mΩ from 100 kHz to 5 MHz, only. This is due to the very low voltage that is measured across the short circuit. As source power is increased, the short is calibrated/ compensated towards less than 10 mΩ up to 5 MHz. In reference to these results, the output power level of the analyser is set to its maximum level of 15 dBm as this higher power maximises the sensitivity in the low-impedance region. Six decades of amplitude range are obtained from 100 kHz to 5 MHz and three decades are maintained up to the maximum frequency of 500 MHz. The estimated gate driver output impedances reach from around 100 mΩ to 10 Ω. Although these values lay below the high accuracy margin (3 %) of the analyser, calibration/ compensation is sufficient to cover this range.

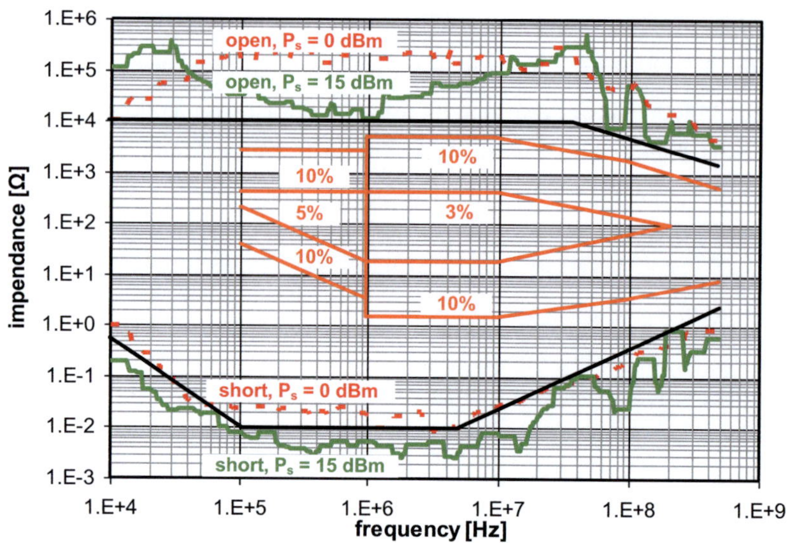

Figure 48: Calibration/ compensation results – shown are sliding MIN and sliding MAX for AC source powers Ps of 0 dBm (red dotted line) and 15 dBm (green solid line); orange: accuracy margins taken from 4395A datasheet.

4.9 Thermal Loss Measurements

In order to verify the electrical efficiencies obtained by electrical measurement and calculation, transient thermal measurements were performed. For this purpose, temperatures of the semiconductor's heat sink, core and winding temperature of L and TR were tracked with an IR-camera and platinum sensors. Knowing the absolute thermal capacity of the respective device and the transient temperature rise, the dissipated power can be calculated according to (4.34).

$$P_{loss} = \frac{\Delta T_{device}}{\Delta t} \cdot C_{th,device} \tag{4.34}$$

The thermal capacity of the heat sink and attached packages has been experimentally determined. For this purpose, the semiconductors have been exchanged by power resistors of the same package style which were then impinged by a certain electrical power. The temperature rise was tracked by platinum temperature sensors attached to the heat sink. In order to reduce the influence of convection, the starting temperature was set equal to the ambient temperature and temperature rise was restricted to less than 10 K. With the specific heat capacity thus obtained, the power loss of the semiconductors in normal operation was determined by performing one temperature measurement prior to and one measurement shortly after operation for a definite time. The measured heat capacitance of the heat sink and of the magnetics, as calculated from available material data, is listed in Table 1.

Table 1: Exemplary specific heat capacity of the measured circuit parts.

part	heat sink	L core	L copper	TR core	TR copper
	fischer LAM4150	RM14		ETD 49	
C_{th} [J/K]	500	74	5.1	127	18.2

As ferrites have rather poor heat conductivity, hot spots are likely to occur. Although the copper of the litz wire has highest conductivity, as windings were set up with insulated strands, turn-to-turn heat conductivity is also poor. Both restrict the accuracy of thermal measurements of the magnetics.

With the given power dissipation in the key elements of the circuit and the precisely measured input power, the circuit efficiency as calculated solely basing on electrical measurements can be validated with the data obtained from the thermal power loss measurements.

4.10 SiC Power Semiconductors for High-Frequency Pulse Operation

The application being the scope of this work puts highest demands on the used power semiconductors. The devices need to operate reliable with lowest conduction and switching loss. Due to the very low duty cycle of the pulsed power systems described in this work, and consequently high transferred peak energies, loss reduction is a major task.

Although SiC had been discovered as early as 1824 (Berzelius 1824), commercial power devices became available as late as in the first years of the 21st century (Kapels 2001). While SiC Schottky diodes had already been matured devices, between 2008 and 2012, different types of SiC power transistors entered the market and are in competition to meet customer demands. A more than tenfold growth in market revenue is expected until 2021 (Eden 2012). Until 2022, a rise by a factor of 18 is forecasted (Research 2013).

Figure 49 visualises the advance of commercial SiC devices compared to Si devices in terms of on-resistance (A) and figure of merit (B) plotted against the maximum operation voltage. While material restrictions of Si lead to an increase of on-resistance of one decade per 600 V operating voltage, their figure of merit (FOM) shrinks by one decade per almost 400 V. Balanced against their Si counterparts, 1200 V SiC power switches offer properties comparable with those of Si devices with a maximum voltage rating of 800 V and lower. The inherent advantages of SiC over Si will boost the SiC power semiconductor market by a compound annual grow rate of approximately 38 % (Coppa 2012) and enable the manufacturers to commercialise devices with even higher voltage ratings. Recent market dynamics as the shutdown of Semisouth Inc. (Hopf 2012) will merely lead to consolidations that leave other players as Cree, Rohm and Infineon strengthened.

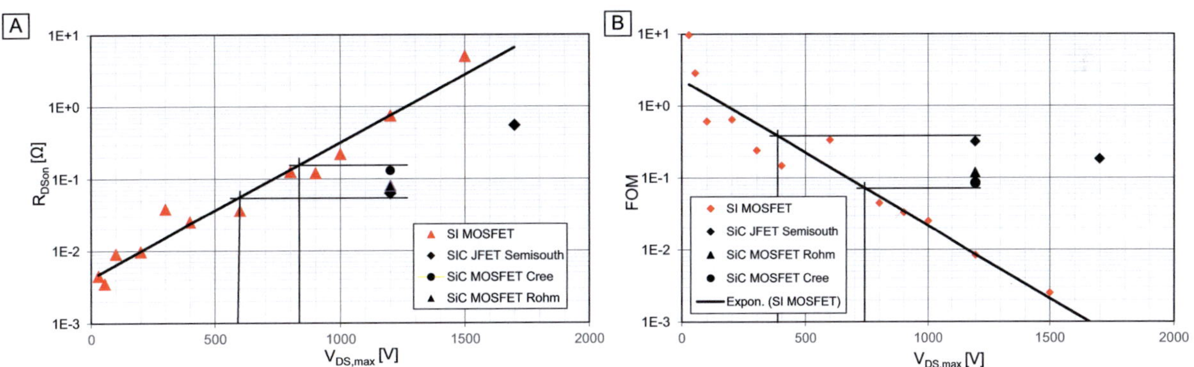

Figure 49: Comparison of 100 fast Si power MOSFETs and SiC power switches. Shown are parameters of best-in-class devices A: On-state resistance versus maximum drain-source voltage of best in class Si MOSFET and commercially available SiC devices. R_{DSon} of Si MOSFETs rises with one decade / 600V. R_{DSon} of 1200 V SiC devices is comparable to 800 V Si devices. B: Figure of merit (FOM = $1/(Q_g*R_{DSon})$) of selected best in class Si MOSFETs and commercially available SiC devices. 200 V and 600 V types comparatively show better performance due to their by far higher market volume and hence their forced development. 1.2 kV SiC device performance is comparable to that of Si MOSFETs with 800 V or lower voltage rating.

If comparing power semiconductor switches intended for high speed switching applications as in resonant converters, relevant key figures are required that characterise the performance of the devices for this application. Indeed, switching speed can either be limited by the input or output characteristic of the transistor or its robustness against faulty turn-on. A common bench mark is the FOM, which is represented here in reciprocal form. The reciprocal of the product of on-state resistance R_{DSon} and total gate charge Q_g indicates the device performance mainly from the view of the gate driver. Conversely, the ratio of continuous drain-source current I_{DS} to output capacitance $C_{DS} = C_{OSS}$ characterises the switching performance related to the output. High I_{DS} is required to charge C_{DS} of the transistor within a short time. As ratio I_{DS}/C_{DS} gets higher, the switch transitions between the high-side switch and the low-side switch can occur faster and possible operating frequencies are higher. It should be noted that C_{OSS} is highly non-linear since its value changes drastically with applied voltage V_{DS}.

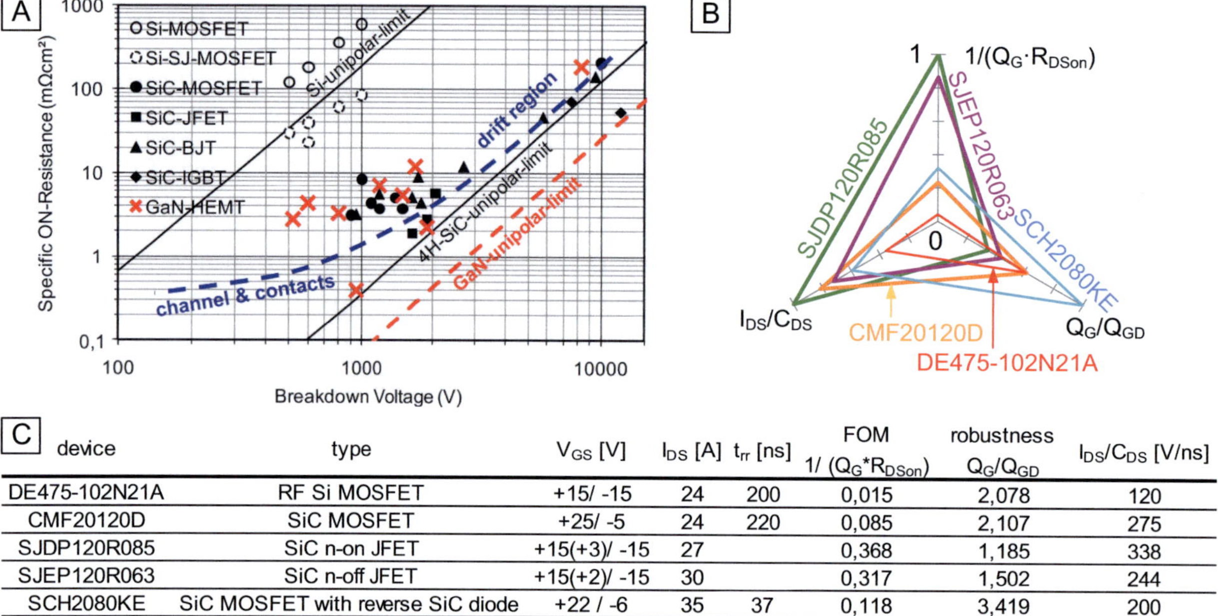

device	type	V_{GS} [V]	I_{DS} [A]	t_{rr} [ns]	FOM 1/ (Q_G*R_{DSon})	robustness Q_G/Q_{GD}	I_{DS}/C_{DS} [V/ns]
DE475-102N21A	RF Si MOSFET	+15/ -15	24	200	0,015	2,078	120
CMF20120D	SiC MOSFET	+25/ -5	24	220	0,085	2,107	275
SJDP120R085	SiC n-on JFET	+15(+3)/ -15	27		0,368	1,185	338
SJEP120R063	SiC n-off JFET	+15(+2)/ -15	30		0,317	1,502	244
SCH2080KE	SiC MOSFET with reverse SiC diode	+22 / -6	35	37	0,118	3,419	200

Figure 50: A: Specific on-resistance of different power semiconductor materials (Kaminski 2011). Theoretically, SiC devices should attain approximately same specific on-resistance at ten-fold voltage rating compared to their Si counterparts. B: Spider chart visualising normalised key figures of different representative power transistors. C: Comparison table with datasheet values and calculated key benchmark figures of these transistors.

The two former parameters indicate that low input and output capacitances are of benefit in any case. However, when it comes to fast switching action, for instance within a half-bridge configuration, the ratio of gate to miller-capacitance Q_G/Q_{GD} is of high importance as it is an indication for the robustness against faulty turn-on due to miller-capacitance feedback. This faulty turn-on can especially occur in bridge configurations where it is triggered by body reverse recovery surge and mostly ends up in destruction of the bridge.

In the table of Figure 50C, commercially available devices with comparable current rating are listed. Figure 50B visualises the comparison of these devices concerning the three mentioned key parameters. All values were extracted from the respective datasheet and may slightly differ in the conditions they were measured. Although the Si RF MOSFET has a package with lowest lead inductance, the chip itself shows lowest performance in terms of input and output switching speed. However, its robustness is high. The SiC MOSFET shows much better output and input switching performance than the Si RF MOSFET while maintaining high robustness. The normally-on JFET has best input and output switching performance but worst robustness. The normally-on characteristic complicates the power stage design as in case of power-up or fault, either off-state of the JFETs needs to be ensured or other switches need to be placed in series in order to prevent short circuit. Balanced against the normally-on device, the normally-off JFET has better robustness but lower input and output performance. To conclude, the SiC normally-off JFET is a good choice for high-frequency bridge applications as long as the gate drive maintains its maximum negative voltage during the transistor off-state in order to prevent a faulty turn-on of the device. The different complexity of the necessary gate drive circuits that have to meet the requirements concerning gate-source voltage levels of the respective devices is not included in this comparison. Before implementing promising devices, their static and dynamic electrical behaviour should be studied (Abuishmais 2011; Haehre 2012a). Above all, the devices should be characterised under application-specific conditions as heat-sink temperature, PCB layout and gate drive. The following chapter presents a test rig capable of that.

Further improvement of switching speed is possible by combining advancing materials (SiC, GaN) with suitable HF-grade packaging technology. Today's SiC power devices could reach highest switching speeds, but still suffer from implementation in standard power semiconductor packages as the TO247. A possible alternative pathway is presented in Chapter 4.7 by means of low-inductive all-SiC modules. It is also referred to publications dealing with the potentials and realms of SiC power semiconductor devices (Stevanovic 2010; Agarwal 2011; Biela 2011; Haehre 2012a; Springett 2012).

4.11 Power Semiconductor Characterisation Rig

The use of novel power semiconductors bears the risk of unknown device behaviour and insufficient data provided in datasheets. Moreover, datasheets do not provide real-world data regarding the ohmic losses of the devices at typical heat-sink temperatures. For these reasons, a characterisation rig was set up which allows for static measurements superior to standard pulsed characterisation. This rig was used, among others, for earlier publications (Haehre 2012a; Haehre 2012b). A block diagram is depicted in Figure 51 and the laboratory set-up is pictured in Figure 52.

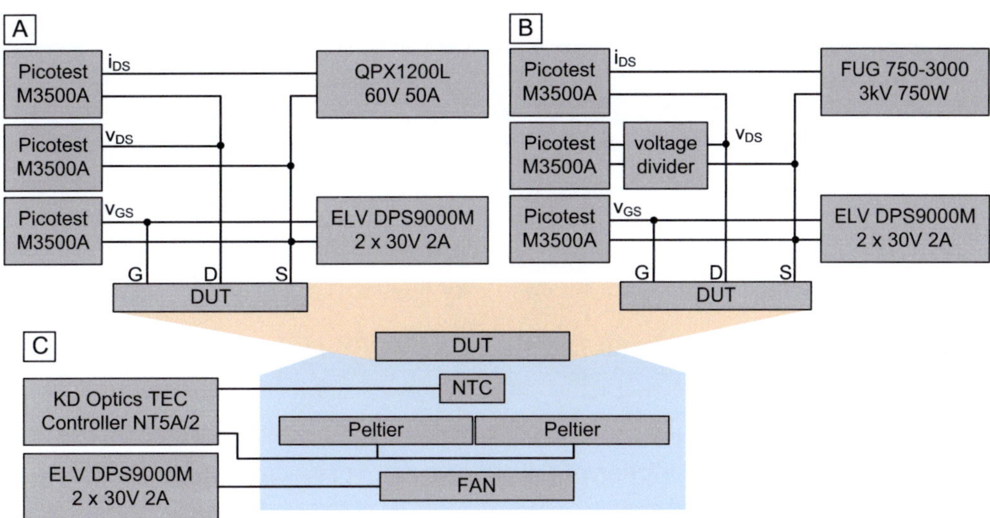

Figure 51: Block diagrams of the static measurement set-up. A: Electrical set-up for static-on characterisation. The DUT is impinged with continuous currents up to 50 A. B: Electrical set-up for static-off characterisation. Instruments, cables and connectors to the DUT are rated for a maximum DC voltage of 2 kV. A customised and calibrated voltage divider is placed between DUT and precision voltage meter. C: Set-up for thermal control. The DUT is heated or cooled to the set heat-sink temperature.

While the thermal management block (Figure 51C) stays the same for both static-on and static-off measurements, the connector PCB to the DUT needs to be exchanged since it is either designed for high current or high voltage.

In contrast to the common pulse measurement technique, the devices are continuously operated during the static-on characterisation according to Figure 51A and C. By doing that, also properties of the internal thermal path are included in the measurement. While the temperature of the device's thermal pad is kept constant by a thermo electric controller, the die temperature rises in dependence of the dissipated power and the thermal path between die and heat-sink.

In other words, the negative effect of the finite thermal conductivity of the die attach, the base plate, etc., influences the measurement in the same way as it does in the real-world application. Consequently, this measurement technique provides more accurate data regarding generated losses as possible by pulsed measurements which do not heat up the die.

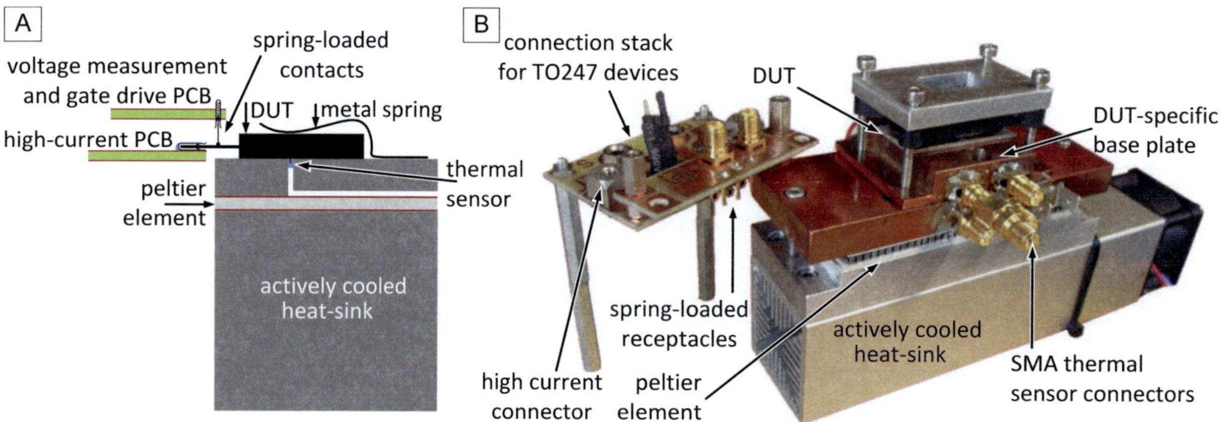

Figure 52: Power semiconductor characterisation rig for static and dynamic measurements. A: Principal set-up of the characterisation rig. The DUT is mounted on a thermally controlled baseplate. The fan and the peltier element are controlled in order to stabilise the set temperature measured at the thermal pad of the DUT. The electrical connection in case of static-on measurements consists of the depicted two PCBs, each responsible for either the high current path (drain-source) or the voltage measurements. The ohmic influence of the high-current spring-loaded contacts is avoided by additional spring-loaded contacts measuring voltage, only (Kelvin connection). In case of dynamic measurements, the PCBs also carry the gate drive and the double-pulse power circuit. B: Picture of the characterisation rig. Left: Connection PCBs for TO247 interconnection. Right: Characterisation rig configured for characterisation of DCB-based modules according to Figure 45A.

Static-off characteristics are measured with the set-up according to Figure 51B and C. Here, a voltage source and connection scheme capable of safe operation up to 2 kV is used.

Both measurements are automatically controlled via a LabVIEW™ environment which includes an over-temperature shut-down. The dynamic characterisation of power semiconductor switches was performed with the set-up of Figure 51C for thermal control and a double-pulse circuit described by Haehre (2012a). It is referred to this publication for a description of the measurement procedure and experimental results.

4.12 Power Switch Gate Drives

Different kinds of power semiconductor switches were implemented in the topologies presented in this work. Obviously, each of the respective power switches needs to be operated with the specified voltage levels and peak currents and therefore requires its very own gate drive circuit. The gate drives used in this work will be described as follows.

4.12.1 Gate Drive Supply

Although bootstrap supply using fast SiC diodes is suggested up to 600 V DC-link level and switching rates of up to 2 MHz (Strydom 2004), in this work, a transformer insulated gate drive supply had been chosen as standard for all drivers used. This is to reduce capacitive coupling and providing a continuous power transfer to the drivers which permits flexible driving signals to be transmitted. The gate drive supply consists of a laboratory DC-source, a 100 kHz full-bridge chopper and an insulation transformer implemented on each gate drive PCB plus a connected fast rectifier. The transformers have a spilt secondary permitting a symmetric supply of the gate drive IC. The transformer's centre tap is connected to gate drive circuit ground and in most designs also directly to the source of the power semiconductor switch. In order to reduce the commutation loop area between gate drive IC and the power semiconductor switch, ceramic bypass capacitors between V+ and ground as well as V- and ground are placed closely to the source connection.

Commercially available gate supply transformers are rated up to 5 KV (at 50 Hz) insulation voltage. As of this writing, no comprehensive de-rating guidelines for gate drive-transformers operated in fast switching power electronics systems are available. Voltage ratings are mostly solely provided for 50 Hz primary to secondary swing. Regarding de-rating from 50 Hz to HF no better than vague hints such as 43 % at 50 kHz for polypropylene (Khachen 1990) and generally 25 % (Bilodeau 1989) can be found.

If high voltage levels are to be insulated or if insulation reliability and safety is of importance, multiple transformers should be stacked. For instance, double galvanic insulation transformers (DGITs) (Brehaut 2006) can be used to insulate 10 kV rated bridge gate driver controlling SiC MOSFETs as presented by Galai (2012).

4.12.2 Signal Transmission

The gate drive signals were generated by Quantum Composer pulse generators and transformed to an optical signal by a proprietary TTL-to-FOC transducer. One millimetre polymer optical fibre (POF) connected the transducer with the individual gate drive board. On the board, the received signal is conditioned by a fast comparator with Schmitt-trigger input. This is especially necessary if fast FOC receivers are used which do not incorporate a TTL-compatible output but have a low-amplitude output signal superimposed by an offset. Digital couplers having a common mode immunity of 50 kV/µs were implemented in the gate drivers of the matrix switch presented in Chapter 6.3.1.

4.12.3 Gate Drive Topologies and Performance Evaluation

In contrast to standard Si power switches, SiC switches require unique driving voltage levels. Compared to Si MOSFETs, the margin between the threshold voltage and the maximum or minimum voltage is by factors of three to eight lower. This bares the risk of unattended switching or operation in the linear region. In order to avoid these malfunctions the following gate drive circuits had been developed. All gate drive circuits were characterised with the impedance characterisation set-up given in Figure 46A of Chapter 4.8. The impedance measurements were conducted at the gate and source terminals of the respective gate drive circuit. Using the DC-blocker presented in Chapter 4.8, the drivers could be characterised for high and low output voltage level providing information about the parasitic inductance of the respective commutation mesh. The power transistors were not connected as their total input capacitance C_{oss} changes dynamically as either gate drive voltage V_{GS} or drain-source voltage V_{DS} changes. The measurement results highlight the operation restrictions that come from components and layout. Drivers only work effectively within a defined frequency range that is marked by low impedance thanks to sufficient bypassing, ground plane and return path design. The lower border of this range is defined by the value of the power supply bypass capacitors (C_{DC}). The upper border is defined by the parasitic inductance of the driver output stage, gate resistor and bypass network.

SiC MOSFETs were favourably used in the HV-SPSP-topology (see Chapter 5.2.4) as the reverse recovery time of their body diodes is not relevant for energy recovery since they can be bypassed by a fast SiC Schottky diode. The SiC MOSFET must be driven by a highly asymmetric gate voltage. The maximum positive gate drive amplitude of the SiC MOSFETs should attain 20 V in order to turn on the device safely while the negative voltage is restricted to 5 V. Figure 53A shows a SiC MOSFET gate drive circuit where the symmetric power supply voltage is level shifted by series capacitor C_{shift} to meet the voltage level requirements of the MOSFET. The Zener diodes ZD1 - 3 are required to restrict the gate voltage accordingly. The Zener diode ZD1 ensures that the -3 V offset during off-state is fixed no matter how long off-phase takes (ensures DC-off-state). Although the MOSFET supports even faster switching, the chosen gate resistor placed in the return path R_B restricts the switching time to about 20 ns. The SIC MOSFET driver is equipped with large value C_{DC} capacitors and therefore, according to the results provided in Figure 53B, only parasitic inductances restrict driver performance in the measured frequency range. As the driver-IC is mounted in a DIP-package, inductances are two- to threefold higher than those of the following driving circuits that use driver-ICs in SOIC or RF packages. The two parallel resonances at 100 MHz and 300 MHz occur between the parasitic inductance (mainly of the driver-IC package) and parallel capacitances as those of the Zener diodes. Experiments showed that the gate oxide of the used MOSFETs does not tolerate any voltage spike that may exceed the negative voltage boundary of -5 V. Therefore, in later experiments a Zener diode combination of +22 V and -1V was used.

The driver layout displayed in Figure 53C and D had been enhanced towards the driver depicted in Figure 54A mainly by the exchange of the DIP package with a DFN package and the use of a four-layer PCB. Comparison of Figure 53B and Figure 54B shows that the inductance could almost be halved.

The drive of SiC JFETs requires a different attempt. As the JFETs contain a forward biased diode between gate and source, during turn-on and the on-time, the driver has to act as current source. The adaptive driver circuit as shown in Figure 55A contains two individual gate drivers and is designed for two-current-level drive of two parallel connected SiC normally-off JFETs. The drive signal (IN, ENB) is received by the POF-link and transmitted to both driver stages.

An on-board logic generates a 100 – 200 ns delay of the enable signal (ENA) in order to switch between the high current high state at switching action and low current high state in static on-state. The different current levels for fast switching and low-R_{DSon} in on-state are provided by a two-output driver and corresponding gate drive resistors R_{HC} and R_{LC}. In low-state and during switching, the driver provides high current through R_{HC}, while in on-state, the current is restricted by R_{LC}. The provided low impedance during the switch-off transition is especially crucial in order to restrict turn-off losses of the low-side switch in the half-bridge configuration and the ERS in the full-bridge configuration.

The source return path is also equipped with one resistor per power transistor in order to prevent drain-source current to flow through the driver circuit in case of two paralleled power devices. The driver ground is generated by the supply transformer centre tap. The loop impedance is reduced by paralleled ceramic capacitors that form a short HF current path from the power transistor source to the driver supply rails.

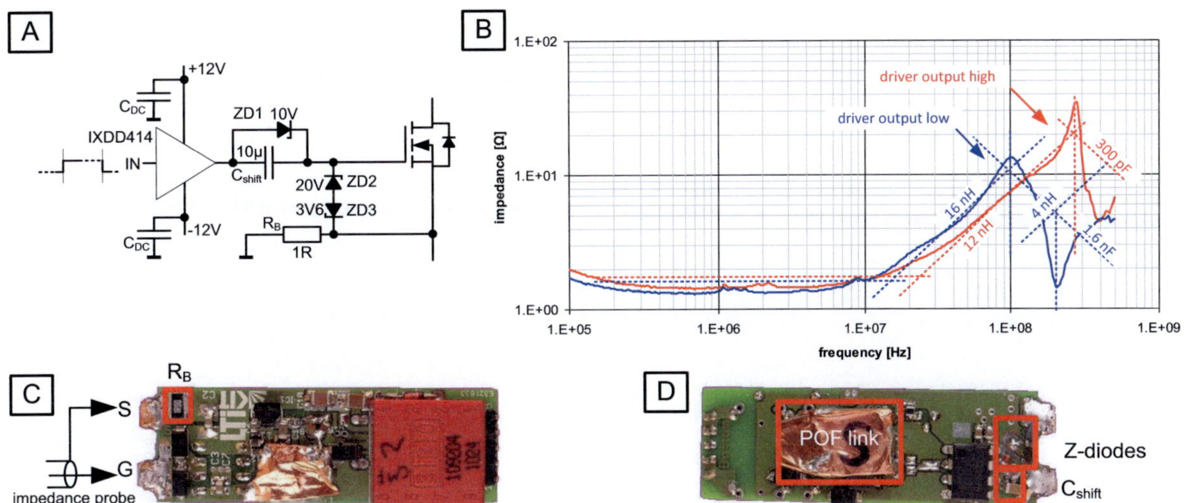

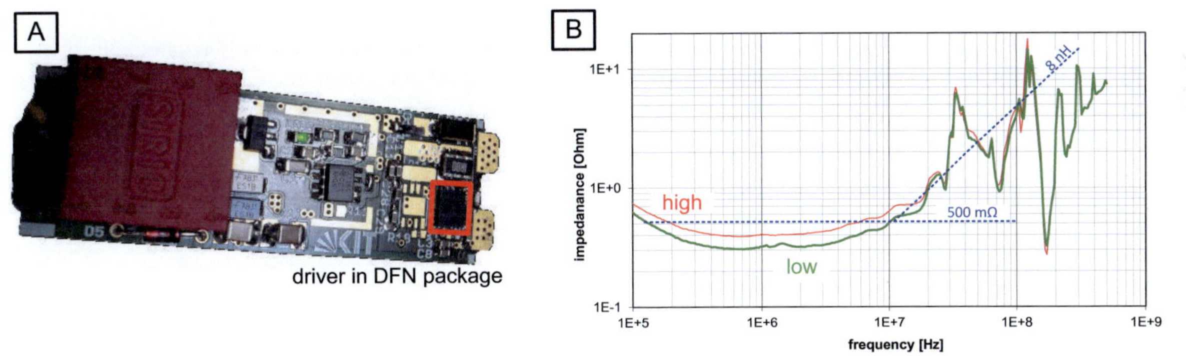

Figure 53: SiC MOSFET gate drive circuit with level-shift capacitor. A: Schematic. B: Impedance measurement results for high and low state; inductive impedance rise 10 – 100 (200) MHz, multiple resonances above 80 (200) MHz. C & D: Top (C) and bottom (D) photographs of the PCB and indicated probe tip connection.

Figure 54: Gate driver with topology similar to that of Figure 53. A: Four-layer PCB with attached driver-IC in DFN package. B: Impedance measurement results for high and low output state. Compared to Figure 53B, parasitic inductance is halved.

As Figure 55B indicates, the negative supply buffer capacitor is of too low value as impedance rises below 700 kHz. The parasitic inductance, which has a value of 8 nH in low and 5.3 nH in high output state, is significantly lower compared to the SiC MOSFET driving circuit. The gate drive circuit depicted in Figure 56 is part of a RF half-bridge (Meisser 2012b). Both driver and RF MOSFET are implemented in RF packages that add low inductance to the circuitry.

Additionally, no gate resistor is implemented in order to guarantee best performance. This results in the lowest measured inductance of 4.8 nH in on-state and 3.2 nH in off-state. The resistance of 80 mΩ is below the datasheet value of 400 mΩ (@ V_{CC} = 15 V) which may be due to the operation at V_{CC} = 24 V. With rising frequency, the impedance rises slowly, partly supported by skin effect influence, until it approaches the level of inductive impedance slope. Above 20 MHz, several resonance points can be identified in the impedance spectrum. Despite of the resonances, the impedance stays below 5 Ω at any frequency. The reason for the multiple resonance points is with high probability the low inductance gate driver IC and the missing gate resistor. Both ensure the best performance below 20 MHz but on the other hand support the occurrence of high-frequency resonances. The measured values above approximately 50 MHz are too close to the calibration/ compensation accuracy boundary of the measurement setup.

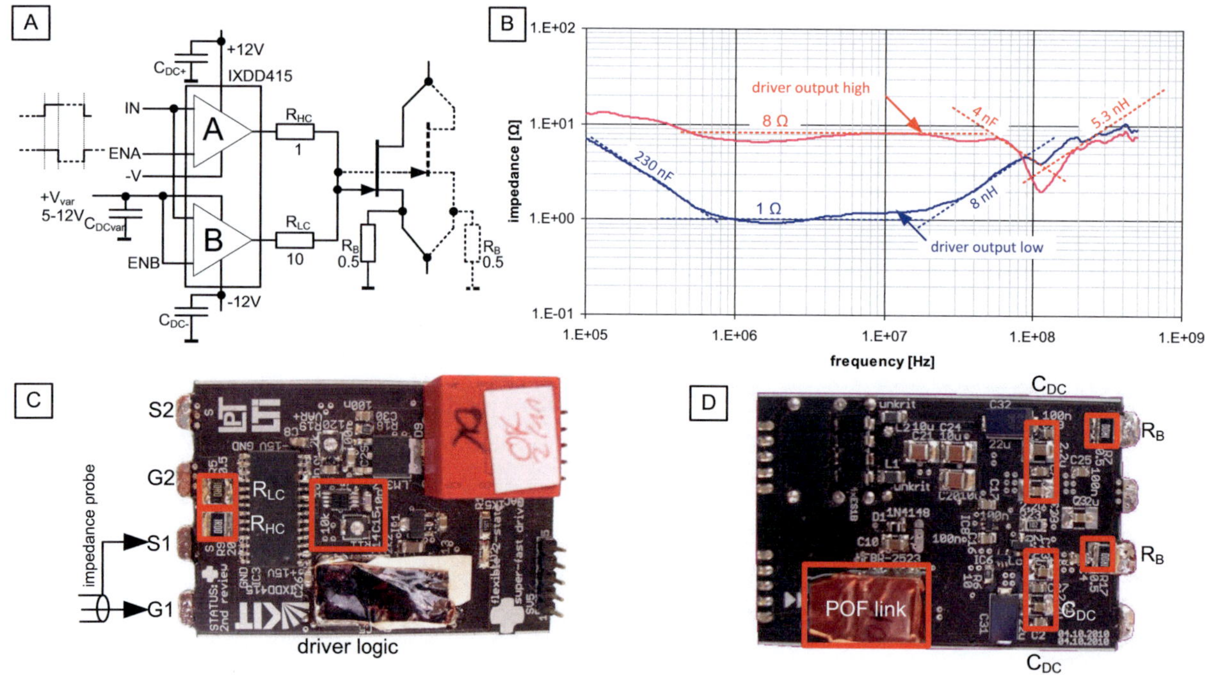

Figure 55: Multi-current-level JFET gate drive circuit consisting of two independent driver outputs and control logic. A: Schematic. B: Impedance measurement results for low current high and high current low state; low inductance due to RF driver package and enhanced layout. C & D: Photographs of PCB layout and indicated probe tip connection.

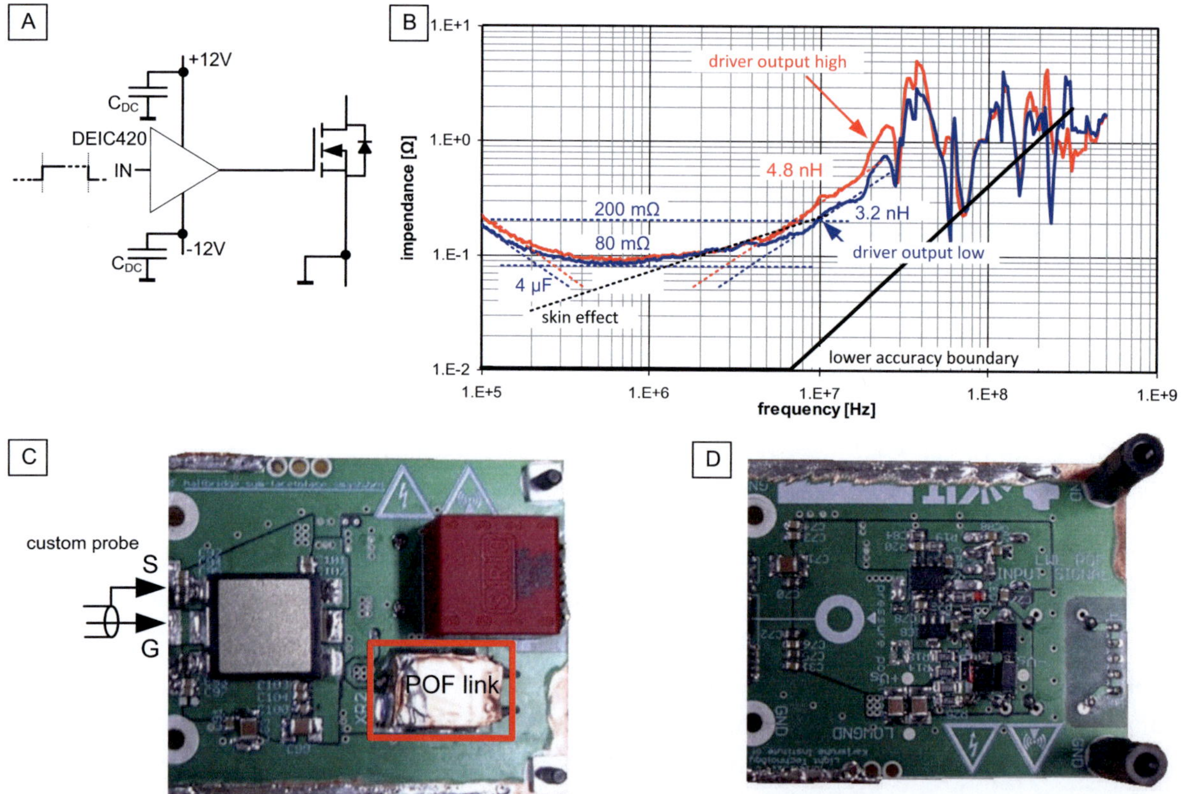

Figure 56: RF-MOSFET driver for use in RF-half-bridge configuration. A: Schematic, symmetrical 12 V supply, low impedance connection of MOSFET source to supply rails by various ceramic capacitors. B: Impedance spectra, very low impedance up to 20 MHz. C & D: Photographs of PCB layout and indicated probe tip connection.

Therefore, safe characterisation of this very low impedance design is restricted to that frequency. In order to enhance the measurement capabilities in the low impedance high-frequency range, enlarging the source power is necessary. An external broadband low impedance power amplifier connected between AC source output and impedance test kit might enhance the high-frequency signal level and thereby could provide better results. Beside suitable HF gate driver packages and careful PCB layout, special emphasis was taken on the necessary gate resistors. If resistor values below one ohm need to be implemented, it is worthwhile to replace conventional resistors with reverse geometry (RG) types.

As measurements depicted in Figure 57 prove, RG resistors can reduce parasitic inductance by a factor of three. This becomes relevant if resistances below one ohm need to be implemented.

The measurements performed allowed to identify and quantify parasitic circuit elements and supported the advance of the gate drive designs. The work underlines the impact of gate driver IC packages and circuit complexity on the parasitic inductance of the driver. Limitations of the impedance spectrum analysis approach come mainly from the restricted source power of the used impedance analyser and the resulting reduced accuracy at very low impedances at elevated frequencies (see Figure 48).

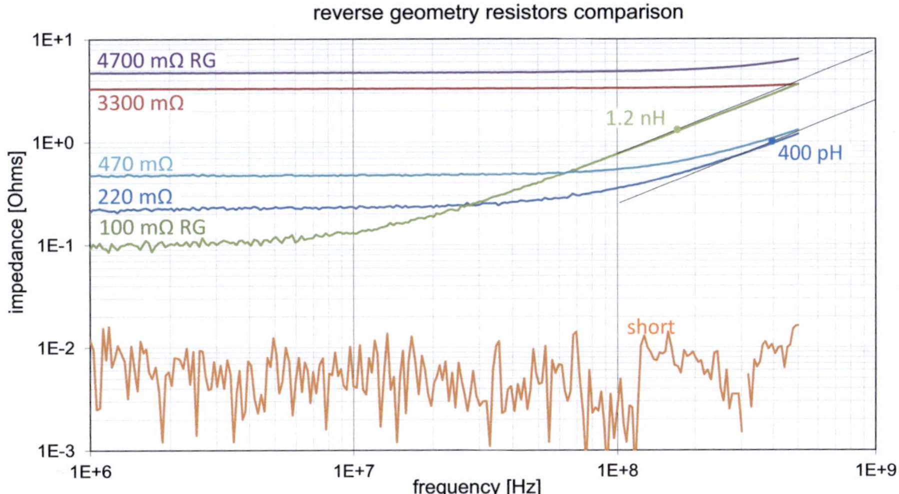

Figure 57: Impedance spectra of reverse geometry 2010 3/4 W and conventional geometry 2512 1 W SMD power resistors. Parasitic inductance of RG resistors is threefold lower.

4.13 Magnetics for High-Frequency and High Peak Power

Beside capacitors which can easily take up one third of volume in power electronic modules (Perret 2009), also magnetic devices (short: magnetics) can be voluminous. This chapter describes the efforts that were made within the scope of this thesis to restrict the volume of magnetics, especially the magnetic storage element. Below some 100 kHz, magnetic designs are saturation limited, whereas at elevated frequencies magnetic's losses are the primary limitation factor (Waffenschmidt 2001). Accordingly, at first the loss mechanisms in magnetic devices will be shortly mentioned.

The inductor and transformer windings are composed of copper conductors that have limited conductance and hence generate resistive losses. At DC or low frequency operation, resistive losses are not frequency-dependent but only rely on the properties of the inductor as length l_W and cross-sectional area A_{CU}:

$$R_{DC} = \rho_{CU} \frac{l_W}{A_{CU}} \tag{4.35}$$

However, for the high operation frequencies of DBD pulse generators AC losses become dominant. This is because at elevated frequencies, current is forced to flow through smaller parts of the conductor's cross-sectional area due to repulsive forces of AC currents of same orientation. While the skin effect is only depending on conductor properties and operating frequency (Formula (4.36)) the proximity effect also highly depends on the relative position of multiple conductors and should be quantified by simulation.

$$\delta \approx \frac{1}{\sqrt{\pi \cdot f \cdot \mu \cdot \sigma}} \tag{4.36}$$

For good conductors, both skin and proximity effect become relevant at skin depths $\delta < r_W$. Skin depth shrinks by one decade per two decades of frequency increase. The AC resistance of a single round wire including skin and proximity effects is defined by Formula (4.37) including skin depth (4.36) and the factor K (4.38) (Kaiser 2004).

$$R_{AC} \approx R_{DC} \frac{K \cdot r_w}{2 \cdot \delta} \qquad (4.37) \qquad \text{with} \qquad K = \frac{1}{\sqrt{1 - 4\left(\frac{r_w}{d_{ww}}\right)^2}} \qquad (4.38)$$

In Formula (4.38) ratio K is defined by the wire radius r_w, and the centre-to-centre wire distance d_{ww}. Formula (4.37) directly refers to the skin effect impact. K adds the influence of wire proximity. For all magnetics manufactured for this work, litz wire with $r_W = 0.05$ mm individual strand radius was used. At 2 MHz, according to Formula (4.36), δ equals the chosen strand radius and therefore the skin effect starts to deteriorate wire conductivity. However, R_{AC} could already increase at frequencies below due to the proximity effect.

The Steinmetz-formula (Formula (4.39)) gives a first-order indication of core losses depending on the actual frequency and the peak flux density. The factors P_0, e_f and e_B are given for the respective material or can be derived from the given core loss graphs.

$$P_{loss} = P_0 \cdot f^{e_f} \cdot B_{max}{}^{e_B} \tag{4.39}$$

The formula is applicable for toroid cores with constant cross-sectional area. However, common power electronics core geometries contain some area variation that lead to a non-uniform local flux density distribution. Additionally, every air gap dramatically changes the distribution of the magnetic field and core loss. For the intended frequency range, both for transformers and inductors, the Mn-Zn material N87 (EPCOS) was mainly used in this work. In fact, comparison of magnetic materials based on the performance factor $f{\cdot}B_{max}$ according to Figure 58 unveils that from terms of efficiency this material is somewhat less than perfect. However, more suited materials for frequencies exceeding one MHz as 3F45 or 3F5 have bad availability. In order to be able to test and characterise a variety of transformer and inductor designs, N87 seemed to be a meaningful choice. As the performance of topologies evolved it is suggested to base future work on Ferroxcube materials up to the Ni-Zn material 4F1.

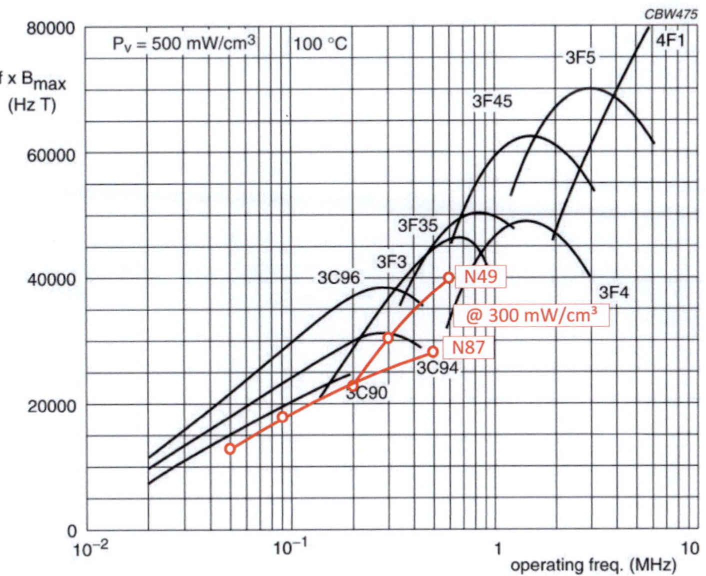

Figure 58: Comparison of ferrite materials regarding their frequency-dependent performance factor being a figure of merit for power magnetic cores. Curves in black are for 500 W/cm² loss and red curves represent data based on 300 W/cm² loss. Chart bases on Ferroxcube and Epcos datasheets.

General guidelines for magnetics design and construction are found in (Maniktala 2004a; Kazimierczuk 2009). It may also be referred to the excellent works of Schülting (1993) and Brockmeyer (1997) in which advanced design guidelines are presented.

4.13.1 Inductor Design and Characterisation

Inductors used in resonant topologies presented in this work are AC chokes used for energy storage. Together with the DBD lamp capacitance they form the main resonant circuit. Energy cycles between the electrical field in the DBD capacitance and the magnetic field in the inductor. The inductance L of the inductor therewith determines the resonance frequency (4.1) and the quality factor (4.10) of the resonant circuit. The inductance of an inductor is given by Formula (4.40):

$$L = N^2 \cdot A_L \tag{4.40}$$

L is thus depending on the turns in square and the popular A_L value. The latter is provided by the vendor for common cores in un-gapped and special gapped versions. To calculate A_L for a core containing a customised air gap Formula (4.41) can be used.

$$A_L = \frac{1}{z}\left(\frac{\mu_r \cdot \mu_0 \cdot A_e}{l_e}\right) \tag{4.41}$$

In this Formula z is the air gap factor introduced by Maniktala (2004b) which will be defined later. The formula further includes geometric parameters such as the average length of the magnetic path within the core l_e and the cross-sectional area of the magnetic core A_e and the core material property μ_r. Besides the inductance, the energy level that can be efficiently stored in the inductor is of high importance. In first order, the required storage capability of the choke can be determined from the parameters of the DBD lamp as peak voltage and DBD lamp capacitance.

$$E_{L,pk} = E_{DBD,pk} = \frac{1}{2} V_{DBD,pk}{}^2 C_{DBD} \tag{4.42}$$

Ferrite cores seem to be the best choice for the frequency range from hundreds of kHz to tens of MHz. Suiting core sets can be chosen by their energy handling capability (4.42) if the vendor supports respective data. Otherwise, the energy density of gapped cores, being the ratio of storable energy E_L to core volume V_C, can be calculated as it is mainly determined by the air gap width expressed as z-factor.

$$\frac{E_L}{V_C} = \frac{B^2 \cdot z}{2 \cdot \mu_c} \tag{4.43}$$

Factor z exceeds one in case the magnetic circuit includes an air gap. μ_c is the core permeability without an air gap. The importance of the air gap will be visualised by the following example: The common N87 material achieves roughly 20 J/m^3 at B = 330 mT. The 0.2 m^2 3.18 nF flat Planilum® DBD lamp typically stores a maximum energy of 6 mJ. Therefore, a core volume of 300 cm^3 would be required, which is equal to 20 RM14 cores. This huge volume is not to be tolerated and can be drastically reduced by introducing an air gap in the magnetic path. Referring to (4.43) the air gap factor z is set to 20. Formula (4.44) then defines the required total air gap length l_g:

$$l_g = \frac{(z-1) \cdot l_e}{\mu_r} \tag{4.44}$$

Regarding the mentioned example, a gap length of l_g = 0.6 mm allows to store the required energy using a single RM14 core. Independent of the core material constant μ_r, the necessary magnetic core volume V_C with respect to factor z is given by:

$$V_C = \frac{2 \cdot E_L \cdot l_e \cdot L}{B^2 \cdot N^2 \cdot A_e \cdot z} \tag{4.45}$$

The formula reveals that the core volume can be reduced by increasing the air gap which is expressed by the air gap factor z being the quotient of the permeability of the core without gap μ_C and the effective permeability of core and gap μ_{eff} as rearranging of Formula (4.44) yields:

$$z = 1 + \frac{\mu_r \cdot l_g}{l_e} = \frac{\mu_C}{\mu_{eff}} = \frac{L_{ungapped}}{L_{gapped}} \tag{4.46}$$

Another way to decrease the core volume is to reduce the average magnetic flux path length. Planar magnetics benefit from the fact that the energy handling capability is not affected by that. This is because the reduction of winding window breadth not only reduces l_e but also leads to a reduction of the overall volume:

$$V_C \propto l_e \cdot A_e \tag{4.47}$$

It can be concluded that for a given core geometry, the energy storage capability of a gapped core is significantly enhanced compared to an un-gapped core. Setting $l_g = l_e$ which mimics an open I-shape core Formula (4.46) reveals $z \approx \mu_r$. Therefore, the energy storage capability would be enhanced by a factor of μ_r.

Besides that, air gaps flatten the B-H-curve and therewith help to maintain the intended inductance independent of the current level.

However, there are hard facts that speak against a too wide air gap. Ferrite cores are needed to guide the H-field and keep stray flux low which could be the source of intense electromagnetic interference. The ferrite core is also needed to concentrate the H-field in air gaps where the highest magneto-motive force is generated and therefore the power density is highest. Because of the high field strength in the gap, especially large gaps as those of the inductor design depicted in Figure 59A and B, produce excessive fringing fields penetrating surrounding windings. The eddy currents generated by the fringing field enhance copper losses in the winding.

One way to defeat these losses is to ensure sufficient space between winding and fringing field. A rule of thumb is to space the innermost winding layer apart from the air gap by two times the gap width. This ensures that the winding loss in the gap proximity is roughly 25 % higher than without a gap. The spacing is done by "bolstering up" the winding near the gap as shown in Figure 59F.

Hence, in this work, multiple distributed gaps of up to one centimetre total width were used to improve the power density of the inductor. Additionally, layers of ferrite foil and stacks of ferrite foil and plastic foil have been used in this work to tame fringing flux. To minimise the stray field and respective electromagnetic radiation the gap was distributed by cutting the centre leg of commercial N87 RM14 cores according to Figure 59C, D and F. By using a specialised air-cooled diamond-blade cutting machine set-up and a developed cutting technique, the outer legs could be kept solid while the centre leg was carefully removed. Concentrating the air gap in the centre leg dramatically reduces the stray field as the windings and the outer core legs efficiently shield the fringing field. The stray field expansion is further reduced by introducing multiple air gaps being placed between the cut core drum pieces (Figure 59D). Non-magnetic spacers keep the distance for all gaps equal. The winding keeps space to the centre leg in order to reduce eddy current losses. Ideally, the required gap is placed at the ends of the centre leg where the bobbin naturally separates the winding from the gap (Figure 59C). Alternatively, the bobbin can be fitted with additional spacers as illustrated in Figure 59F and G. These spacers ensure the required distance of the winding to the individual gap. The inductor pictured in Figure 59F and G only presents the concept and would require larger wire cross-sectional area in order to handle high currents efficiently. Figure 59A and B show an inductor design for transformer-equipped topologies requiring low inductance and medium voltage handling capability. In order to minimise copper loss, the copper filling factor of the winding window was maximised. Hence, low-count windings were made of thick litz wires insulated with natural silk. Due to the operating frequency range, litz wires with single strand diameter of 0.1 mm had been chosen to minimise HF copper losses. The winding was then stabilised by additional lacquer. Since the other cores illustrated in Figure 59 exhibit gapped centre cores and solid legs, metal clamps could be used to press the core halves together. Though, eddy currents would lead to excessive losses if metal clamps would be attached to the inductor of Figure 59A and B. The massive gap of 7.8 mm in total required multiple layers of Kapton® tape for core fixture instead.

An aspect that becomes important for high-voltage inductors exemplarily shown in Figure 59D and E is the careful insulation of the individual turn and especially from winding layer to winding layer. Depending on the required voltage rating, multiple layers of 65 µm Kapton® tape were used for insulation. Special care was also taken to prevent flashovers from winding to core. Impregnation of the winding additionally helped to prevent micro-cavity discharges.

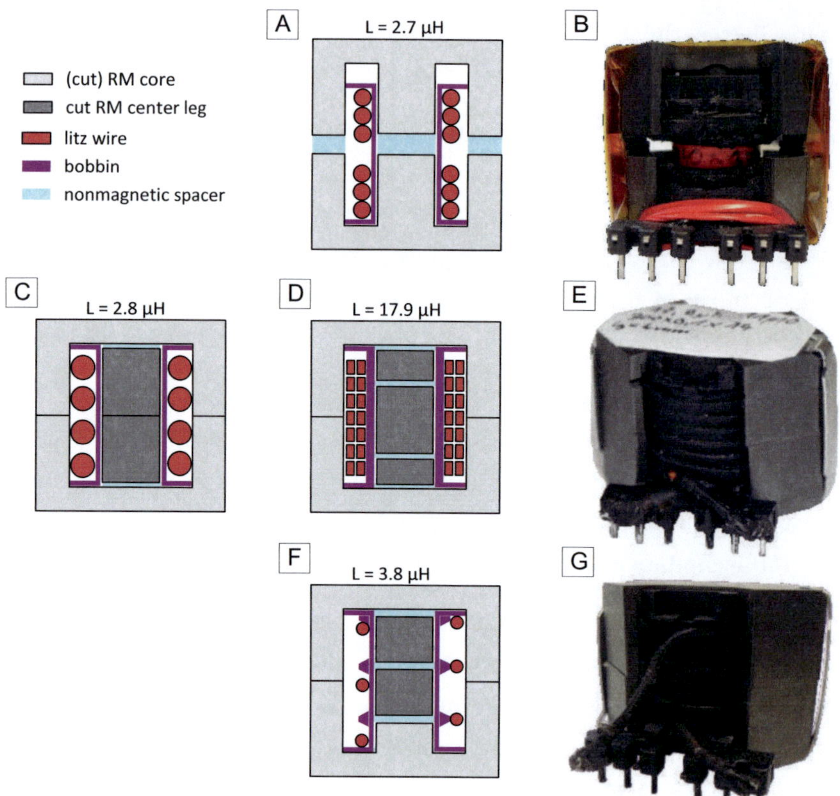

Figure 59: Schematic cross sections of the used RM14 N87 core inductors. A and B: Inductor with excessive air gap in order to handle high energy levels and 6 turns 840x0.071 mm litz wire winding. C: Inductor with cut centre leg in order to minimise stray field and EMI while keeping high energy handling capability and 4 turns 900x0.1 mm litz wire winding. D and E: High inductance high-voltage inductor with further distributed air gap and 14 turns 300x0.1 mm litz wire winding. F and G: Low inductance low eddy current loss concept inductor with 3 turns 175x0.1 mm litz wire being separated from 3 air gaps by extra spacer.

In any case, the gap length needed to be precisely adjusted to obtain the required inductance. This is since stray fields increase the stray inductance of windings depending on their distance to the core centre. Therewith, depending on the location of winding and air gap(s) the inductance may differ from calculated values basing on Formula (4.40) (Voss 2001).

Magnetic devices are also characterised by their resonance frequency. Inductors and transformers are useful up to 90 % of their intrinsic resonance frequency where they still show inductive behaviour. Above the resonance frequency, their response is capacitive. Therefore, it is of high concern to increase the resonance frequency of the magnetic devices according to the requirements.

The inductors designs described and illustrated in Figure 59 were characterised by small signal impedance measurements using the impedance characterisation set-up presented in Chapter 4.8. The results are depicted in Figure 60. The inductance values calculated from the measured impedance are a good estimation of the inductance obtained under high current operation since the considerable air gaps significantly linearize the B-H characteristic. To increase the resonance frequency, theoretically it may be advantageous to increase the number of turns while keeping inductance equal by an increased air gap. Hence, winding capacitance is reduced as more turn-to-turn capacitances are connected in series. A larger air gap favourably enhances the power handling capability of the inductor, too. As a drawback, the available cross-sectional area per turn diminishes and thus, copper loss increases.

De facto, resonance frequency is only slightly enhanced shifting from 4 (C) to 6 (A) turns in order to achieve 2.7 µH. Clearly to see is that high inductance is opposing high resonance frequency. This is also due to the likely increase in parasitic winding capacitance caused by necessary multi-layer windings. The two-layer winding also causes the multiple resonances at higher frequencies (Figure 60 right).

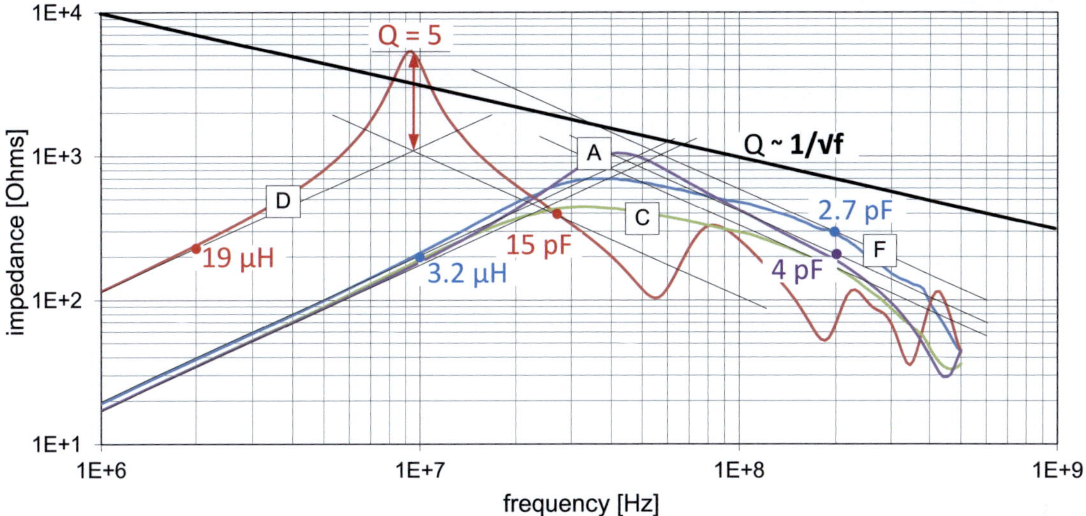

Figure 60: Impedance spectra of the different inductors described and depicted in Figure 59. The letters marking graphs here refer to those of Figure 59. As clearly to see, high achieved inductance is tantamount to reduced resonance frequency. This is partly caused by increased winding capacitance due to necessary multi-layer winding structures.

4.13.2 Transformer Design and Characterisation

As long as the resonant circuit is not directly driven by high-voltage power semiconductors, a transformer is required to step-up the voltage. Transformers are broadly used for pulsed power applications if high output voltages (hundreds to ten-thousands of volts) or galvanic insulation are required, such as to drive DBD lamps. Although the transformer permits a less complex structure of the power semiconductor circuit, transformers bear significant drawbacks and design challenges.

Transformers:

- restrict flexibility, as the volt-seconds product must not exceed a maximum value given by the maximum magnetic flux density,
- add parasitic elements as stray inductance and parallel (winding) capacitance, which add parasitic resonances to the system,
- add losses like copper and core losses, and
- require good magnetic coupling, especially if low mutual inductance as well as high energy storage capability is required.

Therefore, one scope of this work was to eliminate the transformer leading to transformer-less variations of the presented topologies. Nevertheless, the following paragraphs analyse the requirements to the transformer and discuss characterisation and optimisation of this central component.

The main function of a transformer is to match voltages, currents and impedances to different requirements on its primary and secondary side. In case of resonant pulse topologies for the drive of DBDs the transformer is at first used to transform voltage by its transformer ratio r. The device matches the voltage rating of the used semiconductors on the primary side to the higher required voltage to drive the DBD on the secondary side. In this case, the transformation of current (4.48) and capacitive reactance (4.49) performed by the transformer are both unwanted effects.

$$i_{prim} = r \cdot i_{sec} \quad (4.48) \qquad C_{DBD,prim} = r^2 \cdot C_{DBD,sec} \quad (4.49)$$

The inversely transformed secondary current leads to increased conduction losses in the power semiconductors connected to the primary while the impedance transformation restricts the maximum pulse frequency of the series resonant circuit depicted in Figure 61 with C_{DBD} and L being the main resonance partners:

$$f_{res} = \frac{1}{2\pi}\sqrt{\frac{1}{LC} - \frac{R}{4L}} \quad (4.50) \quad \text{with} \quad \begin{cases} C = C_{DBD} \cdot r^2 & (4.51) \\ L = L + L_{stray,Tr} + L_{stray,DBD}/r^2 & (4.52) \\ R = R_{Cu,L} + R_{Cu,Tr} + R_{Cu,DBD}/r^2 & (4.53) \end{cases}$$

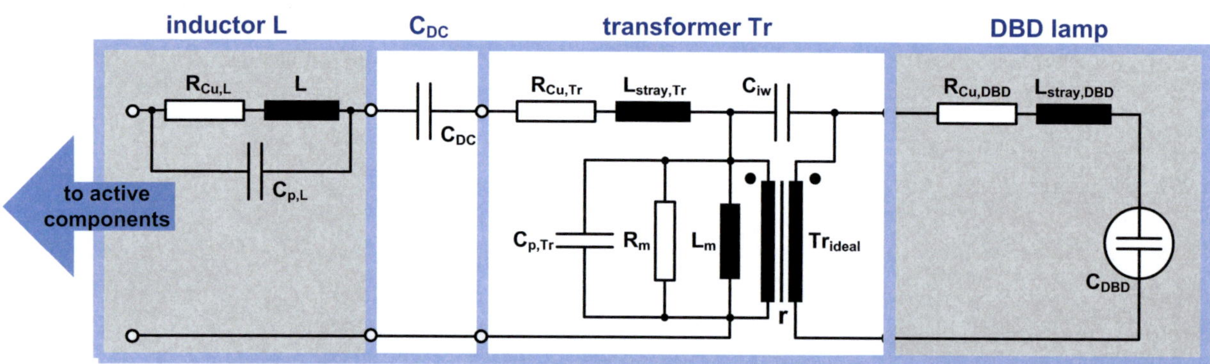

Figure 61: Equivalent circuit of a transformer-equipped, damped series resonant circuit. Beside this totally insulated variant, the transformer can be equipped with a secondary centre-tap to be connected to primary ground. Wammes (1999b) claims that this can significantly reduce EMR.

The circuit outlined in Figure 61 contains three main components: inductor L, transformer Tr and the DBD lamp. In that circuit, the ideal transformer Tr_{ideal} with transformer ratio r is surrounded by the following parasitic elements:

$R_{Cu,Tr}$	resistor representing copper losses of primary and secondary, both DC and AC parts
R_m	resistor representing magnetic core losses
L_m	mutual inductance, reactive part, generated field links primary and secondary side
$L_{stray,Tr}$	inductance of magnetic stray field that does not link primary and secondary
$C_{p,Tr}$	parallel capacitance, turn to turn capacitance
C_{iw}	inter-winding capacitance, capacitance between primary and secondary windings

While the resistive parasitics of the transformer lead to losses and may damp oscillations, the reactive parasitics add additional resonances to the system. The resulting main resonances are depicted in Figure 62.

Transformers implemented in the pulse generator topologies presented in this work need to support broadband operation. Hence, parasitic elements need to be tuned in order to transfer both the pulse (MHz-range) and the pulse repetition (tens to hundreds of kHz) which equals a bandwidth of two to three decades.

The mutual inductance L_m of the transformer is the reason for the appearance of the parasitic parallel resonance with resonance frequency f_{pp} that is comprehensively discussed in Chapter 5.1.3. From terms of power loss, the value of L_m should be high in order to minimise current needed to maintain the magnetic link between primary and secondary. On the contrary, from terms of parasitic resonance prevention, its value and thereby its energy storing capability should be minimised. The latter was done in Chapter 5.1.3 by inserting an air gap. This reduces L_m while likewise enhancing the maximum saturation flux density B_{SAT} making the system more reliable.

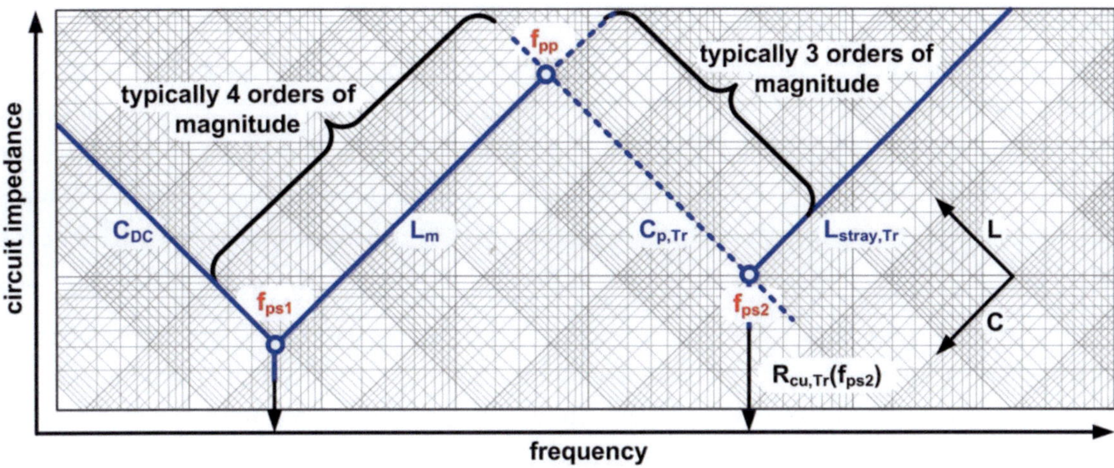

Figure 62: Impedance spectrum showing multi-resonant system of the transformer with main resonances of equivalent circuit of Figure 61. Equivalent circuit and impedance chart are valid for series topologies. Given are the typical relative orders of magnitude that were identified characterising the designed transformers.

The stray inductance L_{stray} of the transformer, however, does not saturate as its magnetic field lines propagate mainly through air. By tuning L_{stray} by means of winding structure it can favourably be utilised as resonance inductor in series topologies (Chapter 4.7). This can effectively save one bulky component and increases the power density. However, L_{stray} is an unwanted parasitic especially in mixed mode topologies (Chapter 7). This is because it leads to parasitic resonances being excited by the current step provided by the pre-charged inductor. To significantly reduce the stray inductance, the primary and secondary windings are favourably be decollated into sections and then arranged in an interleaved manner. This can account for significant reductions of L_{stray} by factors of ten to up to 30. Besides that, the air area enfolded by windings and contact leads needs to be minimised by suitable contact configuration.

The parallel capacitance, also called winding capacitance or end-to-end capacitance, together with the transformer's mutual inductance defines the parallel resonance of the component. In any case, $C_{p,Tr}$ is unwanted and can be minimised by single layer windings and bank winding arrangement if multiple layers are required.

The inter-winding capacitance C_{iw} comes into focus if the electromagnetic radiation of the circuit is considered. It is the parasitic element that transfers CM noise and could be minimised by keeping distance between primary and secondary winding. Evidently, reduction of C_{iw} and $C_{p,TR}$ has contrary effect on the inductive parasitic of the transformer and therefore needs consideration.

Beside the minimisation of transformer parasitics, design also has to ensure that the transformer is able to handle the voltage time product that is defined by the resonant circuit. Refering to Formula (5.5), the voltage-time product of the series resonant circuit step response until current approaches zero amplitude is given as:

$$\Delta v_{DBD} \cdot \Delta t = V_S \frac{\pi}{\omega_d} \approx V_S \cdot \pi \cdot \sqrt{C_{DBD} \cdot L} \tag{4.54}$$

Insertion in the transformer equation yields:

$$A_e \cdot N_{prim} = \frac{V_S \cdot \pi \cdot \sqrt{C_{DBD} \cdot L}}{B_{max}} \tag{4.55}$$

Hence, the design and optimisation procedure includes finding the optimum relation of the cross-sectional area A_e of the magnetic core to the number of primary winding turns N_{prim}. In order to allow for systematic progress, ETD ferrite core shapes were used in this work, exclusively. ETD cores are preferable since the round centre leg has advantages compared to their square (E) or rectangle (EFD, ELP) shaped counterparts in terms of winding loss and leakage inductance (Marinos 2010). For most designs, an ETD 49 core was chosen.

Figure 63A visualises the general transformer structure. The primary high current winding requires a large cross-sectional area and multiple-strand conductors whose bending radii are limited. Additionally, a lot of winding breadth is consumed if a connection between different layers needs to be established. With standard bobbins, the litz-wire diameter is often too large to fit between the pin slots of the bobbin. Therefore, the primary winding is best placed as outer winding or in the middle if an interleaved winding arrangement is desired. The secondary starts with the high-voltage end close to the centre of the bobbin and ends with the low voltage end (ground) at the insulation layer between the primary and the secondary. Advantageously, this minimises the capacitive coupling and reduces the insulation strength required between primary and secondary. However, the insulation strength of the bobbin needs to be further strengthened as the ferrite core needs to be threaten as conductor in that respect.

According to the maximum allowed flux density (4.55), the minimum number of turns for the primary winding is determined. The number of turns for the secondary is then given by:

$$N_{sec} = \frac{V_{sec}}{V_{prim}} N_{prim} \tag{4.56}$$

In order to unify primary and secondary DC-losses, the available winding window area A_w is distributed to primary and secondary windings. As the winding breadth is assumed to be equal for all windings, the respective winding width is the parameter to be determined. Figure 63 shows the geometrical set-up. The average length of one turn is set equal for primary and secondary. Then, the areas for primary and secondary windings are equalised as well. This is a good approximation for large cores with a high ratio of inner bobbin radius r_{inner} to winding area width w. However, the characteristic ratio of cores for power electronic applications is around one.

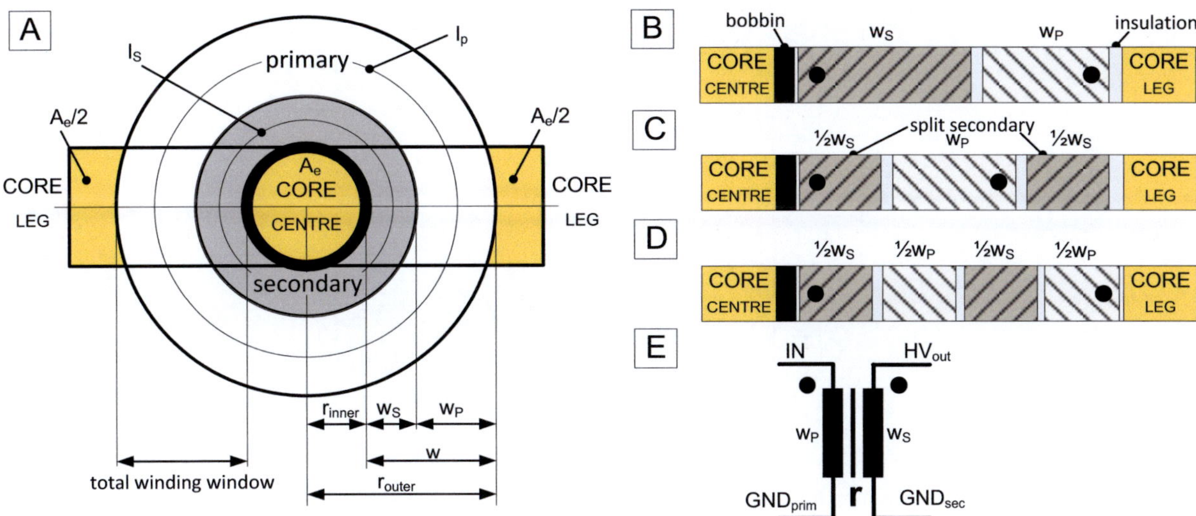

Figure 63: A: Geometry of common magnetic cores (ETD core shape) with all geometric dimensions. B: Cross-section of a winding structure consisting of a secondary close to the core centre and a primary wound on top of the secondary. C: Interleaved winding structure where the primary is surrounded by a split secondary. D: Interleaved winding with split primary and secondary. E: Schematic with the relevant electrical parameters.

To obtain equal DC resistance and yet uniform power loss in primary and secondary, the width for the secondary winding w_s has to be lower than ½ w and is given by:

$$w_s = \sqrt{r_{inner}(w + r_{inner})} - r_{inner} \tag{4.57}$$

The formula is applicable to core geometries with a round inner leg. The ratio of inner winding radius to winding window area width is a characteristic parameter of a bobbin-core combination. Based on Formula (4.57), the standardised winding area width of the secondary in relation to this specific parameter is plotted in Figure 64. However, this allocation of winding areas is only valid for the two-winding standard design according to Figure 63B. In case of an interleaved winding structure as shown in Figure 63C and D, the copper cross-sectional area would have to be adjusted for every layer. However, due to economic reasons, this is not feasible. Interleaving plays an important role where the stray inductance needs to be minimised as in the case of flyback coupled inductors.

After w_s and w_p are found, the DC power loss ratio between primary and secondary needs to be defined:

$$F_{Pdc} = \frac{P_{loss,prim}}{P_{loss,sec}} \tag{4.58}$$

The primary loss should be chosen to be higher than the secondary loss since the heat emission capability of the outer primary winding is by far higher. The ratio of primary to secondary winding wire cross-sectional area is then defined to:

$$\frac{A_{cu,prim}}{A_{cu,sec}} = r \cdot F_{Pdc} \frac{r_{inner} + w_s + \frac{1}{2} w_p}{r_{inner} + \frac{1}{2} w_s} \tag{4.59}$$

This is the product of the transformer ratio, the DC power loss ratio F_{Pdc} and the ratio of the average winding length per turn of primary and secondary.

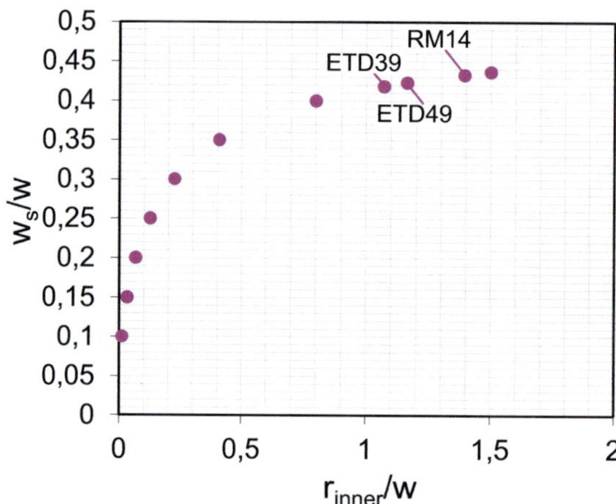

Figure 64: Winding area distribution for equal DC-copper losses in primary and secondary: ratio of winding area width of secondary to total winding area width in relation to bobbin-parameter.

A major task in transformer design and manufacture is the guarantee of a safe and reliable insulation between low-voltage primary and high-voltage secondary. The task is exacerbated by high-frequency operation of the transformer which makes partial discharges likely to occur. The underlying rationale for this behaviour is that resistors determine the voltage sharing between series connected circuit elements for DC as opposed to the capacitors being the determinator for the voltage sharing in the case of AC. Consequently, with an insulation structure composed of a solid insulator and an air gap, the electrical field strength in the air is enhanced due to the high ε_r of the solid dielectric. Hence, an electrical breakdown of air and a discharge that punctures the solid may be the consequences. In order to prevent partial discharges while keeping the conductor distance equal, either low-ε_r materials such as Teflon are used as insulator or a dielectric casting ensures a void-free insulation path. Within this work, the transformers were impregnated with PLASTIC 70 (CRC Kontakt Chemie) and multiple KAPTON® foil layers insulated the primary from the secondary windings.

A special case is presented by flyback transformers, which are actually coupled inductors. In that respect, they are a combination of the purposefully energy storing inductor and the transformer which adapts voltages and currents. In order to enhance the energy storing capability of the core, like inductors, flyback transformers are equipped with air gaps. Being implemented in a flyback topology to drive DBDs as outlined in Chapter 6.1, the transformer is impinged by high di/dt since current flows not continuously but is chopped between primary and secondary. Compared to other magnetics, this dramatically increases winding (skin and proximity effect) and core losses. The negative impact of stray inductance is very severe in flyback designs. However, while for off-line power supplies intense interleaving of primary and secondary is possible, the requirements regarding winding insulation restrict interleaving to first order interleaving according to Figure 63C. Reduction of stray inductance always goes in hand with enhanced winding to winding capacitance C_{iw} and higher dielectric stress to the insulation.

4.14 State-of-the-Art of DBD Pulse Generators

The previous chapters gave an insight into the different pulse inverter components and its principal mode of operation. Following up on this, the objective of the actual chapter is to exemplarily present state-of-the-art topology concepts in order to provide evidence for the novelty of the developed topologies. In the Chapters 5 - 7, known and novel topologies for the pulsed operation of DBD lamps will be described and characterised. Additionally, following the intent of this work, the topologies and driving schemes are classified according to the classification scheme presented in Chapter 4.5. Finally, novel topologies will be compared and rated regarding their key parameters in Chapter 8.

Kyrberg (2007b) presented a topology that utilises the DBD capacitance to store energy during the pulse pause. Beside pulsed operation, also square-wave operation is possible. The also patented topology family (Kyrberg 2005) consists of unipolar switches made of a power semiconductor switch and a series connected diode placed in half-bridge or full-bridge configuration as well as an inductor which may be separated into two devices for reasons of symmetry. The basic topology is shown in Figure 65A. In contrast to the common resonant half-bridge, the unipolar switches prevent any recovery of energy back into the supply rail, which is indeed not intended. Instead, the operation bases on energy feed-in only. After a run-up of the system energy level, the energy drawn from the supply rail only has to compensate circuit losses and energy consumed by the lamp. Run-up is initiated by enabling S1. According to Figure 65D, the lamp voltage ramps up following a negative cosine wave-shape while the current wave-shape is a sinusoidal half-wave conforming to Figure 39A. The lamp voltage plateau amplitude ideally is twice the supply voltage. Closing the low-side switch S2, the positive plateau can be transferred into a negative plateau of the same absolute amplitude if zero loss is assumed. A further switch transition from S2 to S1 quadruples the lamp voltage amplitude as compared to the supply voltage level. The scheme can be continued until the ignition voltage of the lamp is reached. The next switch transition will then lead to an inverted voltage applied to the lamp. Depending on the chosen inductance the voltage slew rate is determined for a given lamp capacitance. A circuit variation according to Figure 65C allows for independent slew rates for positive and negative slopes and also supports ZCS conditions by preventing capacitive short circuits in the bridge-leg.

Figure 65E shows a full-bridge configuration with symmetric inductor placement enabling for energy feed-in both during positive and negative voltage slope. Since every energy feed-in also leads to a voltage amplification the topology could, in principle, be supplied by a low supply voltage.

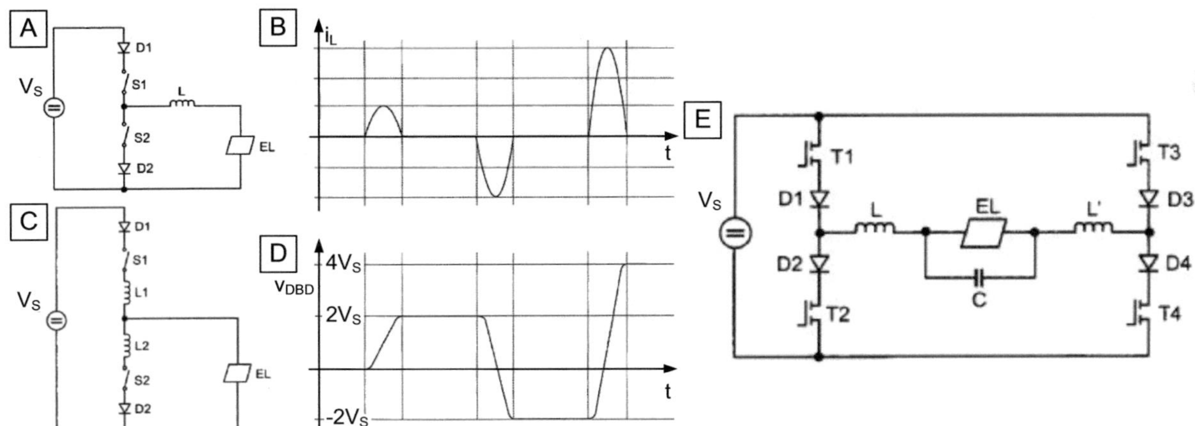

Figure 65: Converter with unipolar switches for DBD drive with resonant charge of the DBD capacitance (Kyrberg 2007b). A, C and E: Circuit variations. B and D: Voltage and current waveforms of circuit A. Pictures modified for clarity.

Despite of its flexibility, the circuit is not without drawbacks. As compared with the conventional half-bridge which is made of bipolar switches that inherently limit the voltage level, the use of unipolar switches for commutating inductively-shaped current bears the risk of over-voltages leading to the destruction of the power semiconductors. This is due to the missing freewheeling path. Since freewheeling is not possible, the current flow shaped by the resonance cannot be interrupted, for instance in order to modify the transferred energy level as is possible with current cut-off. The topologies that will be presented in Chapters 5.2 and 5.3

share the advantages of this approach but prevent its drawbacks. Figure 66A depicts a quasi-resonant converter that generates unipolar high-voltage pulses with a claimed electrical efficiency of 55 % (Pemen 2012). The voltage is amplified by a transformer with a ratio of 18 and resonant overshoot. The rising voltage slope is a negative cosine while the current is of sine-shape as is common with series resonant converters. The converter includes an actively controlled input filter and a complex pre-magnetisation and de-magnetisation circuit controlled by power switch M1. The transformer's mutual inductance freewheels into C3 in order to compensate magnetisation built up by the pulse. In advance of the pulse, M1 may be turned-on to pre-magnetise the transformer core in order to allow for a greater magnetic flux-swing. As resonance tank, C_{SDBD} and the transformer's stray inductance define the pulse frequency.

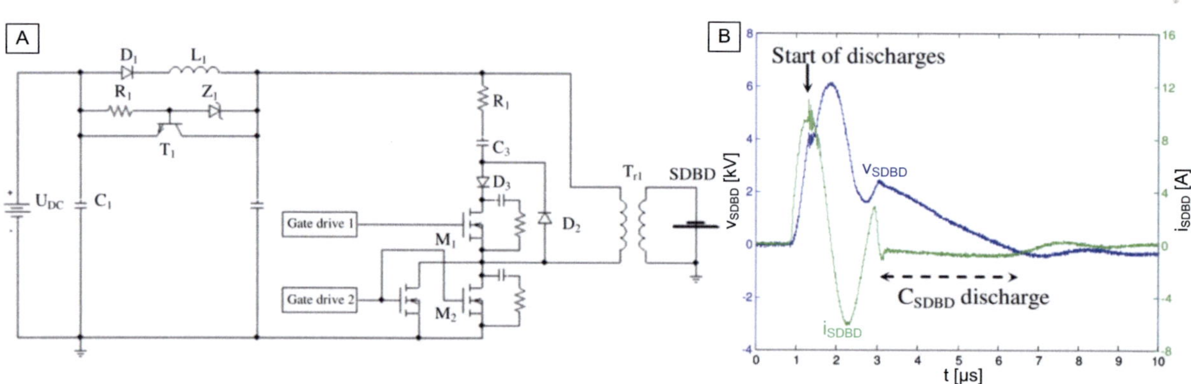

Figure 66: Quasi-resonant pulse generator with 55 % efficiency claimed and transformer pre-magnetisation and de-magnetisation (Pemen 2012). A: schematic. B: Voltage and current of DBD. Pictures modified for clarity.

The unipolar pulse having a voltage tail is not advantageous as it does not support the advantageous second plasma ignition. Both the magnetisation circuit and the active filter include dissipative components (T1, R1) while energy recovery is not intended. Both account for the poor efficiency of this approach. In contrast, the less-complex SP-topology presented in Chapter 5.1 efficiently generates bipolar pulses and allows for energy recovery. The voltage tail is prevented by the topology depicted in Figure 67.

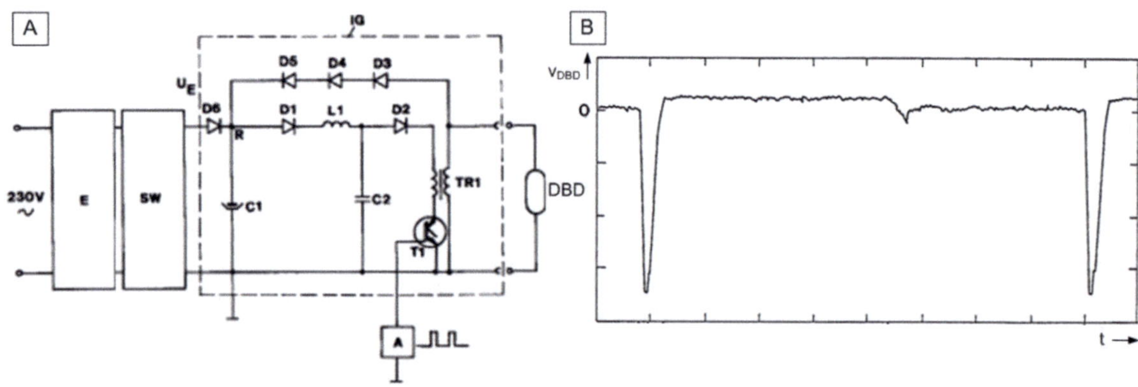

Figure 67: Resonant flyback derived topology (Huber 1996) with secondary energy recovery. A: Suggested circuit. B: Typical DBD waveform. Pictures modified for clarity.

In his European patent Huber (1996) suggests a resonant converter with secondary voltage clamping by series connected diodes D3 – D5 that is to say energy recovery. At the first glance, the circuit appears to be a flyback-derived topology. However, the patent description discloses a differing operation mode. The lamp capacitance is charged by enabling T1. The subsequent resonant current backswing discharges the lamp capacitance via the high-voltage diodes back into the supply rail insuring immediate pulse termination. L1 and C2 only act as filter and D2 prevents any reverse current. As T1 is turned off under ZCS condition, IGBT switches could be implemented. It is assumed that the transformer stray inductance shapes the pulse waveform and that this concept could achieve fairly high efficiencies as long as compact lamps are driven. On the contrary, the large number of components could make this concept bulky if scaled up for high power DBDs.

Figure 68 depicts a topology presented in a patent of Huang (2007) which is intended to generate bursts of square-waves. The claimed invention uses the mesh T-M2-M4 to short-circuit the transformer T between the bursts and therefore prevents the parasitic oscillations observed in the prior art. Unfortunately, this technique virtually freezes the current through the mutual inductance of T, which is a highly lossy process. Energy freezing is further addressed in Chapters 6.1.3 and 7.3.2.

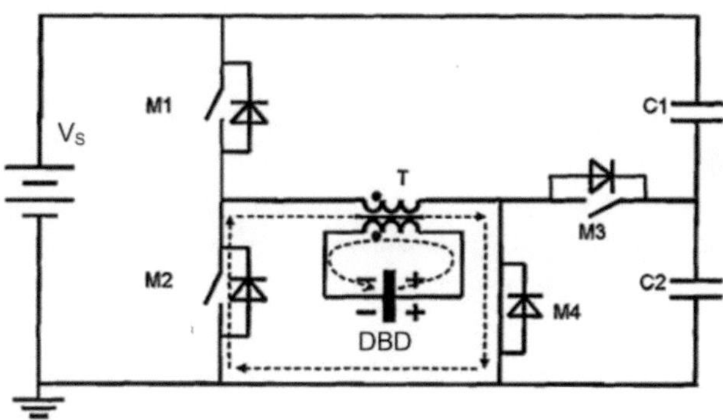

Figure 68: Topology of Huang (2007) for diming of continuously driven DBDs including clamping of transformer in idle time. Schematic modified for clarity.

A resonant full-bridge topology according to Figure 69A that generates bipolar pulses with alternating polarity used to drive a water-cooled DBD ozone reactor is presented by Shin (2006). Experimental data is given for the generation of 200 kHz pulses with a repetition frequency of one to ten kilohertz.

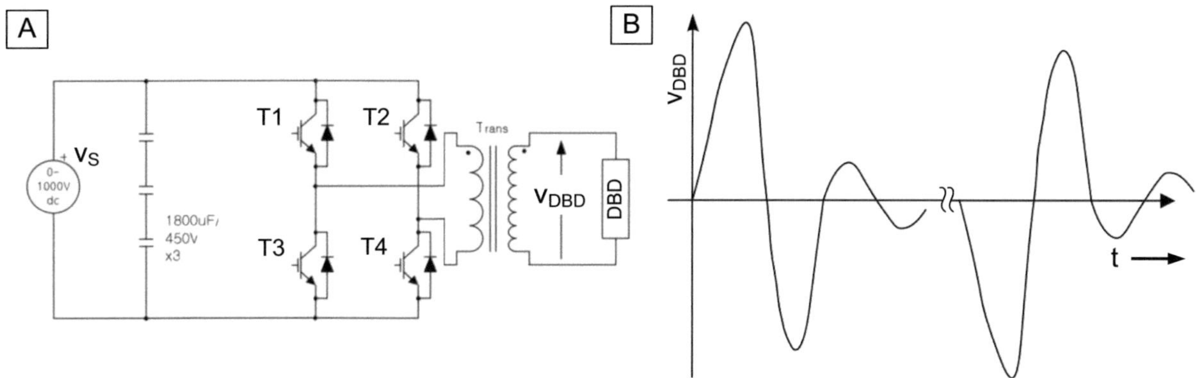

Figure 69: Resonant full-bridge topology generating bipolar pulses (Shin 2006). A: Schematic hiding the resonant inductor. B: Voltage pulse-shape applied to the DBD reactor. Pictures modified for clarity.

The generated waveform resembles those created by the SP-topology discussed in Chapter 5.1 except for the alternating pulse polarity suggested by Shin. Assuming two ignitions per pulse, the suggested alternating polarity involves a non-alternating ignition polarity. According to Chapter 3.4.3 this does not support homogeneity and efficiency of the discharge, especially in case of high repetition frequency. A modified resonant flyback converter operated in CCM is depicted in Figure 70A. In contrast to common resonant flybacks that are intrinsically limited to unipolar pulse generation, the use of a unipolar switch enables the design to provide bipolar pulses. The additional resonant capacitor C_r ensures ZCS turn-off of the switch while resonance leads to ZVS. The duty cycle (pre-charge time of mutual transformer inductace L_m) determines the amount of energy transferred to the DBD lamp. Figure 70B shows CCM simulation and experimental results obtained by the authors. Although this is a very interesting approach due to its simplicity, it has limited flexibility regarding the variation of the repetition rate. However, this is provided by the traditional resonant flyback.

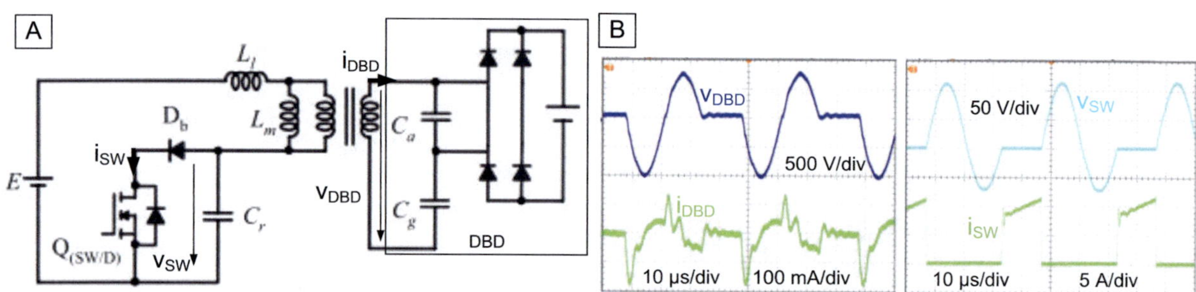

Figure 70: Bipolar sinusoidal pulse generated by a series resonant flyback derived topology having a unipolar switch (Sugimura 2006). Schematic (A) and experimental data (B) of switch voltage v_{SW}, switch current i_{SW}, DBD voltage v_{DBD} and DBD current i_{DBD}. Pictures modified for clarity.

These six examples of state-of-the-art topologies show that primarily bridge-derived or flyback-derived topologies are comonly used for pulsed DBD operation. The topologies dynamically differ in the generated pulse wave-shape and their flexibility to vary pulse parameters.

The following chapters give an insight into topologies which were newly developed as part of this work.

5 OPERATION MODES AND RESONANCE BEHAVIOUR OF SERIAL TOPOLOGIES

The following chapters present serial topologies that have been developed or improved in this work. So far, parallel and mixed topologies attracted a broader international interest. Thus, they were already investigated thoroughly. In contrast, the presented serial topologies and operation modes were preponderantly developed and investigated at the KIT's Light Technology Institute. Theoretically, serial topologies offer a higher efficiency than their parallel counterparts as disclosed in Chapter 4.5.

The sinusoidal pulse (SP-) topology (Paravia 2008a) was comprehensively investigated. Its properties are discussed in Chapter 5.1 which includes its resonance behaviour and optimisation steps.

Resulting from the SP-topology the novel single-period sinusoidal pulse topology (Meisser 2010c) presented in Chapter 5.2 improves the wave-shape and renders it more suitable for the requirements of the DBD lamp.

Other variations bear the possibility to go without the otherwise necessary transformer. The high-voltage half-bridge topology of Chapter 5.1.8, the high-voltage SPSP-topology (Meisser 2011a) of Chapter 5.2.4 and the novel upswing full-bridge topology (Meisser 2010d) of Chapter 5.3 are transformer-less.

5.1 HALF-BRIDGE EXCITED SERIES RESONANT SINUSOIDAL PULSE TOPOLOGY

The genesis of this topology goes back to investigations of Paravia (2008a). To start with, the basic topology will be described by means of schematic and principal mode of operation. It follows the investigation of the major parasitic resonances in Chapter 5.1.3. Chapters 5.1.5 and 5.1.6 discuss pulse wave-shape modifications by tailored switching patterns.

The last two sub-chapters deal with ways to enhance the maximum operating frequency margin valid for the basic topology. First, Chapter 5.1.7 discusses principal design considerations regarding PCB layout and power semiconductors. Chapter 5.1.8 discusses the transformer-less high-voltage topology variation that operates at high supply voltage exceeding ½ $v_{DBD,max}$.

5.1.1 SCHEMATIC AND PRINCIPAL MODE OF OPERATION

The sinusoidal pulse shown in Figure 71B driving the excimer DBD lamp is generated by a half-bridge and an asymmetrically connected series resonant circuit as depicted in Figure 71A.

The half-bridge is supplied with a voltage of up to 400 V complying with standard PFC output levels and permitting the use of fast, reliable and cost-effective Coolmos power switches for the low-side switch LS and the high-side switch HS. The transformer Tr matches the primary voltage generated by the resonant tank of inductor L and primary-transferred lamp capacitance C_{DBD}' to the secondary high-voltage required to ignite the lamp. DC-blocking capacitor C_{DC} provides a freewheeling path and thereby prevents the transformer from saturation due to the possibly unsymmetrical pulse shape. However, the offset given by C_{DC} limits the voltage step amplitude and thereby the maximum level of v_{DBD}'. Figure 71A also depicts main parasitic elements. These transformer parasitics are the stray inductance L_{stray} (primary-transferred) and the primary mutual inductance L_m. To ensure high efficiency of the implemented energy recovery mechanism a silicon carbide diode D_2 is used to bypass the body diode of HS and thereby significantly reduce reverse recovery time. While the low voltage silicon Schottky diode D_1 prevents conduction of the HS's body diode, D_2 takes over carrying the reverse current. In the experimental set-up, two discrete devices were connected in parallel for each power semiconductor displayed in Figure 71A in order to ensure high circuit efficiency when operating large DBD lamps.

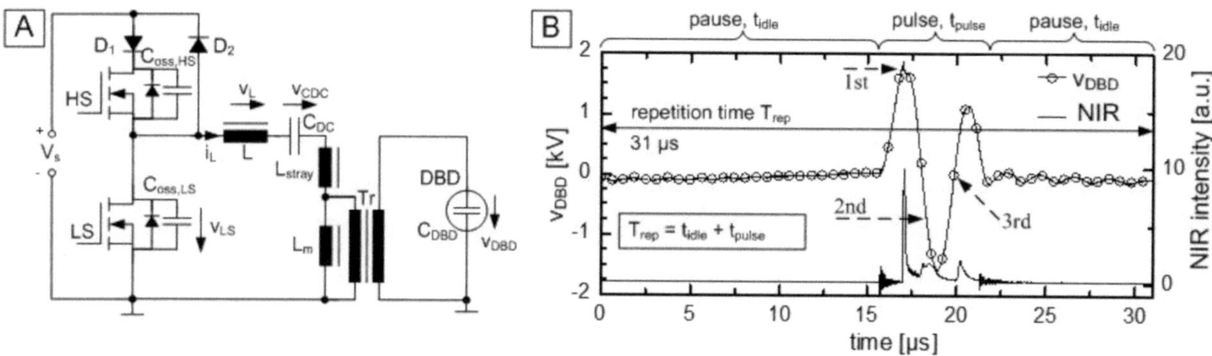

Figure 71: A: Extended schematic of the SP-topology: L_{stray} is the condensed stray inductance of transformer Tr. B: Typical waveform of DBD voltage v_{DBD} and corresponding NIR emission due to three ignitions per pulse.

A typical sinusoidal pulse shape with corresponding NIR emission is shown in Figure 71B. A pulse consists of three sine half-waves and typically leads to three ignitions per pulse which are indicated by NIR peaks (see Chapter 2.2). The circuit of Figure 71A is capable of driving a flat Planilum® lamp ($A_{active} = 0.2\ m^2$, $C_{DBD} = 3.18$ nF) with an electrical efficiency of up to 87 %. The lamp absorbs $P_{DBD} = 100$ W at a PF of 12 % and shows homogeneous appearance.

5.1.2 Time-Domain Behaviour

In Figure 72A the three main time-domains of a sinusoidal pulse are depicted. To generate a pulse HS is switched on. C_{DBD} is then charged through HS while series inductor L limits the initial current. Switching on HS is therefore of ZCS type. As the slope of v_{LS} at HS's turn-on is several orders of magnitude higher than the maximum slope of the main series resonance, the resonance behaviour is not affected. The lamp ignites during the positive voltage slope. In addition to the aforementioned NIR peak the ignition is also indicated by the break-down of v_{DBD}. At almost maximum lamp voltage and still positive current HS is switched off to induce a ZVS commutation between HS and LS. After commutation to the LS body diode the current swings back resonantly. The lamp voltage falls and the DBD is charged with negative voltage leading to a second ignition. While LS is still closed, a further backswing charges C_{DBD} positively inducing a third ignition. At a definite current level LS is opened and the magnetised inductor L forces a current that recovers most of the energy stored in the resonance circuit through D_2 back to the supply rail. The DBD is then left idle. Preparatory to the next pulse a pause with almost zero slope and DBD lamp voltage amplitude is maintained to allow the charges inside the DBD to recombine.

The respective basic equivalent circuit with corresponding initial values of voltages and currents is depicted in Figure 72B. For reasons of clarity the resonance behaviour of the circuit in combination with the typical switching scheme is analysed with the following simplifications:

- a simplified circuit as depicted in Figure 72B without transformer is considered
- the influence of semiconductor output capacitances C_{OSS} is neglected
- the initial v_{DBD}-level is assumed to be zero volts

The response to the initial voltage step V_{step}, generated by the half-bridge, is a resonant overshoot of the primary-transferred lamp voltage v_{DBD}' which depends on the quality factor Q_s of the series resonant circuit. The peak factor as ratio between V_{step} and $V_{DBD,pk}'$ solely depends on the quality factor Q_S of the resonant circuit as already defined in Formula (4.15) which is now modified to:

$$pkf = \frac{V_{DBD\,pk}'}{V_{step}} = 1 + e^{-\frac{\pi}{2 \cdot Q_s}} \qquad \text{with} \qquad Q_S = \frac{1}{R_S} \cdot \sqrt{\frac{L}{C_{DBD}'}} \tag{5.1}$$

Formula (5.1) is valid for $C_{DC} \gg C_{DBD}'$; $C_{OSS,HS}$, $C_{OSS,LS} \ll C_{DBD}'$; $L \ll L_m$.

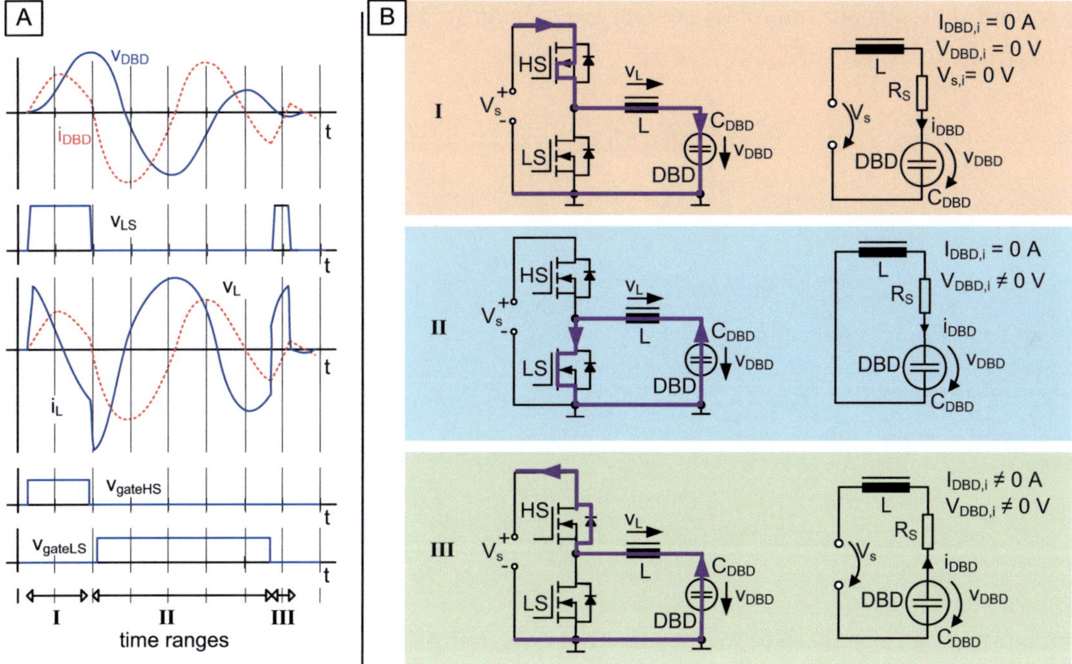

Figure 72: Basic operation mode of the SP-topology. A: Time ranges and related time-dependent waveforms. B: Simplified equivalent circuits for the three main time-domains. All values are primary-transferred with respect to the transformer.

By using the Laplace transformation the relevant time-dependent electrical functions can be calculated. At first, voltage and current are obtained from the simplified circuit of Figure 72B(I) for the first time-domain:

$$v_{DBD}(t)_{D1} = V_s\left[1 - \left[\cos(\omega_d \cdot t) + \frac{\alpha}{\omega_d}\sin(\omega_d \cdot t)\right] e^{-\alpha t}\right] \tag{5.2}$$

$$i_{DBD}(t)_{D1} = \frac{V_s \cdot \sin(\omega_d \cdot t) \cdot e^{-\alpha t}}{L \cdot \omega_d} \tag{5.3}$$

In order to obtain a simple expression for the maximum current in the first time-domain, an undamped circuit is assumed. Therewith, the peak current solely depends on the supply voltage and the characteristic impedance of the circuit:

$$i_{DBD,pk,D1} = \frac{V_S}{L \cdot \omega_0} = \frac{V_S}{Z_0} \tag{5.4}$$

Regarding the transformer design, it is important to determine the voltage-time (v-t) product that is applied to the primary winding. The first positive voltage half-wave generates the largest v-t product of the whole pulse. The v-t product can be calculated by integrating Formula (5.2) from $t = 0$ to $t = \pi/\omega_d$ which yields:

$$(v \cdot t)_{D1} = V_S \frac{\pi}{\omega_d} \tag{5.5}$$

For the second time-domain the initial current is assumed to be zero. We therefore gain for the changed equivalent circuit according to Figure 72B(II):

$$v_{DBD}(t)_{D2} = V_{DBD,i} \left[\cos(\omega_d t) + \frac{1}{\sqrt{4Q_S^2 - 1}} \sin(\omega_d t) \right] e^{-\alpha t} \tag{5.6}$$

$$i_{DBD}(t)_{D2} = \frac{V_{DBD,i} \cdot \sin(\omega_d t)}{L \cdot \omega_d} e^{-\alpha t} \tag{5.7}$$

The derivative of the DBD current (5.7) has zeros at times defined by Formula (5.8) which are identical to the zero-crossings of the DBD voltage (5.6):

$$t[v_{DBD}(t)_{D2} = 0] = \frac{1}{\omega_d} \arctan\left(\frac{\alpha}{\omega_d} \right) \tag{5.8}$$

However, integrating Formula (5.6) from $t = 0$ to the time defined by Formula (5.8) does not deliver concise information. Instead, the wave-shape is approximated to be a pure cosine quarter-wave. This approximation is valid as long as a high quality factor is ensured. The voltage-time product enclosed by the DBD voltage in the second time-domain up to the zero crossing defined in Formula (5.8) is then:

$$(v \cdot t)_{D2} = \frac{V_{pk}}{\omega_d} = 2\pi \frac{V_{pk}}{T_d} = \frac{\pi \cdot V_{pk}}{2 \cdot t[v_{DBD}(t)_{D2} = 0]} \tag{5.9}$$

With the peak voltage from Formula (4.14) and the v-t product of the first domain according to Formula (5.5) the voltage-time product of the first voltage half-wave is:

$$(v \cdot t)_{hw1} = \frac{V_S \cdot \pi + V_{pk}}{\omega_d} = \frac{V_S}{\omega_d}\left(\pi + \left(1 + e^{-\alpha\frac{\pi}{\omega_d}}\right)\right) \tag{5.10}$$

Finally, the complex characterisation of time-domain (III) is expressed by Formulae (5.11) and (5.12), both having sinusoidal and cosine terms.

$$v_{DBD}(t)_{D3} = V_s + \left[\left[\alpha V_s + \frac{V_{DBD,i}(R_1 - \alpha L)}{L} - \frac{I_{DBD,i}}{C_{DBD}}\right] \cdot \frac{\sin(\omega_d t)}{\omega_d} + (V_{DBD,i} - V_s)\cos(\omega_d t)\right] e^{-\alpha t} \tag{5.11}$$

$$i_{DBD}(t)_{D3} = \left[\left[\frac{V_s - V_{DBD,i}}{L} + \alpha I_{DBD,i}\right] \frac{\sin(\omega_d t)}{\omega_d} - I_{DBD,i}\cos(\omega_d t)\right] e^{-\alpha t} \tag{5.12}$$

5.1.3 Parasitic Resonances and Their Damping

Beside the intended main resonance that is excited by the half-bridge in order to generate and shape the pulse that drives the DBD lamp, unwanted parasitic oscillations occur in DBD pulse generators and are major aspects in terms of reliability and efficiency.

Reliability may be affected since resonance phenomena can lead to violations of the SOA of the semiconductors as a result of magnetic saturation or excessive apparent power that increases switching and conduction losses. Handling of the parasitic resonances is possible by sufficient damping and other actions that restrict their occurrence.

The wave-shape of v_{LS} (see Figure 71A) is suited to give an overview of the occurring parasitic resonances. Figure 73 exemplarily shows the wave-shape of v_{LS} during a full repetition period.

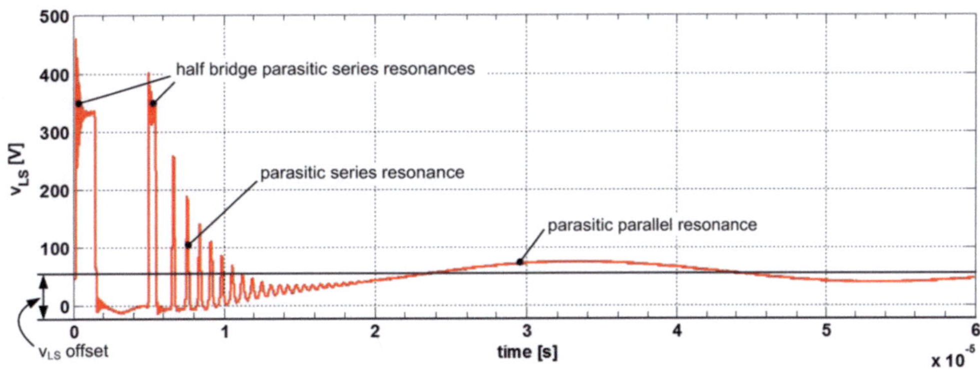

Figure 73: Parasitic resonances within a full repetition period. Visualised is v_{LS} at T_{rep} = 6 µs, V_s = 350 V, $L_{m,prim}$ = 550 µH and C_{DC} = 10 µF.

The fast turn-on of HS initiates the pulse but beside that the generated steep voltage step also excites the output capacitance $C_{OSS,LS}$ of LS to resonate with the parasitic bridge stray inductance. A comparable resonance occurs when energy is recovered through anti-parallel diode D_2 (see Figure 71A). However, the ZVS-commutation from HS to LS shows nothing but little ringing which is due to the met ZVS condition.

After energy recovery, the energy left in the circuit cycles again. But since the energy level is too small, v_{LS} does not reach V_s level any more. Instead, the v_{LS} peak levels decay. Due to the non-linear, highly voltage-dependent nature of the switch output capacitances, the wave-shape of the parasitic series resonance is highly non-symmetric. As can be seen from Figure 73, the short-time average value of v_{LS} rises continuously between 10 µs and 30 µs. Then, the voltage cycles around the v_{LS} offset value. This offset value represents the steady state voltage level necessary to keep the average voltage-time product applied to the transformer at zero. In other words, i_{Lm} recharges C_{DC} during every idle time. The subsequent parasitic parallel resonance also charges $C_{OSS,LS}$ to an offset visible in Figure 73.

Parasitic resonances also occur in circuits that are not intended to show any resonance such as hard switched bridge topologies. Huang (2007) shows a way of dampening oscillations that occur subsequently to the burst of pulses generated by the presented topology. The parasitic resonance is dampened by short-circuiting of the transformer's primary winding. Here, different attempts will be discussed.

After pulse generation and energy recovery, typically 10 % of the energy originally fed into the circuit remains stored in the circuit's components as remaining energy E_{rem}. This energy is the basis for parasitic resonance excitation. Figure 74 displays an equivalent circuit containing the relevant components for the time between the pulses (t_{idle}). $C_{OSS,HS}$ represents the equivalent capacitance of D_1, D_2 and HS.

In Figure 74, an equivalent circuit, including the components interfering in the parasitic series resonance, is depicted.

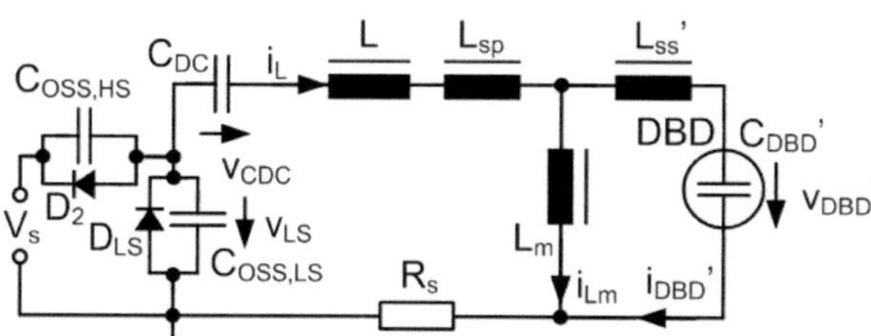

Figure 74: Equivalent circuit for time range between pulses t_{idle}. Note the multi-resonant structure with a main series (f_{PS}) and a main parallel resonance (f_{PP}). The DBD capacitance, the DBD current and voltage as well as L_{ss} are primary-transferred.

Figure 75 shows key waveforms of a 250 kHz pulse with a repetition period of 25 µs. It can be seen that when LS is opened, energy recovery occurs around t = 5 µs by means of the negative current i_L, which forces V_{LS} to increase. After i_L is back to zero again, the remaining charge on $C_{OSS,LS}$ forces i_L to increase, which charges C_{DC} and C_{DBD}'. The anti-parallel diode of LS now shortens $C_{OSS,LS}$. This explains why at first V_{LS} shows an asymmetric waveform. It is not until the DC-level of V_{LS} rises that the resonance becomes symmetric. Whenever v_{LS} crosses zero, resonance parameters, such as quality Q_{ps} and natural frequency f_{ps}, change. Formulae (5.13) to (5.15) express this change which is also visible from the waveform of v_L in Figure 75. Equation (5.15) confirms that the change in quality is solely negligible if the semiconductor capacitances are large compared to the product of C_{DC} and C_{DBD}'. As long as $C_{OSS,LS}$ and $C_{OSS,HS}$ are small, period and peak voltage amplitudes of the resonance significantly change depending on v_{LS} amplitude. $Q_{ps,VLS>0V}$ and $f_{ps,VLS>0V}$ are by principle higher than the parameters at v_{LS} = 0 V due to the extra series capacitance given by $C_{OSS,LS}$ and $C_{OSS,HS}$.

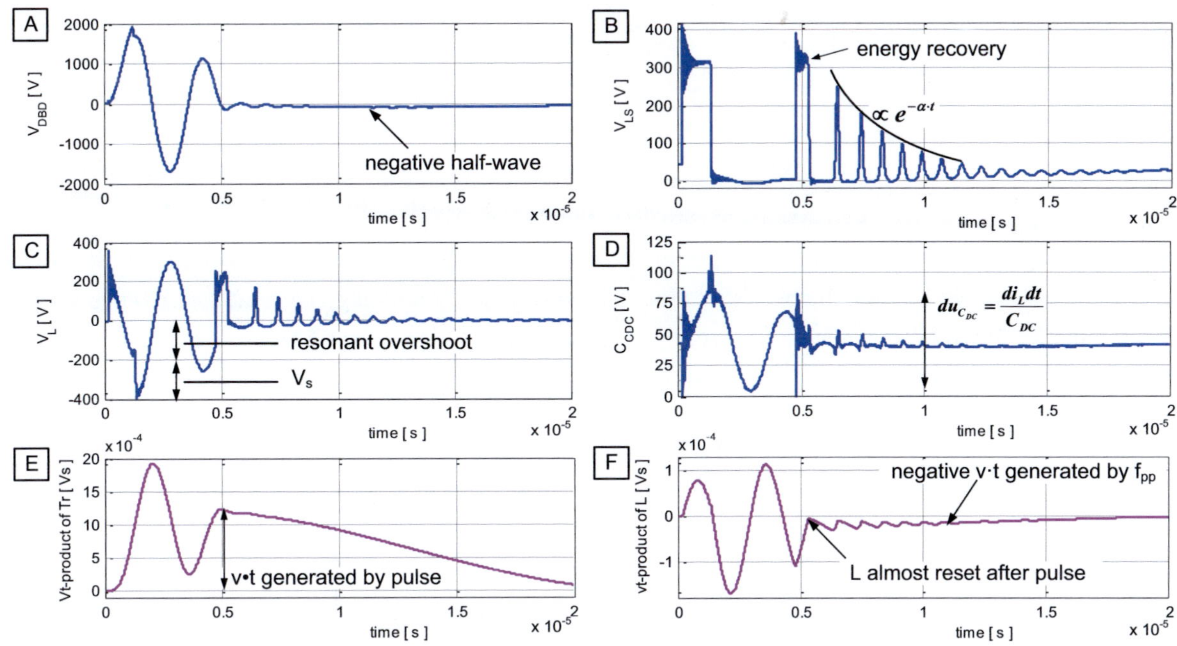

Figure 75: Key waveforms for a pulse with T_{rep} = 25 μs and t_{pulse} = 250 kHz. Note that the inductor is left with almost zero dv/dt while the transformer needs to compensate almost 2/3 of the maximum dv/dt by freewheeling during t_{idle}.

$$f_{\substack{ps \\ V_{LS}=0}} = \frac{1}{2\pi\sqrt{L_s \cdot \left(\frac{C_{DC} \cdot C_{DBD}'}{C_{DC}+C_{DBD}'}\right) - \left(\frac{R_s}{2L_s}\right)^2}} \tag{5.13}$$

$$f_{\substack{ps \\ V_{LS}>0}} = \frac{1}{2\pi\sqrt{\frac{L_s}{\frac{1}{C_{DC}}+\frac{1}{C_{DBD}'}+\frac{1}{C_{OSS,LS}+C_{OSS,HS}}} - \left(\frac{R_S}{2L_s}\right)^2}} \quad \text{with} \quad L_s = L + L_{sp} + L_{ss} \tag{5.14}$$

$$\frac{Q_{ps,V_{LS}>0}}{Q_{ps,V_{LS}=0}} = \sqrt{1+\frac{C_{DC} \cdot C_{DBD}'}{(C_{OSS,HS}+C_{OSS,LS})(C_{DC}+C_{DBD}')}} \tag{5.15}$$

The parasitic series resonance frequency f_{ps} for v_{LS} = 0 V, as given in Formula (5.13), equals the natural resonance frequency of the main series resonance that is used for operating the lamp. Since the resonant current transports charges to all capacitors in the circuit, the amplitude of the voltage shape is inversely proportional to the respective capacitance. In conclusion, for v_{LS} > 0 V, the frequency is high and the amplitude is restricted by D_2, while for v_{LS} = 0 V, the frequency as well as the quality factor are low. Therefore, the parasitic series resonance is not capable of saturating one of the magnetic components. However, this resonance is a significant source of electromagnetic radiation. Furthermore, Figure 75 confirms that, although the volt-seconds of L after energy recovery already have reached zero, the series resonance at first builds up another negative volt-second level. This is compensated by means of i_L which freewheels while charging $C_{OSS,LS}$ positively. Finally, due to its high capacitance, C_{DC} acts as a level shifter that shifts the amplitude of the low frequency parallel resonance waveform and thereby provides a symmetric waveform shape across LS. At f_{ps}, L_m shows rather high impedance and therefore has almost no influence on the resonance parameters.

PARASITIC PARALLEL RESONANCE

Beside L, also the transformer needs to be reset after the pulse. In Figure 75 a second positive peak of volt-seconds after energy recovery is finished (t = 5 µs) can be found. This is decreased by the freewheeling current i_{Lm}, which continuously charges C_{DBD} with negative polarity, although it is lower than the freewheeling current of L. Freewheeling is finished when v_{DBD} crosses zero (Figure 75, t = 25 µs). If the repetition period is enlarged beyond this point in time, a parallel resonance occurs between the mutual transformer inductance L_m and all parallel connected capacitors. Calculating the total capacitance which is connected in parallel to L_m as depicted in Figure 76, C_{pp} is defined by Formula (5.16).

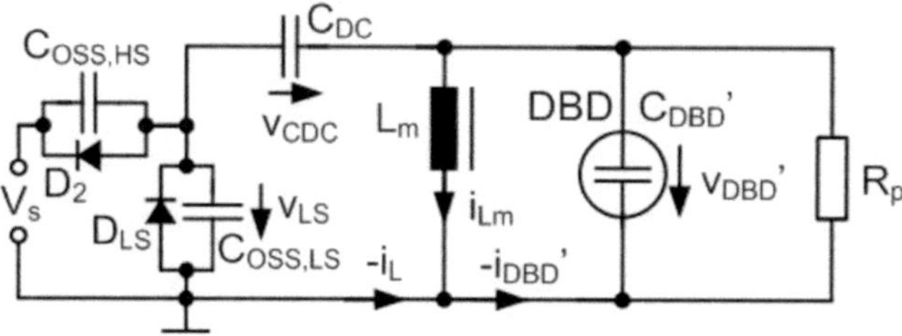

Figure 76: Basic equivalent circuit relevant to describe the parasitic parallel resonance.

$$Q_{pp} = R_p \cdot \sqrt{\frac{C_{pp}}{L_m}} \quad \text{with } C_{pp} = \left(C_{DBD}' + \frac{C_{DC} \cdot (C_{HS} + C_{LS})}{C_{DC} + C_{HS} + C_{LS}} \right) \tag{5.16}$$

The parasitic parallel resonance is characterised by its natural frequency f_{pp} and quality factor Q_{pp}. The primary capacitances play a minor role because their combined value is much lower than the value of the primary-transferred lamp capacitance C_{DBD}'. Because of the parallel connection, v_{LS} and v_{DBD}' show equal polarity at free resonance. The series inductances L, L_{sp} and L_{ss} show low impedance at f_{pp} which is typically at least one order of magnitude lower than f_{pulse}. However, f_{pp}, as given in Formula (5.17), is high enough for C_{DC} to show also low impedance.

$$f_{pp} = \frac{1}{2\pi \sqrt{L_m \cdot C_{pp} - \left(\frac{1}{2 C_{pp} \cdot R_p} \right)^2}} \tag{5.17}$$

The corresponding quality factor Q_{pp} and equivalent capacitance C_{pp} are given in Formula (5.16). A voltage offset across LS and C_{DBD} limits the initial voltage step and thereby also limits the maximum lamp voltage. Low resonance frequency and high voltage amplitudes produce large volt-seconds and therefore can force the transformer into saturation. With the assumptions $(C_{OSS,HS} + C_{OSS,LS}) \ll C_{DBD}'$ and $R_p \to \infty$, the maximum volt-seconds generated by the parasitic resonance can be calculated. Combining Thomson's oscillation formula and the definitions of the area of a sine half-wave and the capacitor's energy, Expression (5.18) is achieved.

$$\Delta v \Delta t = \sqrt{2 \cdot L_m \cdot E_{rem}} \tag{5.18}$$

If L_m is diminished by means of lower winding count N in order to reduce maximum volt-seconds, the voltage time product will change by the factor N. However, reduction of the magnetic core area will linearly reduce the volt-second rating of the transformer, but will reduce the actual volt-seconds generated by the resonance merely by square-root. A key to restricting the resonance is to lower the remaining energy E_{rem}. The problem is that E_{rem} changes significantly with variation of T_{rep} because of an intrinsic feedback. The resonance waveform also determines the voltage level of v_{LS} at pulse initialisation. A high v_{LS} voltage level reduces the initial step height. This leads to a variation of the pulse shape and thereby a changed E_{rem} after pulse.

Managing Parasitic Resonances

As has been found, the level of v_{LS} at the point of pulse initialisation has direct impact on the pulse shape and the resonance appearance. It can be inferred from Figure 75 that in the worst case, the negative volt-seconds are at least in the same range as the positive ones, created by the actual pulse. The optimum, characterised by resetting the core without generating negative volt-seconds, is found here at a repetition time of 25 µs. It is achieved at the time when v_{DBD} returns from the first negative half-wave and crosses zero volt. Referring to Figure 77, by variation of L_m, f_{pp} can be chosen to fulfil Formula (5.19) with n = 0, 1, 2, ..., determining the number of additional sine periods.

$$f_{pp} \approx \frac{1+2n}{2(T_{rep} - t_{pulse})} \tag{5.19}$$

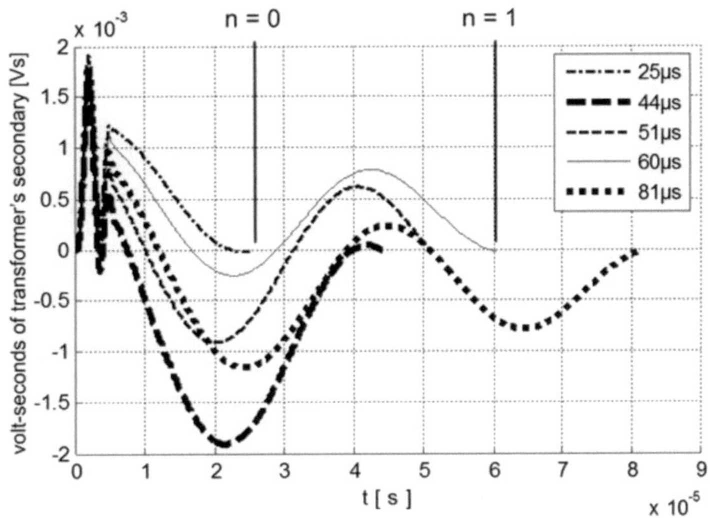

Figure 77: Variation of T_{rep} with standard pulse shape and constant f_{pp}. Volt-seconds seen by the transformer's secondary are displayed.

Furthermore, for a given f_{pp}, T_{rep} and t_{pulse} can be adjusted accordingly. This will achieve that the lamp voltage at the beginning of the new pulse is still negative and v_{LS} is minimal while freewheeling of i_{Lm} is just finished. Another possibility could be the variation of the mutual inductance. According to Formula (5.16), a possibility to reduce Q_{pp} is to significantly increase the transformer's mutual inductance. This will also reduce f_{pp} and thereby, referring to Formula (5.19), allow adjustment to a certain T_{rep}. However, the maximum achievable L_m is restricted by the power dissipation of the transformer windings if winding count N is increased. If the transformer core cross section is increased, limitations come from volume restrictions.

In Figure 78, the opposite had been done: L_m was reduced by inserting an air gap between the core halves. The experiment verifies that with reduced L_m, Q_{pp} as well as volt-seconds of the voltage half-waves increase significantly. This relation suggests that, according to Formula (5.18), E_{rem} must have increased.

Daub (2010) suggests a C_{DC} value of at least three orders of magnitude higher than the primary-transferred DBD lamp capacitance C_{DBD}' in order to minimise the voltage drop across C_{DC}.

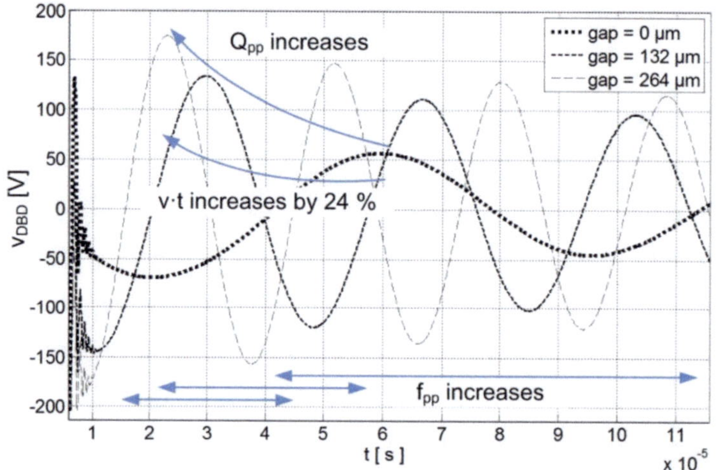

Figure 78: L_m reduced by insertion of an air gap. Q_{pp} and volt-seconds of transformer increase significantly.

As shown in Figure 79, due to its direct impact on all resonances, C_{DC} must be carefully selected. A value too small will not support the plasma current at the moments of ignition as it virtually reduces the barrier capacitance C_b to:

$$C_{b,tot} = \frac{C_b \cdot {}^{C_{DC}}\!/_{r^2}}{C_b + {}^{C_{DC}}\!/_{r^2}} \tag{5.20}$$

As can be seen from Figure 79A, for too low capacitance values the dynamic voltage drop across C_{DC} decreases the voltage level of v_{DBD} directly after ignition and therefore restricts a further energy transmission into the lamp. Consequently, as capacitance lowers, the AC voltage amplitude across C_{DC} and thus also the necessary supply voltage V_s rise. However, a small C_{DC} also reduces Q_{pp} and E_{rem} because it contributes less energy to the resonance as shown in Figure 79B. Because this effect is negligible compared to the disadvantage of increased voltage drop, reduction of C_{DC} capacitance is not beneficial.

Again, the SP-topology does not allow for a complete removal of C_{DC} either. Otherwise i_{Lm} may contain a significant DC component that leads to ohmic losses in the copper wire and conducting semiconductors (LS body diode). Additionally, turn-on of HS no longer meets ZCS condition. The loss can be calculated by comparing the energy in mutual inductance E_{Lm} after and before the pulse. Figure 80 shows waveforms of a SP-topology driving a 33 pF load. The transformer has a ratio of r = 8 and a primary mutual inductance (small signal measurement) of 48 μH. As can be seen, due to the unbalanced voltage-time product of v_{LS}, L_m forces a DC current of initially 2.5 A through the circuit. The ohmic resistance of the circuit and the forward-voltage across the anti-parallel body diode of LS dissipate power that reduces E_{Lm} by a factor of four during idle time. Although the parasitic parallel resonance is effectively circumvented if C_{DC} is missing. Conversely, the prevention of excessive ohmic losses or even catastrophic transformer saturation cannot be ensured.

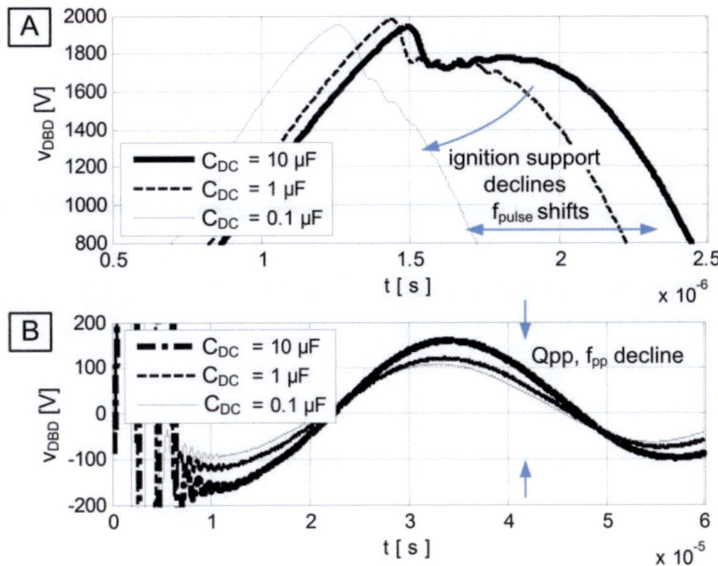

Figure 79: Influence of C_{DC}, details of v_{DBD} waveform. A: Detail of v_{DBD} at first pulse peak - decline of ignition support. B: Changes of parasitic parallel resonance during T_{pause}.

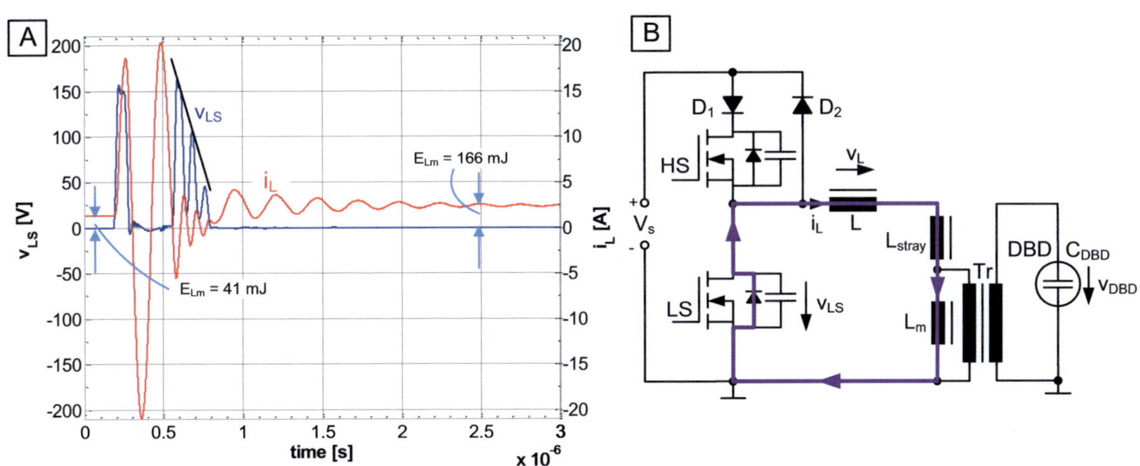

Figure 80: A: Effect of insufficient freewheeling – DC component of i_{Lm}; 4.8 W of wasted power at f_{rep} = 40 kHz, SP-topology without C_{DC}, V_s = 150 V. B: SP-topology without C_{DC} and marked i_{Lm} freewheeling path.

A final option is to modify the switching pattern of the power semiconductor switches. Coming back to the original topology of Figure 71, to obtain a maximum initial step height, it is suggested to briefly switch LS on before HS is activated. If at this time v_{LS} is positive, by switching on LS, the stored charges of C_{DC} and C_{DBD}' are used to load L. If LS is then opened, similarly to the energy recovery mechanism, L forces current to charge $C_{OSS,LS}$. Depending on the charge time, the transferred charge is enough to load $C_{OSS,LS}$ up to V_s level. This enables HS to turn on with ZVS, reducing ringing and EMI significantly. The initial step height is thereby optimised and is equal to V_s level. This method is a milestone since the actual v_{LS} level in advance of the actual pulse cannot influence the pulse shape any more.

This technique, which is also used in combination with other topologies, works as long as Inequality (5.21) is fulfilled at the point in time immediately before LS is switched on.

$$(C_{OSS,HS} + C_{DC}) \cdot V_S^2 < C_{DBD}' \cdot V_{DBD}'^2 + C_{DC} \cdot V_{CDC}^2 \tag{5.21}$$

However, energy stored in the semiconductor capacitances is shorted by LS and therefore lost. Different switching patterns may be used to enhance the shape of the waveforms between the pulses. LS can be switched on during the whole pause time or even multiple on-times can be initiated as demonstrated in Figure 81. The aim here is to reduce the resonance frequency f_{pp} for adaptation to different T_{rep} according to equation (5.19) as well as to provide ZVS for pulse initialisation. Due to turned-on LS, solely L_m and C_{DC} define f_{pp}. Figure 81 shows that, although the volt-seconds of the transformer increase shortly after the pulse, the total volt-second swing is reduced. Inverter efficiency is enhanced by 2 % because of reduced volt-seconds and reduced switching losses due to ZVS turn-on of HS. The linear drop of the vt-product in Figure 81 occurs due to the almost constant voltage level of the high-value C_{DC}.

A similar switching pattern is introduced in Chapter 5.2.3 on page 139 where it helps to manage the parasitic resonances of the SPSP-topology.

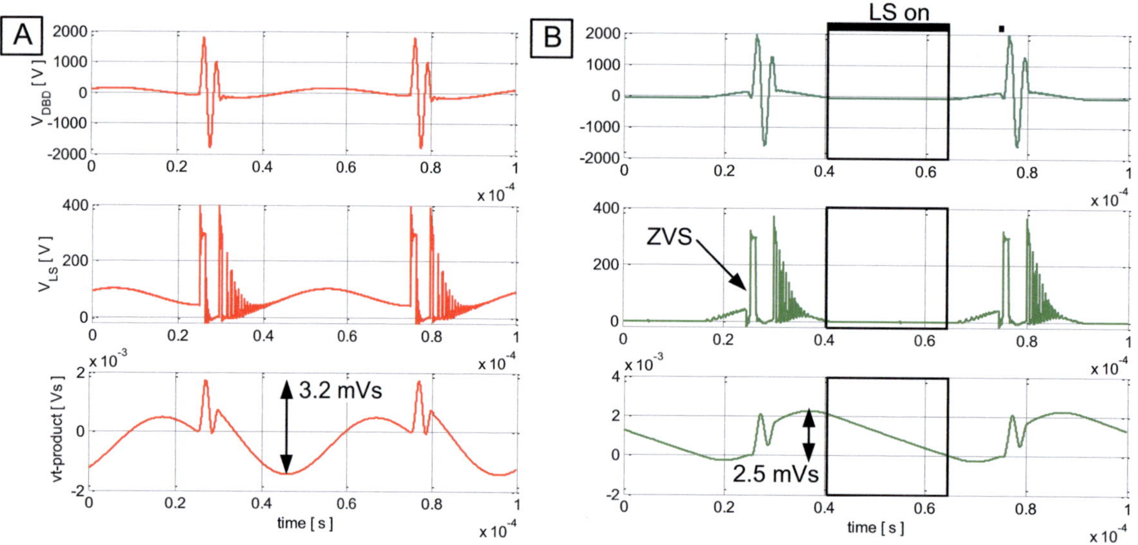

Figure 81: Comparison of SP-topology without and with modified switching pattern, wave-shapes of v_{DBD}, v_{LS} and volt-seconds of the transformer. A: Standard switching pattern as given in Figure 72A. B: Enhanced switching pattern with LS additionally enabled between pulses for linear transformer reset and briefly directly before pulse to achieve ZVS turn-on of HS.

5.1.4 Inductor Placement in Relation to Transformer

At first glance, relating to the SP-topology schematic it should make no difference whether the inductor is placed on the primary or secondary side of the transformer. Required inductance, voltage and current rating are transformed in the way that the energy storing requirement stays the same. On the primary side, the inductor faces high current and low voltage and needs to have comparatively low inductance while on the secondary side, the opposite is the case. However, a closer look reveals that a placement close to the square-wave output (half- or full-bridge) is to be preferred due to the following reasons:

First of all, the copper filling factor of a secondarily placed inductor is much lower since insulation requirements are stronger. Moreover, instead of a multi-layer winding, a single layer winding is sufficient due to lower voltage-time product easing achievement of high self-resonance frequency.

Secondly, the inductor filters high-frequency components and thereby reduces the HF losses of the more complex transformer if connected to the primary side. For the same reason, the insulation of the transformer's secondary winding is less stressed because dv/dt is reduced.

Equally important, electromagnetic interference (EMI) is drastically reduced as the area of fastest di/dt and dv/dt is minimised.

An alternative is to include the inductor in the transformer by utilising its stray inductance.

5.1.5 Unipolar Pulse Generation by Current Cut-off

The SP-topology is also suited to generate unipolar pulses that mimic the wave-shapes generated by parallel topologies presented in Chapter 6. However, compared to the resonant flyback topology (see chapter 6.1) based on inductor pre-charge (Figure 40B), here the DBD capacitance is charged resonantly (Figure 40A). Consequently, the more efficient energy transfer provided by serial topologies presumably results in lower losses in the driving circuit. In order to shape unipolar pulses using SP-topology hardware, the standard switching pattern of the power semiconductor switches needs to be modified.

A possibility that allows for a variety of pulse parameter changes is the cut-off of inductor current. This can happen in two ways. The one used for energy recovery opens the resonant circuit in order to include the supply voltage source (V_S) in which the commutated current transfers the circuit's energy. This is done by opening LS while inductor current is negative. The experimental data gained by that is depicted in Figure 82. Another possibility is to restrict the energy flow from the voltage supply source to the resonant circuit by opening HS while positive current flows through it. This current cut-off scheme is used in Chapter 5.2.3 in order to enhance the pulse frequency. The principles of current cut-off were also suggested in a patent by Park (2008) and are the base of the optimum pulse wave-shape suggested in Chapter 3.4.6.

As depicted in Figure 40C, by opening HS or LS earlier than necessary for the generation of the standard pulse shape, the pulse properties change due to the changed equivalent circuit. This includes a principal reduction of the pulse period t_{pulse} and a significant deviation from the sinusoidal pulse shape. As can be seen in Figure 82A, pulse termination by energy recovery is depending on the content of energy to be recovered. An extreme case is the unipolar pulse shown. For its generation, LS is enabled for comparatively short time (see Figure 82C, green graph) and energy is recovered immediately after that. This unipolar pulse does not include a second or third half-wave. However, as the second half-wave grows with increased on-time of LS also the third half-wave appears. Note that the fully established three half-wave sinusoidal pulse leads to three ignitions indicated by the measured NIR radiation depicted in Figure 82D.

As experimentally evaluated in Chapter 5.2, the third ignition is not necessary for achieving discharge homogeneity. In fact, it can worsen system efficiency since both the third discharge may be dominated by glow losses and the topology efficiency is reduced by the ohmic losses generated by the additional apparent power that needs to be transferred. The SP-topology does not permit energy recovery of positive inductor current which is provided by the SPSP-topology presented in Chapter 5.2. To prevent the third ignition using the SP-topology, the negative half-wave needs to be significantly weakened by early energy recovery as done in Figure 82.

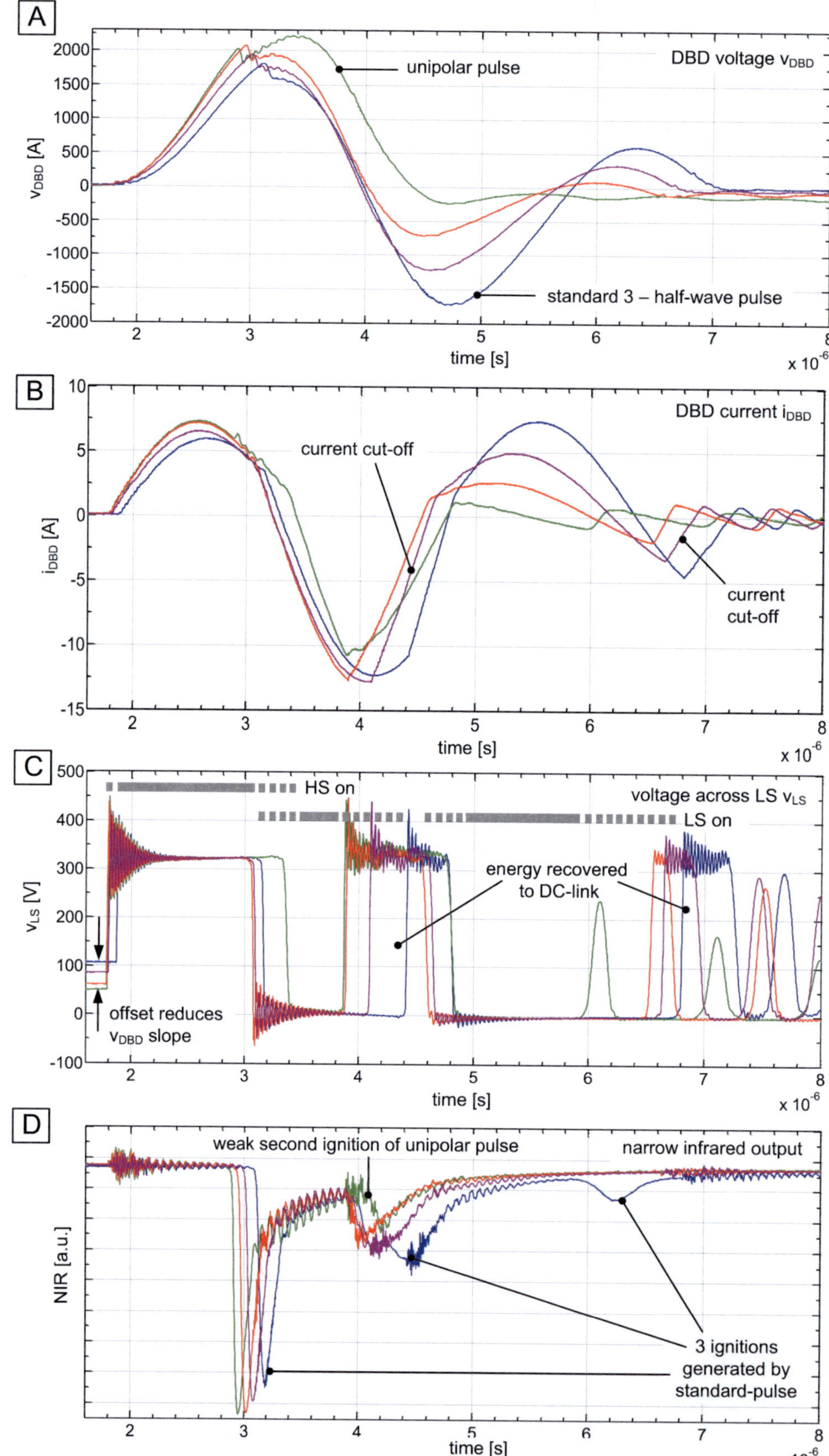

Figure 82: 2.9 nF flat DBD lamp operated with varied pulse shape; current cut-off during first negative current half-wave adjusts characteristics of the second negative v_{DBD} half-wave; variation from unipolar pulse to three half-wave standard pulse shape. f_{rep} = 40 KHz, V_s = 340 V. DBD voltage (A), DBD current (B), voltage across switch LS (C) with indication of times when LS and HS are switched on and NIR signal (D).

5.1.6 GENERATION OF DISTRIBUTED BIPOLAR PULSE

The SP-topology can also be used to generate distributed bipolar pulses. A variation of the delay length was experimentally examined driving a 2.9 nF flat DBD lamp at f_{rep} = 40 kHz. The results obtained are shown in Figure 83. To generate the wave-shape depicted in Figure 83A, HS had been kept on for the whole first current period that ends at the moment 3.9 µs. During that time-domain, the lamp capacitance is charged up to ignition voltage. Afterwards, the back swinging current lowers the amplitude of v_{DBD} to a 450 V offset. Put simply, this offset marks the energy consumed by the DBD and losses in other circuit elements during the first voltage half-wave. This is because an undamped resonant circuit would reach zero volts instead. According to Formula (2.27), this energy amounts to at least 294 µJ. However, the effective DBD capacitance immediately after ignition can vary between the extreme values C_{DBD} (lowest) and $C_b + C_p$ (highest). The latter is the parallel connection of C_b and C_p due to a short-circuited gap. Hence, assuming C_b = 1880 pF and C_p = 2200 pF, the maximum energy amounts to 405 µJ congruent to 16.2 W of dissipated real power.

Figure 83C shows the voltage across LS. Here it can be seen that after turn-off of HS, a low amount of energy still cycles within the circuit creating voltage valleys. For the sake of reduced turn-on loss it is favourable to turn on LS within such a voltage valley. This had been done in the first, third, fifth and seventh valley in order to vary the distance between the v_{DBD} voltage half-waves and thus also between the first and the second DBD ignition.

Paravia (2008b) already used distributed bipolar pulses with variable delay to determine a threshold current density required to achieve homogeneous ignition. He found that with rising delay time, and therewith rising time between first and second ignition, homogeneity is reduced. This is also a qualitative result of the experiment of Figure 83.

However, on the other hand DBD efficiency rises with extended delay time. Therefore, the experiment partially contradicts the statement that an immediate back-ignition after the first ignition enhances the DBD efficiency (see Chapter 3.4.3).

The inverter efficiency slightly scales down which is mainly due to the larger turn-on loss of LS.

The NIR radiation of the first ignition is almost independent from this delay time. However, the second ignition's NIR radiation shows significant variance. While the short-time peak continuously rises with enhanced delay time, the long NIR tail lowers its maximum amplitude. As in Chapter 5.2.3, it is also assumed here that the NIR signal consists of two signal parts having different time constants. They are possibly generated by different kinetic pathways and have possibly slightly differing wavelengths (see Figure 8 on page 36). The respective first peak of the back-ignition's NIR radiation rises with rising DBD efficiency while the part with the obvious larger time constant lowers if the delay of the back-ignition is enhanced.

This finding calls for further research regarding the time constants of the kinetic pathways of the Xe excimer system. Moreover, it calls the doctrine of the bipolar pulse into question.

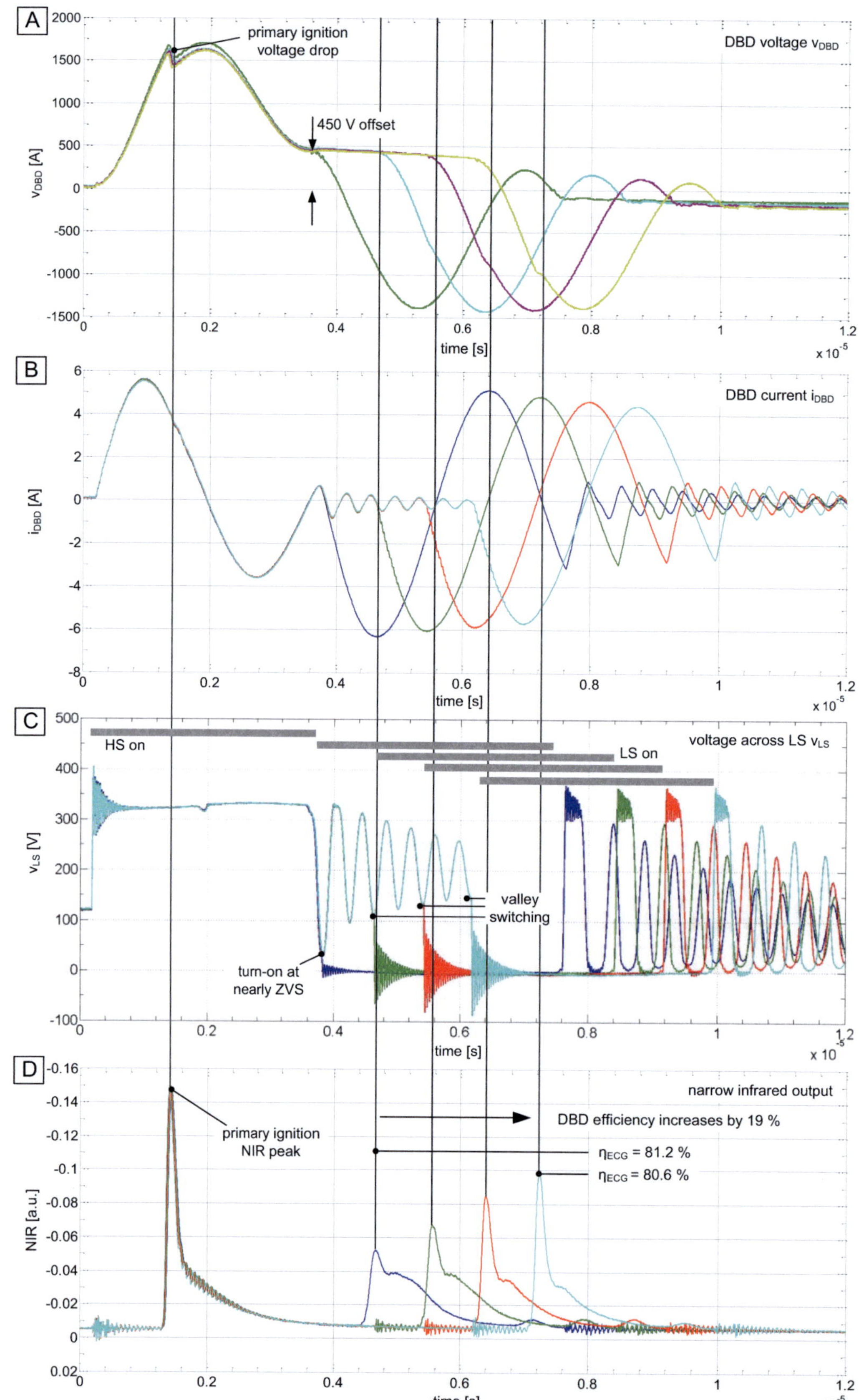

Figure 83: 2.9 nF flat DBD lamp operated with distributed bipolar pulse variation by delayed activation of LS; turn-on of LS in voltage valleys is advantageous as switching loss is reduced; f_{rep} = 40 kHz. DBD voltage (A), DBD current (B), voltage across LS (C) with indication of times in which switches are turned on and NIR signal (D). The author assumes that the NIR signal consists of two signal parts having different time constants.

5.1.7 High-Frequency Operation

As stated in Chapter 3.4, an enhanced pulse frequency improves the general conditions of the DBD discharge. Lamp homogeneity and efficiency can be maximised. The fundamental limitations for high-frequency operation were discussed in Chapter 4.6. The objective of this chapter is to show that as long as the DBD capacitance does not exceed hundreds of pico-farads the basic SP-topology of Chapter 5.1.1 in combination with standard or improved switching schemes (Chapter 5.1.3) is able to operate at pulse frequencies exceeding one MHz. Precondition for that is the enhancement of the three key components: the power semiconductors, the gate drivers and the power stage PCB. A key is to minimise stray inductances both on component and PCB level in order to allow for fast switching and high dv/dt as well as di/dt within the bridge-leg. Implemented power semiconductors need to have high ruggedness against Miller-feedback, a low output capacitance to current rating ratio and low gate charge. Suitable devices are presented in Chapter 4.10.

A first high-frequency design uses silicon RF-MOSFETs (Hartmann 2007). Figure 84A shows the schematic including main parasitic elements. The MOSFETs are mounted between the half-bridge PCB (Figure 84B, right) and a heat-sink. Their internal lead-frame structure as well as their short, wide contact pins which are directly soldered on the PCB reduce the bridge-leg inductance. Both RF half-bridge and resonant circuit PCB (Figure 84B, left) exhibit extensive interconnected ground areas in order to minimise ground impedance. Also, by means of PCB-trace layout the mesh C_{link}–HS–LS is optimised for minimal inductance in order to prevent any ringing and overshoots of V_s or $v_{DS,LS}$.

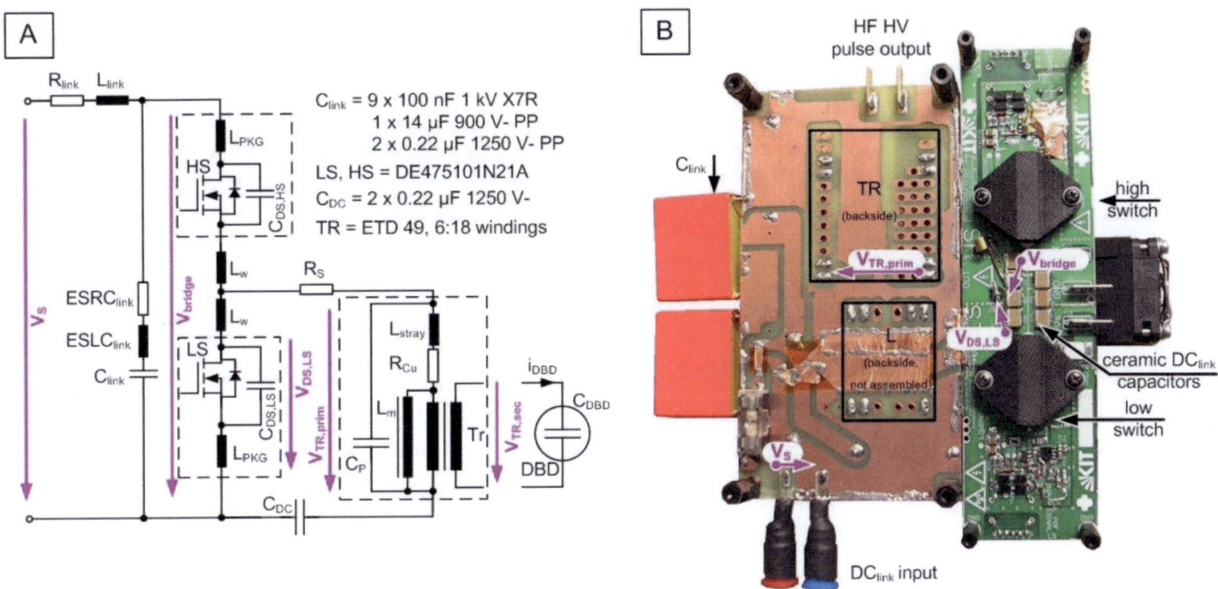

Figure 84: A: Schematic of the RF half-bridge with connected series resonant circuit. Parasitic circuit elements as well as impedance measure points are indicated. B: Picture of bottom side of RF half-bridge PCB (right) with connected series resonant circuit (left). The RF MOSFETs (DE475102N21A) and their drivers are pressed on a backside heat sink.

The switching scheme used to generate the pulse depicted in Figure 85 consists of a short on-time of LS in advance of activation of HS as suggested in Chapter 5.1.3 (Meisser 2010b). Thereby, HS can be turned on under partial ZVS condition further minimising ringing of $v_{DS,LS}$. The bulk capacitors merged in C_{link} are a collection of three different types of different form factors and materials. Ceramic capacitors (X7R) are placed directly opposite the MOSFETs on the PCB while foil capacitors (MKP) are placed on the application-specific resonance circuit PCB. This combination ensures low impedance broadband decoupling of supply voltage V_s and V_{bridge}. The pulse depicted in Figure 85 is mainly shaped by the series resonance of C_{DBD} and L_{stray}. An additional inductor is not required as the transformer's stray inductance was intentionally chosen to define the resonance frequency. 2.5 MHz are achieved driving a Planon® flat DBD lamp. However, the power switches exhibit excessive losses. Partly these are ohmic losses visualised by a significant voltage drop that occurs across LS at around 350 ns in Figure 85. Device characterisation revealed a significant increase of the on-resistance at elevated temperatures which is presented in Chapter 4.7. A second loss contribution has its origin in switching losses principally at LS turn-off.

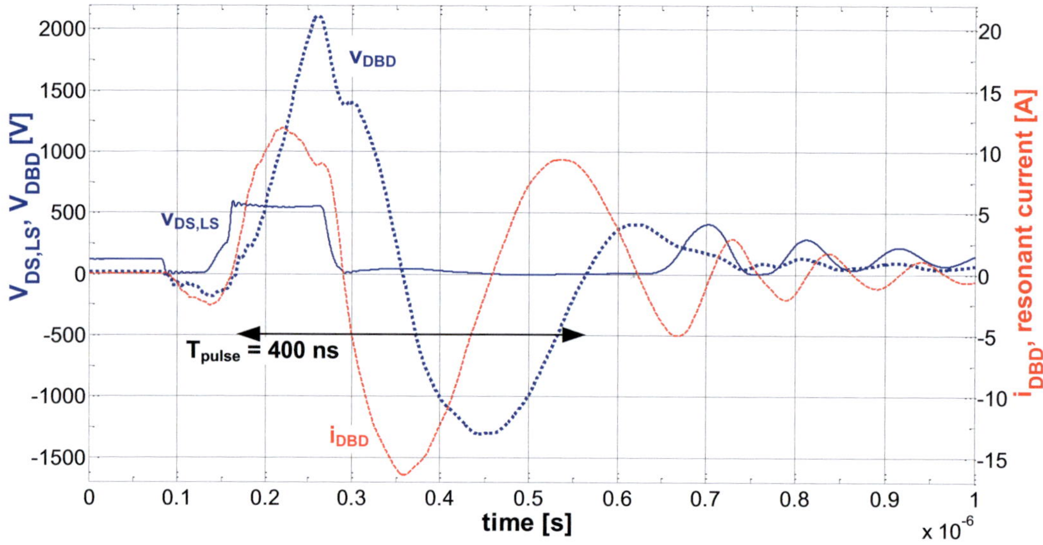

Figure 85: Waveforms obtained with the RF resonant pulse generator. f_{pulse} = 2.5 MHz, V_S = 500 V. Note that HS is switched on under partial ZVS condition due to previous LS turn-on. $V_{DS,LS}$ indicates excessive losses due to a high R_{DSon} during the main negative current half-wave.

The parasitic elements contained in the set-up were investigated by impedance spectroscopy as explained in Chapter 4.8. Measurements were carried out at zero DC voltage at different mesh points of the circuit as marked in Figure 84B. The results are depicted in Figure 86. Depending on the measured mesh point, the equivalent circuit applied to the measurement system changes. Consequently, comprehensive information regarding the parasitic element values is gained. Due to the missing DC-offset, $C_{OSS,LS}$ is almost an order of magnitude greater than the primary-reflected lamp capacitance. Having this in mind, the round-edged wave-shape of v_{LS} around 275 ns in Figure 85 can be explained. The main resonance between L and C_{DBD} is not visible which is also due to the large $C_{OSS,LS}$ at low voltage. The low impedance of $C_{OSS,LS}$ superimposes this resonance.

Two further impedance measurements were performed under DC bias. For both, the DC-blocker presented in Chapter 4.8 was used. The output capacitance C_{oss} of the MOSFETs is strongly voltage-dependent (IXYS 2009) and thus significantly influences commutation. It has an impact on the current through the MOSFET at turn-on and on v_{DS} at turn-off. C_{oss}'s value peaks at low v_{DS} amplitude and falls significantly if v_{DS} exceeds 10 % of $V_{DS,max}$. In the aforementioned half-bridge configuration, the change of C_{oss} is of high importance for the dynamic drain source voltage v_{DS}. If the ratio C_{oss}/I_{DS} is high, ΔC_{oss} has significant impact on the wave-shape of v_{DS}. Figure 87 compares $C_{oss}(V_{DS,LS})$ taken from the datasheet with the measured data. For this measurement, the RF half-bridge was supplied via a feed-in resistor with V_s = 600 V, while $V_{DS,LS}$ was varied between zero and 600 V. The impedance analyser was connected to the presented DC-blocker. The capacitance value was derived from the capacitive reactance value measured at 1 MHz. The measured data is a good match to the calculation of the total capacitance of the half-bridge basing on the datasheet value of $C_{oss}(V_{DS})$ and assuming V_s to be short circuited. Energy of up to 1 mJ is stored in the transistor output capacitances that would be dissipated at hard turn-on.

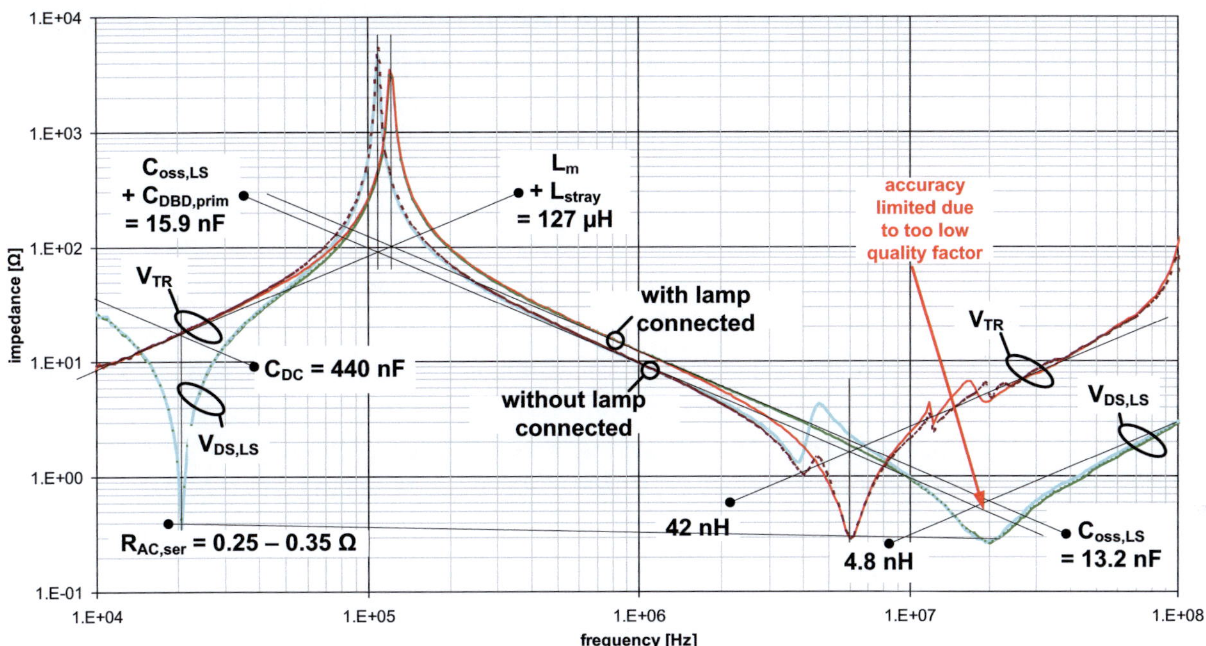

Figure 86: RF half-bridge from Figure 84B with connected resonant circuit. Impedance spectra are shown for different measurement points and configurations. Main parasitic parallel resonance is visible around 100 kHz.

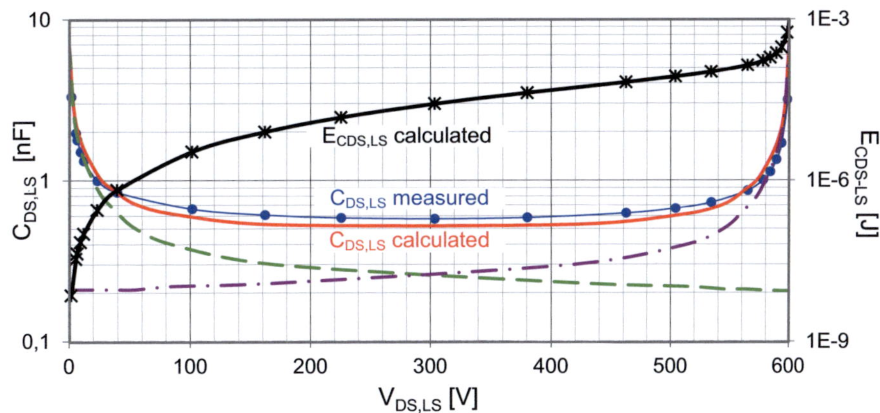

Figure 87: Measurement results and datasheet values of output capacitance C_{oss} and stored energy of a half-bridge configuration of two DE475102N21A RF MOSFETs versus drain-source voltage. The datasheet values related to the respective drain-source voltage of HS and LS were added to obtain the total capacitance which corresponds well with the measurement. Note that the extensive energy level of 1 mJ is stored within the total capacitance.

A third measurement concerned the filter performance of the DC-link capacitor group. The requirements for capacitors used in power electronic circuitry and especially those implemented for broadband supply rail stabilisation/ decoupling, e.g. bulk capacitors, include low parasitic inductance (ESL) and stable capacitance, especially at operation voltage. As long as no magnetic core is included, inductance is stable and independent from current and voltage. However, especially X7R ceramic capacitors have significant capacitance drift with changing voltage. Lower capacitance at elevated voltage is tantamount to reduced available charge, e.g. energy. Consequently, stabilisation properties deteriorate. Therefore, it is important to evaluate the high-voltage performance of the bulk capacitor arrangement in order to make sure that HF currents are still provided at high-voltage. Hence, impedance measurements with high DC-bias voltage have to be performed. For that measurement, the measurement probe was directly connected to the half-bridge, namely drain of HS and source of LS. The high-voltage measurement revealed that under applied supply voltage the decoupling is worse than under zero voltage as can be seen in Figure 88. This is because parts of the bulk capacitor arrangement, namely the MLCCs, significantly lose capacitance with rising voltage. Part of the decoupling capacitance is lowered from 3 μF to 265 nF.

This leads to an impedance peak at around two MHz. Consequently, peak current support is negatively affected. As a remedy, voltage-stable NP0 ceramic capacitors could be added in order to keep the capacitance more stable at elevated voltages. In addition to the measurement data also the accuracy margins of the measurement system (std. margin) and the used DC-blocker are depicted in Figure 88. It becomes obvious that evaluation of DC-link capacitors is too inaccurate below 200 kHz. Since the repetition frequency lies in the range of 20 – 200 KHz, it is nevertheless important to gain information about the buffer performance in this frequency range. An attempt that was not followed here could be the use of an impedance analyser supporting a lower frequency range such as the Agilent 4192A, for example.

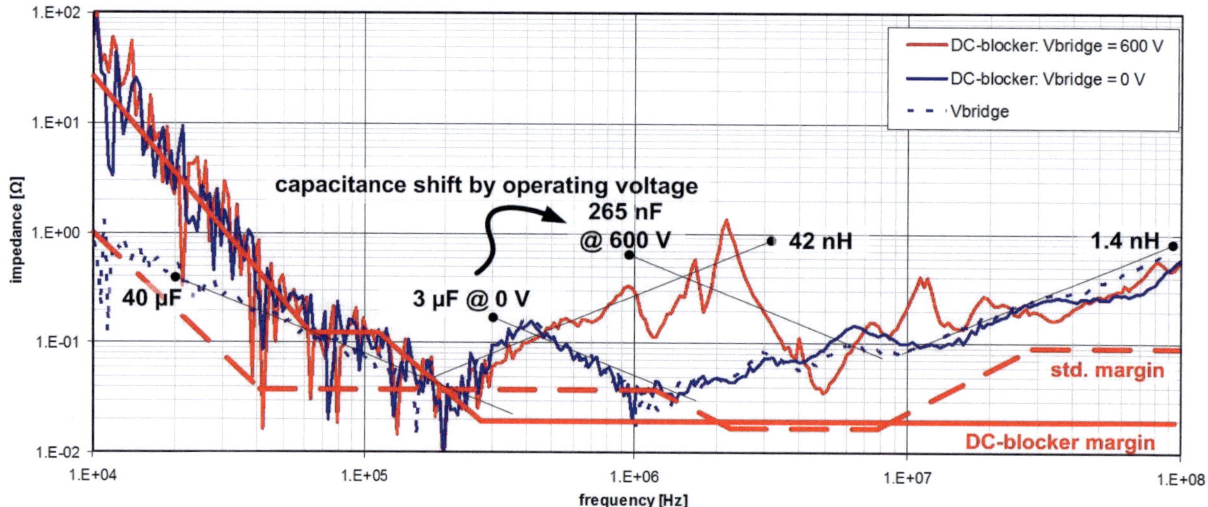

Figure 88: Decoupling performance under DC-bias: Respective lower accuracy margin of measurement system in continuous (attached DC-blocker) and dotted (no DC-blocker) lines.

5.1.8 New Transformer-less High-Voltage Operation

Due to the drawbacks of the transformer and inspired by the work of Kyrberg (2007b), ways to remove this component were investigated. One way is to operate the half-bridge with the required high voltage which is presented here while the alternative resonant upswing is discussed in the chapter 5.3.

Compared to the standard SP-topology incorporating a transformer, the high-voltage (HV) SP-topology exhibits much smaller bridge currents and consequently also lower di/dt but higher dv/dt at the bridge node. Equally important is the fact that capacitive energy storage becomes more important than magnetic energy storage at high voltage. This changes the resonance behaviour compared to transformer-equipped topologies and may lead to a higher electromagnetic interference generated during fast switch transitions. Additionally, losses related to hard switching become more severe and a higher dv/dt immunity of the gate drive galvanic insulation, both of signal and energy transmission, is required. This demand is met by using electromagnetic shielding techniques, POF signal transmission and best-in-class gate drive transformers.

The schematic of the basic transformer-less high-voltage supplied sinusoidal pulse (HV-SP-) topology is depicted in Figure 89A. The resonance circuit containing the DBD capacitance C_{DBD} and the series inductor L is excited by a half-bridge configuration consisting of two SiC JFETs. As the JFETs do not contain body diodes, the additional SiC Schottky diodes D1 and D2 are beneficially connected anti-parallel in order to support ZVS turn-on of the low-side switch LS and energy recovery at pulse termination. A pulse is initialised by enabling high-side switch HS with ZCS commutation. The L-C circuit reacts with a voltage overshoot of an amplitude given in Formula (4.15), being characteristical for the series topologies. The DBD voltage amplitude follows a wave-shape determined by Formulae (4.13) and (4.17) as depicted in time range 2 of Figure 89B. During the last part of this positive voltage slope, the DBD lamp is intended to ignite for the first time. By initiating current commutation from HS to LS, ZVS turn-on of LS is achieved. During time range 3, driven by a negative current, v_{DBD} swings back and leads to an advantageous second ignition of the DBD (Paravia 2007). After a third voltage half-wave, LS is turned off leading to an energy recovery preventing a further lamp ignition. This turn-off occurs under high current. Therefore, fast switching is important to minimise switching loss. C_{DBD} is almost discharged by the negative current i_L during one energy recovery cycle. This is in contrast to topologies that feature an inductor pre-charge and require multiple recovery cycles as presented in Meisser (2011b).

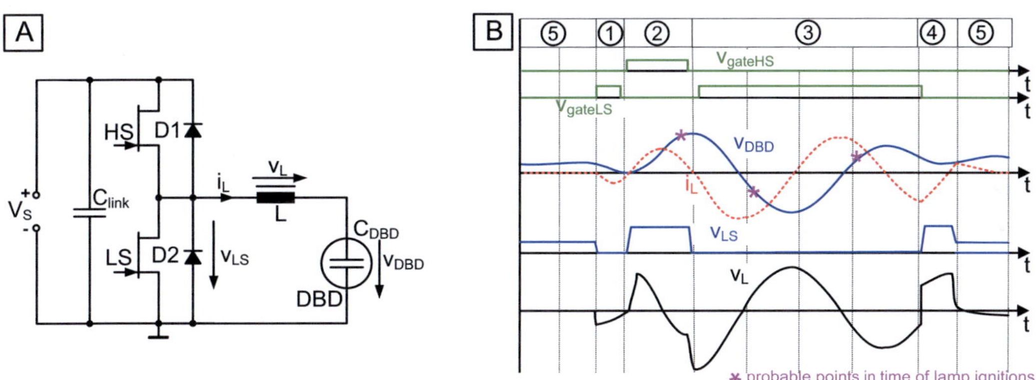

Figure 89: Schematic (A) and waveforms (B) of the HV-SP-topology equipped with SiC JFET power switches. Note that LS is turned on prior to the actual pulse in order to allow for a partly ZVS turn-on of HS. The energy stored in C_{DBD} is utilised for that. Due to the generated three voltage half-waves, three ignitions may occur per pulse.

However, an amount of energy equal to Formula (5.22) is left in the circuit which concentrates in terms of a positive voltage offset of v_{DBD} and v_{LS}.

$$E_{left,rec} = \frac{1}{2}\left[V_s^{\,2}\left(C_{DBD} + C_{LS,oss}\right) + i_L^{\,2} L\right] \qquad (5.22)$$

A unique turn-on of LS in advance of pulse initialisation utilises the energy left in the DBD capacitance in order to store it in inductor L during time range 1. With this, ZVS turn-on of HS is supported as suggested in Meisser (2010b). Time range 5 in Figure 89B is the idle time in which no pulse is generated.

Figure 90 shows the results of the experimental operation of the suggested HVS-SP-topology with a pulse frequency of 580 kHz. The supply voltage V_s has been adjusted to 1.15 kV. As can be seen in Figure 90A, the resonant circuit generated a peak voltage of 2 kV. This gives an overshoot factor of 1.74, which corresponds to an effective quality factor Q_{res} of 7 during the first voltage slope. Also presented is Q_{res} gained by logarithmic decrement calculation.

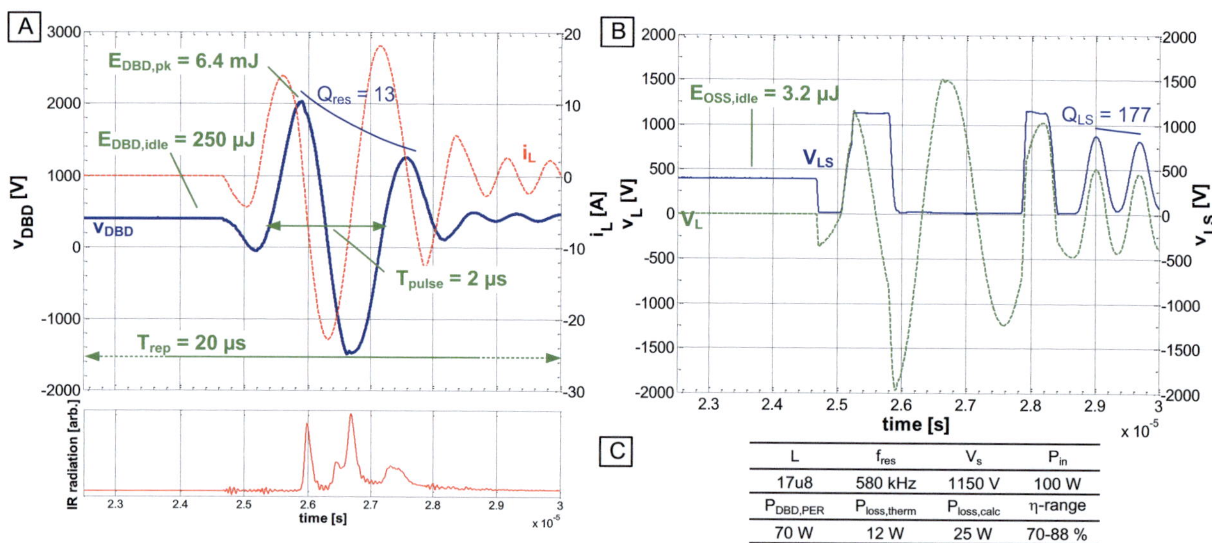

L	f_{res}	V_s	P_{in}
17u8	580 kHz	1150 V	100 W
$P_{DBD,PER}$	$P_{loss,therm}$	$P_{loss,calc}$	η-range
70 W	12 W	25 W	70-88 %

Figure 90: Experimental results operating a 3.18 nF flat DBD lamp. A: DBD voltage v_{DBD}, inductor current i_L and corresponding IR radiation indicating three ignitions; B: low-side switch voltage v_{LS} and inductor voltage v_L. Note that the turn-on of LS before the actual pulse is initiated helps to turn on HS under almost zero voltage. C: Key parameters proving the high efficiency of the transformer-less attempt.

Three ignitions of the DBD are indicated by NIR radiation shown in Figure 90A. After energy recovery, which is finished at 2.84 µs, merely 4 % of the C_{DBD} peak energy remains in the circuit. This energy leads to a parasitic series resonance between the switch output capacitances and L which is damped by the resistive components of the circuit. The maximum current during the pulse occurs during the maximum negative slope of DBD voltage at time 2.63 µs. The amplitude of approximately 23 A corresponds well with Formula (4.29). Although the efficiency is electrically measured to be 70 %, the thermal verification yields an efficiency of up to 88 %. The experience gained with these experiments was successfully used to set up a HV single period sinusoidal pulse topology which will be presented in Chapter 5.2.4.

To conclude, Chapter 5.1 describes different novel operation modes and a transformer-less circuit variant of the known SP-topology (Paravia 2008a). It builds upon the principal concepts of bipolar pulse generators as presented in Chapter 4. Notable outcome of the investigations is the successful handling of parasitic resonances by using a modified switching pattern and the variation of the amplitudes of the negative voltage half-wave in order to modify the second ignition (Chapter 5.1.5). The presented transformer-less variant bears the potential to drive very large DBD lamps with high energy requirements at highest pulse frequency. It is therefore predestined for industrial applications where power density is of higher importance than system price.

5.2 New SPSP-Topology

Rooting in the SP-topology presented in Chapter 5.1 the novel single period sinusoidal pulse (SPSP-) topology aims to remove the third voltage half-wave in order to prevent a third ignition. The third ignition is supposed to be less efficient and involves a comparatively low DBD power factor leading to avoidable losses in the power stage. Genuine single-period sinusoidal pulse waveforms are known from literature (Yukimura 2007) but efficient ways for their generation have not yet been reported. It is believed that the restriction to two ignitions per pulse is mainly relevant at low pulse frequencies which promote thermalisation of the gas. At higher pulse frequencies, even multiple sinusoidal periods forming a burst may be used to increase the power density of the lamp without declining the DBD efficiency (Beleznai 2009).

This objective is approached by providing an alternative path for energy recovery that terminates the pulse already after the second voltage half-wave. Consisting of two bridge-legs driving the DBD, the left leg is responsible for energy feed-in while the right leg is used for energy recovery, only. This distinguishes the SPSP-topology from a full-bridge topology presented by Lee (2008b) which generates alternating polarity sinusoidal half-waves instead of a whole sinusoidal period pulse. Because energy flow paths are insulated, the bridges may be operated with different voltage levels.

The following sub-chapters describe the SPSP-topology and present further switching schemes and circuit variations.

5.2.1 Schematic and Principal Mode of Operation

Figure 91 compares the schematics and typical waveforms of the SP- and the SPSP-topology. Additionally to the half-bridge of the SP-topology (Figure 91A) the SPSP-topology (Figure 91B) is equipped with an additional bridge-leg consisting of the energy recovery switch ERS and the energy recovery diode D_{ER}. This additional right bridge-leg permits to terminate the pulse by discharging L and C_{DBD} while the circulating current i_L is positive. As the comparison of the waveforms depicted in Figure 91A and B shows, it is then possible to terminate the pulse already after the negative voltage half-wave.

If this mode of operation is pursued, the requirements on the power semiconductors change. The energy recovery current no longer passes through the left bridge-leg and therefore fast Schottky diodes D1 and D2 can be omitted. Since the HS body diode does not conduct, a fast recovery property is not required. LS turns on under ZVS condition but is not turned off under high current. Hence, especially if high supply voltage levels are considered an IGBT may be used in order to reduce conduction losses. Energy recovery is performed by opening ERS to allow current to leave the circuit through fast Schottky diode D_{ER}. ERS may be turned on under ZVS or ZCS during the idle time, but generates losses at turn-off. D_{ER} needs to show fast recovery behaviour in order to reduce losses by circulating recovery currents.

For comparison, the SP- and the SPSP-topologies have been used to drive a flat DBD lamp (C_{DBD} = 3.18 nF, $V_{DBD,ign}$ = 2 kV) with a pulse frequency of 500 kHz. The time-domain experimental results are shown in Figure 92A while a comparison of key parameters is depicted in Figure 92B and C. Due to the different pulse shapes, the offset across DC-blocking capacitor C_{DC} varies slightly. Therefore, the supply voltage V_s has been adjusted to obtain homogeneous discharge. As the third ignition is omitted with the SPSP-topology the luminous flux would be lower if repetition frequency was kept constant. Instead, repetition frequency f_{rep} was adjusted for equal luminous flux. Due to excessive primary currents, all switches and diodes were replaced by two paralleled power semiconductors. An IGBT is used as LS instead of a CoolMOS™ transistor. Diodes D_1 and D_2 have been used for both topologies in this experiment.

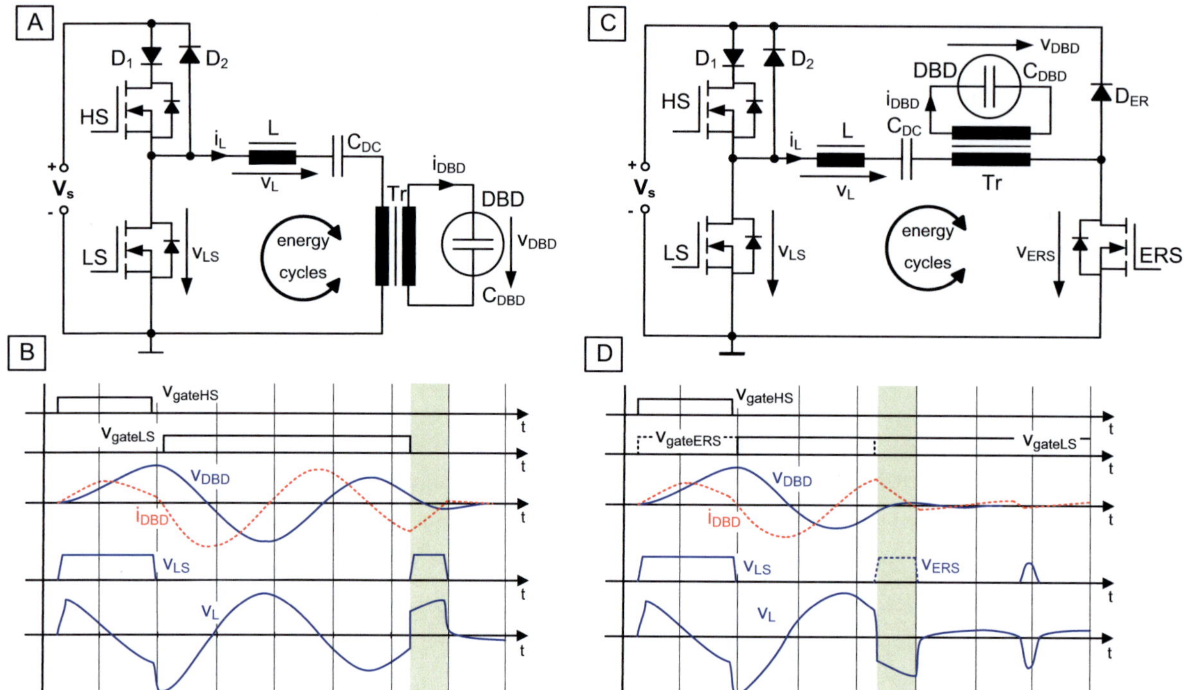

Figure 91: Comparison of schematics of the sinusoidal pulse topology (A) and the single period sinusoidal pulse topology (C) with corresponding qualitative waveforms (B and D). Note that the SPSP-topology allows for earlier energy recovery (green area) and therefore avoids a third ignition of the DBD. This bases on the ability to recover also positive inductor current via D_{ER}.

The SPSP-topology shows a slightly lower electrical efficiency. During energy recovery, the current level through the opening switch ERS is higher, leading to increased switching loss. Additionally, the extra switch leads to higher conduction losses within the circuit. However, under comparable operating conditions, the SPSP-topology leads to a higher efficiency and an enhanced power factor of the DBD lamp by preventing the third ignition. Hence, the system efficiency is slightly improved.

By adjusting switching times, the pulse shape can be further improved in order to meet the requirements of the DBD lamp, as done by current cut-off in Chapter 5.2.3.

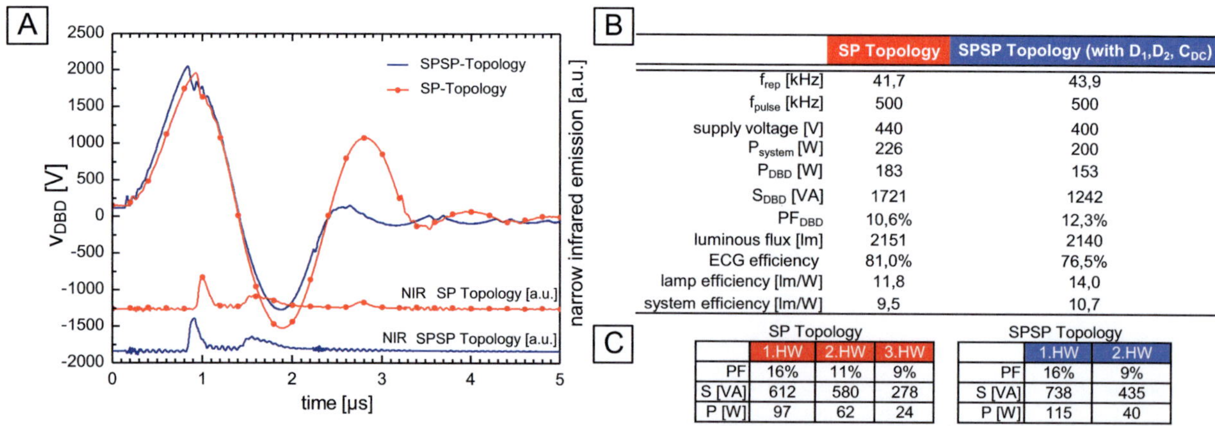

	SP Topology	SPSP Topology (with D_1,D_2, C_{DC})
f_{rep} [kHz]	41,7	43,9
f_{pulse} [kHz]	500	500
supply voltage [V]	440	400
P_{system} [W]	226	200
P_{DBD} [W]	183	153
S_{DBD} [VA]	1721	1242
PF_{DBD}	10,6%	12,3%
luminous flux [lm]	2151	2140
ECG efficiency	81,0%	76,5%
lamp efficiency [lm/W]	11,8	14,0
system efficiency [lm/W]	9,5	10,7

SP Topology

	1.HW	2.HW	3.HW
PF	16%	11%	9%
S [VA]	612	580	278
P [W]	97	62	24

SPSP Topology

	1.HW	2.HW
PF	16%	9%
S [VA]	738	435
P [W]	115	40

Figure 92: Comparison of the SP- and the SPSP-topology by experimental operation of a C_{DBD} = 3.18 nF, $V_{DBD,ign}$ = 2 kV DBD flat lamp at T_{rep} = 22.8 µs, $f_{pulse} \approx$ 500 kHz. A: Time-depending lamp voltage v_{DBD} and NIR radiation. Key parameters (B) and power distribution (C) itemised per half-wave.

5.2.2 Time-Domain Behaviour and Parasitic Resonances

Beside the main resonance used for pulse generation including a voltage overshoot, also in the circuit of the SPSP-topology parasitic resonances can occur. Figure 93 shows an equivalent circuit containing the components that are relevant during circuit idle time and are involved in parasitic resonances.

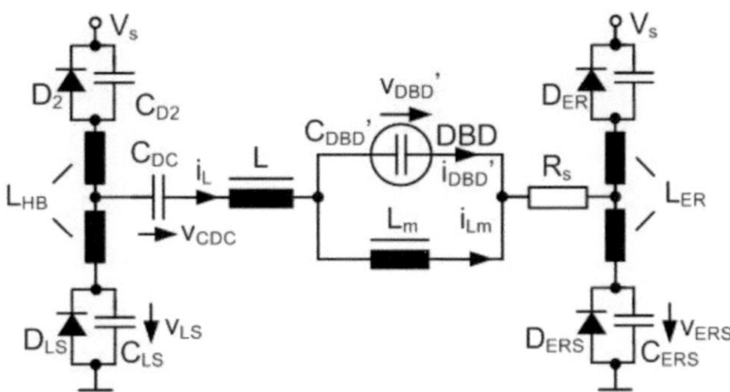

Figure 93: Equivalent circuit of the SPSP-topology for the time range between pulses t_{idle}. Switch body diodes and C_{OSS} and bridge-leg inductances L_{HB} and L_{ER} remain from the bridges. Supply source V_S can be treated as short circuit. The main resonant circuit is depicted with primary-transferred DBD parameters.

A resonance that occurs due to the bridge configuration is excited by high-voltage slopes generated by the switch transitions. The corresponding resonance frequency is given in Formula (5.23). Damping is possible by minimising parasitic stray inductances through short, wide PCB paths. Alternatively, the slope of the switch transitions can be reduced at the cost of increased switching losses. Because of the small values of the corresponding components, the resonance is of low energy but of high frequency. The generated voltage overshoot might be restricted by the avalanche capability of the switches. In any case, the resonance leads to increased electromagnetic radiation.

$$f_{ps} = \frac{1}{2\pi\sqrt{\frac{L_S}{C_S} - \left(\frac{R_S}{2L_S}\right)^2}} \tag{5.23}$$

Another parasitic series resonance occurs when energy recovery is finished. At this time, v_{ERS} is charged to V_s level. Inductor current i_L as well as v_{DBD} are almost zero. Because the inductor is therefore exposed to level $v_L = -(V_s + v_{CDC})$, a negative i_L is forced which then oscillates within the circuit. Depending on the current direction, the parameters of the established resonance change since the involved capacitances change. If LS is kept turned on after energy recovery, the equivalent series capacitance is characterised by Formulae (5.24) and (5.25), depending on the actual current direction which is set by v_{ERS}. Damping of this resonance is possible by incrementing the parasitic capacitance of ERS and D_{ER}. However, this also raises the energy available for resonances. The shape changes further if LS is turned off within the pulse pause. Then, depending on the current direction, either the body diode of LS (Formulae (5.24) and (5.25)) or the body diode of ERS (Formulae (5.24) and (5.26)) conducts. Formula (5.27) is valid if both v_{LS} and v_{ERS} exceed 0 V. Hence, resonance parameters change depending on the different capacitances of the devices.In Formula (5.23) L_s is the total circuit inductance, R_s is the total series resistance and C_S is the equivalent series capacitance depending on v_{LS} and v_{ERS}.

C_s is especially defined for the four cases:

$$C_S^{-1}\Big|_{\substack{V_{LS}=0V\\V_{ERS}=0V}} = \frac{1}{C_{DC}} + \frac{1}{C_{DBD}'} \tag{5.24}$$

$$C_S^{-1}\Big|_{\substack{V_{LS}=0V\\V_{ERS}>0V}} = \frac{1}{C_{DC}} + \frac{1}{C_{DBD}'} + \frac{1}{C_{ERS}+C_{DER}} \tag{5.25}$$

$$C_S^{-1}\Big|_{\substack{V_{LS}>0V\\V_{ERS}=0V}} = \frac{1}{C_{DC}} + \frac{1}{C_{DBD}'} + \frac{1}{C_{LS}+C_{D_2}} \tag{5.26}$$

$$C_S^{-1}\Big|_{\substack{V_{LS}>0\\V_{ERS}>0}} = \frac{1}{C_{DC}} + \frac{1}{C_{DBD}'} + \frac{1}{C_{LS}+C_{D_2}} + \frac{1}{C_{ERS}+C_{DER}} \tag{5.27}$$

C_{DBD}' refers to the primary-reflected DBD capacitance. Besides C_{DBD}, all capacitances included in Formulae (5.24) - (5.27) except for C_{DC} are highly non-linear variables. The switch output capacitances and the diode capacitances show a huge dependence on the voltage level.

However, the parasitic parallel resonance which establishes mainly between C_{DBD}' and L_m within the pulse pause has the highest impact on the reliability of the circuit. Immense v-t products can be generated by this resonance, raising losses and even leading to transformer saturation. The resonance parameters are given in Formulae (5.28) and (5.29) with the corresponding equivalent parallel capacitance given in Formula (5.30).

$$f_{pp} = \frac{1}{2\pi\sqrt{L_m \cdot C_{pp} - \left(\frac{1}{2C_{pp} \cdot R_p}\right)^2}} \tag{5.28}$$

$$Q_{pp} = R_p \cdot \sqrt{\frac{C_{pp}}{L_m}} \tag{5.29}$$

$$C_{PP} = C_{DBD}' + \frac{1}{\frac{1}{C_{DC}} + \frac{1}{C_{LS}+C_{D_2}} + \frac{1}{C_{ERS}+C_{DER}}} \tag{5.30}$$

Management of the parasitic parallel resonance is possible by parameter variations of the involved components and enhanced switching patterns. Corresponding methods are presented in the next chapter.

5.2.3 CIRCUIT MODIFICATIONS

Although the SPSP-topology bases on the principles of the SP-topology, due to the differences in the schematic and operation mode, certain further improvements are possible that are presented here.

REMOVAL OF HS DIODES

The initially used diodes D1 and D2 are not required if recovery is intended to occur via D_{ER}, only. Therefore, these diodes can be removed which reduces the bridge inductance and reduces ringing. Additionally, the losses related to the diode forward drop are avoided.

Removal of C_{DC}

Because energy is fed in solely at the beginning of the pulse, namely during the positive slope of the first voltage half-wave, the negative half-wave will always show a lower maximum amplitude and therefore a lower v-t product. Therefore, the transformer will in any case remain with a positive v-t product excess after the pulse. Due to the prevention of the third half-wave by the SPSP-topology, this excess is lower compared to the SP-topology. In any case, a freewheeling path for the current, stored by the mutual inductance L_m, has to be provided to prevent transformer saturation. For the SP-topology, C_{DC} is the only option. C_{DC} is charged by I_{Lm} and due to its high capacitance, C_{DC} preserves an almost constant voltage offset. This removes any DC component of the voltage seen by the primary transformer winding. However, the offset of C_{DC} reduces the initial step height and also feeds parasitic resonances. In contrast, the SPSP-topology provides an intrinsic freewheeling path. Freewheeling of the positive I_{Lm} occurs right after the pulse termination by means of charging C_{ERS}, while the bypass diode of LS conducts. Therefore, C_{DC} is not necessary and should be removed in order to save volume and component count. The removal of C_{DC} also allows for a direct access to the DBD lamp by the power semiconductor switches giving more grades of freedom of DBD voltage control. For example, after transformer freewheeling, the DBD could be shorted by closing LS and ERS in order to minimise the v_{DBD} RMS voltage.

Keeping The Low-Switch Turned on Within Pulse Pause

If C_{DC} is already removed, a very effective way to damp the parasitic parallel resonance is to leave LS turned on within the pulse pause. By doing that, the primary voltage of the transformer is bound to $v_L + v_{ERS}$. Because the inductor L shows low impedance at this resonance frequency, the voltage drop across L is almost zero. The body diode of ERS prevents a resonant backswing and thereby any resonance. Figure 94 compares the key waveforms for the standard operation mode referring to Figure 91D and the enhanced switching pattern with keeping LS in on-state during the pulse pause. Note that the transformer is not excited by a negative v-t product in the latter case.

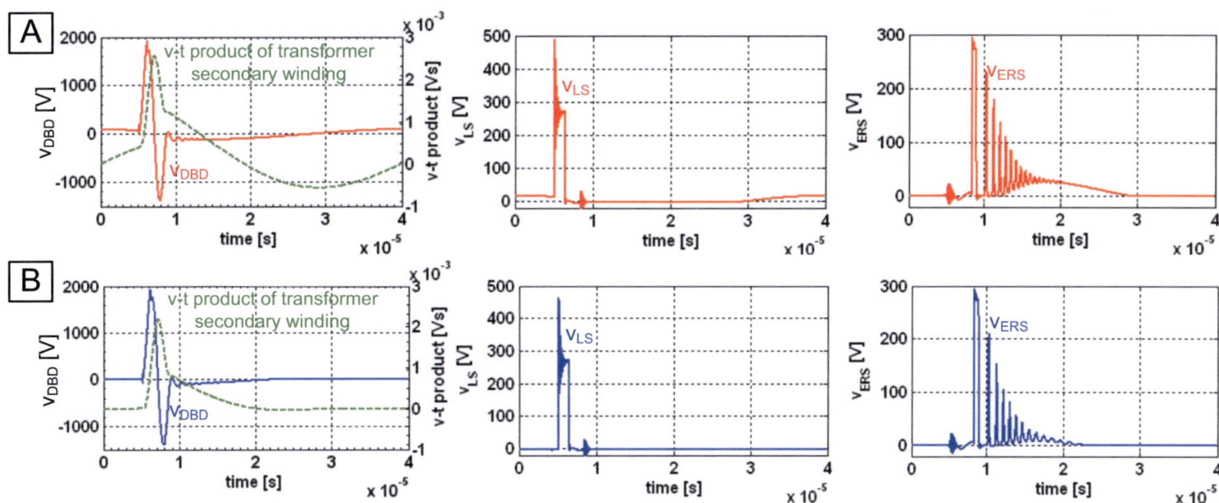

Figure 94: Key waveforms of SPSP-topology, f_{pulse} = 266 kHz, T_{rep} = 40 µs, V_s = 280 V. LS turned off (A) in pulse pause leading to higher v·t product than operation with LS turned on (B) in pulse pause. Both operation modes achieved η_{ECG} = 79 % and η_{DBD} = 13.5 lm/W.

CURRENT CUT-OFF

Figure 95 shows a possibility to further improve the pulse shape according to the suggestions of Chapter 5.1.5. This is done by cutting off the continuous current flow in one or even multiple half-waves of the actual pulse. The adjusted pulse provides high slopes of the lamp voltage at the moments of DBD ignition. Hence, the discharge is externally supported by a high current. Due to the suddenly initiated energy recovery, the discharge glow phase is shorted by the vastly descending current after ignition. Compared to the standard switching scheme, the effective pulse frequency of the adjusted pulse is 20 % higher. In summary, multiple current cut-offs can mimic a pulse created by a circuit with a much higher resonance frequency. In extension, this can significantly enhance lamp efficiency. A disadvantage is the lower electrical efficiency of the circuit. Because inductor energy is partly immediately recovered and not fully utilised to step up voltage, a higher supply voltage or transformer ratio is needed to reach the ignition voltage of the lamp. Furthermore, multiple energy recoveries cause additional switching losses because turn-off transitions are neither of ZVS nor of ZCS type.

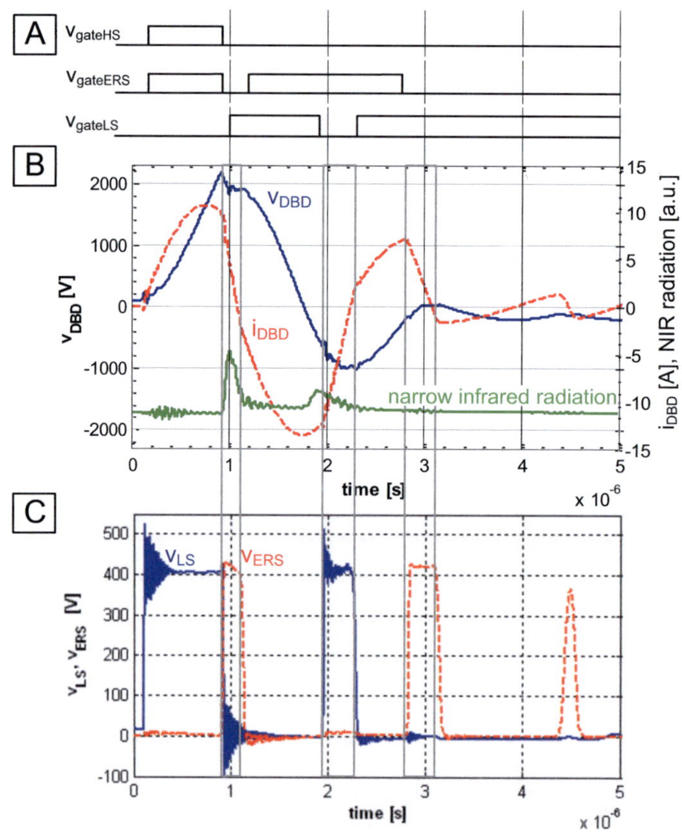

Figure 95: Switching patterns and waveforms from experimental verification of enhanced pulse wave-shape with 3 current cut-offs. A: Switching scheme. B: DBD voltage, current and NIR emission due to occurring DBD ignitions. C: Voltages across LS and ERS. HS diodes are implemented and topology operated at T_{rep} = 40 µs, V_s = 380 V, η_{ECG} = 70 %, η_{DBD} = 14 lm/W.

INCREASED PULSE FREQUENCY

Operation of DBDs with high-frequency pulses is referred to as being highly efficient (Paravia 2008a; Beleznai 2010). However, solely lamps with a small electrode area and therefore low capacitance have been under investigation so far. In contrast, the high-frequency operation of the 3.18 nF flat DBD lamp is investigated here. Due to the significant parasitics of DBD lamp, f_{pulse} could only been increased to 847 kHz by removal of the inductor and enhancement of the transformer in order to minimise it's stray inductance.

Compared with the 476 kHz pulse, at f_{pulse} = 847 kHz the lamp efficiency remained constant while the efficiency of the inverter dropped to 63.9 %.This low inverter efficiency is due to the excessive primary currents with a negative peak at 105 A and positive peak of 75 A at energy recovery initialisation. A comparison of the voltage pulse shapes for frequencies of 476 kHz and 847 kHz with corresponding NIR emission is shown in Figure 96.

This experiment reveals that there are fixed boundaries for a further frequency increase. A minimum Q_s has to be achieved in order to benefit from the voltage amplification and the energy recovery features. Further, the minimum circuit inductance is limited by the stray inductance of the transformer, PCB traces and the lamp connection cable.

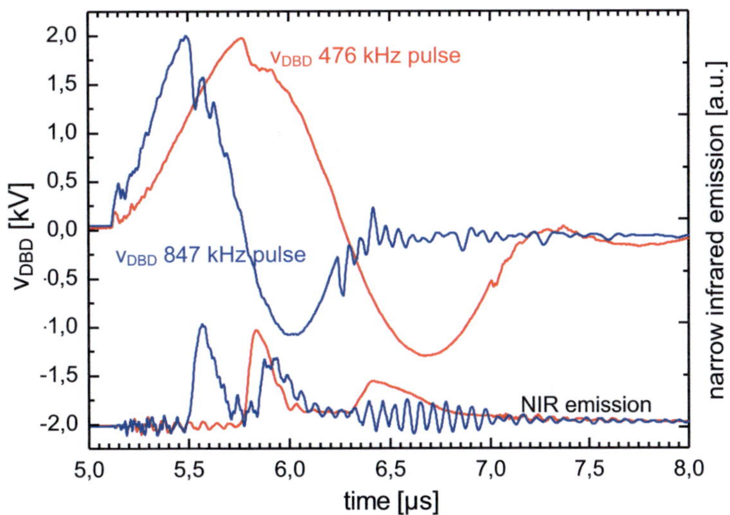

Figure 96: High-frequency pulse comparison of 476 kHz and 847 kHz pulse. Operation without diodes attached at T_{rep} = 40 µs, V_s = 380 V. 847 kHz pulse (blue curves): LS = MOSFET, η_{ECG} = 66.7 %, η_{DBD} = 12.9 lm/W. 476 kHz pulse (red curve): LS = IGBT, η_{ECG} = 80 %, η_{DBD} = 13.3 lm/W.

In order to investigate a further increase of the pulse frequency, a smaller coaxial DBD lamp had been chosen. The experimental results obtained from the high-frequency drive of this lamp are presented in Figure 97. The circuit used here bases on the SPSP-topology but consists of two separate power stage PCBs where the power semiconductor switches are arranged as described in Figure 44 on page 84. SiC JFETs were used as power semiconductor switches in order to provide fast switching and highest flexibility in terms of supply voltage level. The driven lamp has a small calculated capacitance of only 26.5 pF and is filled with xenon at a pressure of 150 mbar. The waveforms presented in Figure 97 show the results of a pulse frequency variation up to 3.1 MHz. In order to change the pulse frequency, different inductors and transformers were implemented into the circuit. The timing of the power semiconductor switches and the supply voltage level was adjusted in order to gain comparable pulse wave-shapes. The peak lamp voltage was adjusted to the smallest value necessary for a homogeneous discharge. According to Chapter 2.6.1, the discharge homogeneity was assessed by means of an ICCD camera. The outer ignition voltage depends significantly on the pulse frequency and rises by about 38 % as visible in Figure 97A. The reason for this is the pre-ionisation of the gas that occurs to a higher extent the lower the v_{DBD} voltage slope, e.g. the longer the time from first change of the electric field to ignition.

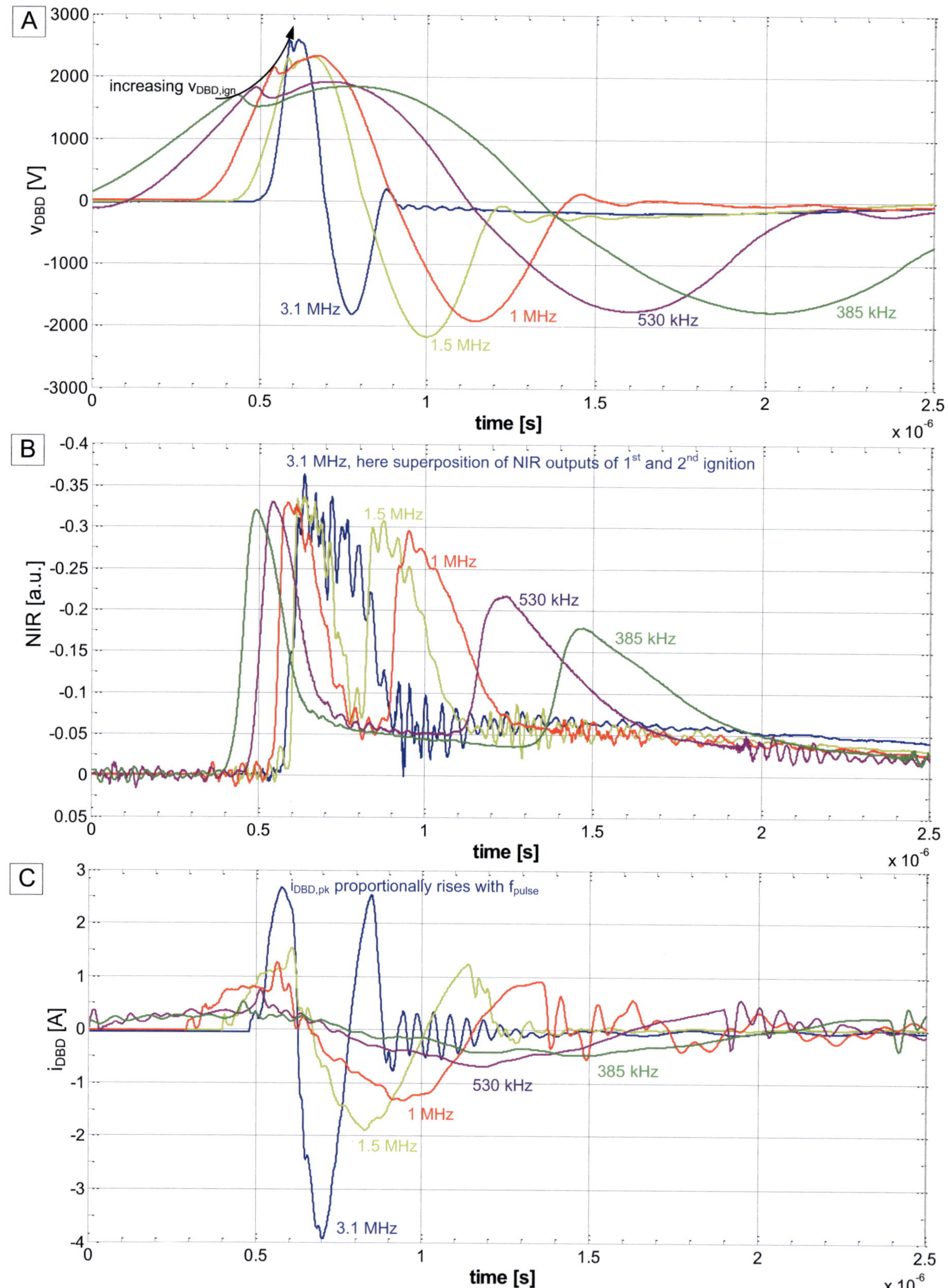

Figure 97: Pulse frequency variation driving a coaxial 26.5 pF $C_g/C_b = 2.4$, $p_{Xe} = 150$ mbar DBD lamp. A: Outer DBD lamp voltage v_{DBD}. B: Relative NIR output. C: DBD lamp current i_{DBD}. Note that the outer ignition voltage increases with rising pulse frequency. At 3.1 MHz pulse frequency, the NIR outputs of the first and second ignition overlap significantly.

Consequently, at low pulse frequencies, the gas conductivity is enhanced which lowers the necessary electrical field strength to ignite the gap. The NIR outputs generated by the respective ignition are plotted in Figure 97B. Evidently, as the pulse frequency rises the time between the NIR peaks shrinks significantly. Finally, at f_{pulse} = 3.1 MHz, the NIR outputs merge. This phenomenon is also known from square-wave pulses with low duty cycle (Carman 2004b). Further, it can be seen that the wave-shapes of the first and second ignition's NIR output differ significantly. While the initial slope of the respective first NIR pulse is almost independent from the pulse frequency, the initial slope of the second NIR pulse rises significantly with rising pulse frequency. Another change is related to the tail of the second NIR pulse, which is shortened with rising pulse frequency.

It is believed that these effects are related to the change in the excitation level due to the lower pre-ionisation and higher discharge current support provided by higher pulse frequencies. Unfortunately, the simplified time-dependent kinetic pathways depicted in Figure 8 (page 36) do not provide any further explanation.

Future investigations regarding the wavelength-resolved time-dependent behaviour of the NIR output could support the assumption that Figure 97B actually shows superimposed NIR emissions with different time constants.

In Figure 98A, the VUV efficiency and the VUV output measured at the different pulse frequencies is plotted. The VUV output shows an almost twofold increase if the pulse frequency is enhanced by a factor of eight. At frequencies above 1 MHz, homogeneous discharge behaviour was proven by ICCD camera recordings. However, at 530 and 385 kHz, stable partial filamentation was obtained as shown in Figure 98B. This explains the 20 % increase in efficiency at higher pulse frequencies.

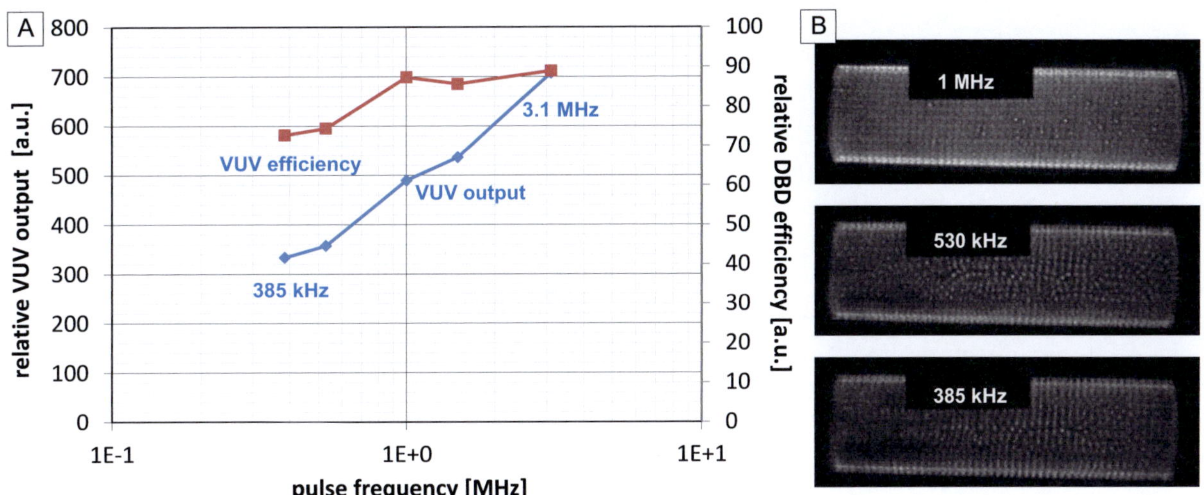

Figure 98: A: Relative VUV efficiency and relative VUV output plotted against pulse frequency f_{pulse}. The data corresponds to the experiments presented in Figure 97. B: Corresponding ICCD camera images characterising discharge homogeneity. At 1 MHz absolute homogeneity is obtained. Bright dots are contaminations on the glass surface. The black shadow is caused by the VUV measurement equipment directly mounted to the lamp.

To conclude, from the lamp's point of view, the operation with pulse frequencies in excess of 1 MHz is favourable since an exceptional power density can be achieved. Further research needs to be conducted in order to investigate the effect of even higher pulse frequencies. This could contribute additional information about up to which point enhancement of the pulse frequency is advisable.

The electrical efficiency of the inverter dropped by 40 % as pulse frequency increased from 385 kHz to 3.1 MHz. However, it needs to be noted that the magnetic components used were not optimised for operation at multiple MHz. It is believed by the author that efficiencies of more than 70 % are achievable using tailored components.

5.2.4 New Transformer-Less High-Voltage Operation

The combination of the single period sinusoidal pulse topology and fast power semiconductors able to operate at voltage levels that exceed the half of the required maximum DBD voltage leads to the high-voltage supplied single period sinusoidal pulse (HV-SPSP-) topology presented here. As long as the DBD ignition voltage stays below two kilovolts, commercially available SiC switches and diodes can favourably be used. Assuming the availability of a high-voltage DC source, the pulse generator set-up is less complex as neither a high-frequency, high-voltage power transformer nor a C_{DC} is required. Beside omitting these bulky components also the problems connected to transformer stray inductance and parallel parasitic resonance (Meisser 2010b) are solved. Furthermore, the DBD voltage level within the idle time is no longer controlled by transformer freewheeling action and therefore stays constant. However, the drawbacks of high-voltage operation as discussed in Chapter 5.1.8 apply. The circuit presented benefits from the possibility to utilise the energy stored in the DBD capacitance for the following pulse. As the transformer ratio no longer enhances the effective DBD capacitance, the highest possible pulse frequency is increased by a factor of this ratio.

The schematic of the proposed HV-SPSP-topology is depicted in Figure 99A. This topology permits to recover energy from the DBD immediately after the second ignition like its transformer-equipped counterpart, the SPSP-topology. By that, a third ignition, claimed to be less efficient (Meisser 2010c), is prevented. The pulse consisting solely of two voltage half-waves is feasible since also positive inductor current can recover energy through D_{ER}. A sketch of time-dependent waveforms is given in Figure 99B. Low-side switch LS and energy recovery switch ERS must be enabled shortly in advance of the pulse in order to maximise the step amplitude of v_{DBD}. By enabling LS, a negative i_L allows a partial ZVS turn-on of ERS. The pulse is then initialised by disabling LS and enabling HS with ZCS-condition in time range 2.

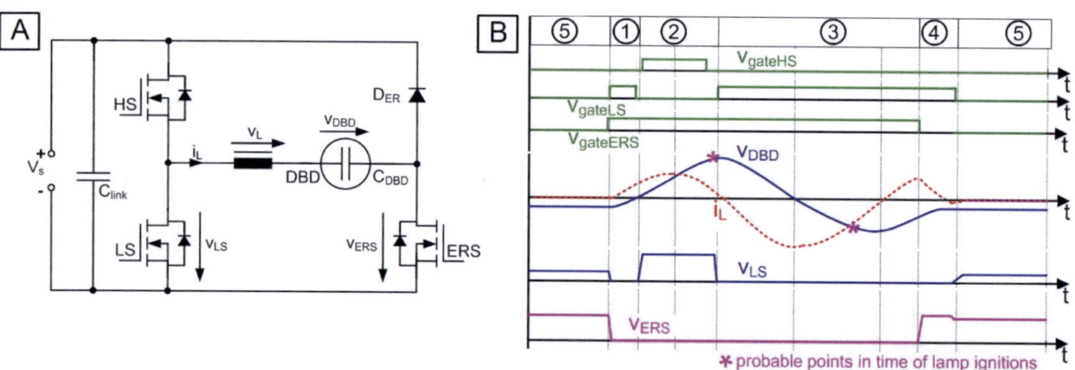

Figure 99: A: Schematic of the proposed HVS-SPSP-topology. B: Respective ideal waveforms with indication of probable moments of ignition. Mind the negative offset of v_{DBD} during t_{idle}.

Then, in time range 3, after the current i_L returns to being positive, energy recovery through diode D_{ER} is initiated by turning off energy recovery switch ERS. Depending on the time at which ERS turns off, a certain amount of energy remains in the capacitances C_{DBD} and C_{OSS} of ERS:

This is roughly one quarter of the maximum C_{DBD} energy. The resulting voltage offset dramatically reduces the relative amplitude of the negative voltage half-wave of the pulse. Thereby, current through the DBD is cut off shortly after the second ignition. Although the negative offset of v_{DBD} may not be favourable for every application, it reduces the required supply voltage amplitude nevertheless as the voltage amplification is enhanced by the resonant upswing.

Although D_{ER} is made of SiC SBDs, a small reverse current flows through L after energy recovery. As $C_{DBD} \gg C_{OSS}$, V_{DBD} stays unaffected by this current. The negative i_L slightly rises V_{LS} above zero, while V_{ERS} falls. During idle time range 5, V_{DBD} is equal to V_{LS} - V_{ERS} and, due to the full-bridge configuration, is restricted to $-V_s$. In contrast to resonant upswing techniques (Meisser 2010d) presented in the following chapter, absolutely no excitation by changes of V_{DBD} occurs in advance of the generated pulses. This is beneficial as it prevents any pre-ionisation of the gas inside the DBD. Consequently, a high degree of homogeneity as well as high DBD lamp efficiency can be achieved (Paravia 2009b). See Chapter 3.4 on page 63 for reference.

For the experimental verification of this topology, for all switches depicted in Figure 99A SiC 1200 V MOSFETs were implemented. The measurement of input power has been performed using a high accuracy power meter. However, because of the high-frequency, high-voltage pulse, the measurement of DBD power has to be performed with an oscilloscope and respective probes. All probes were DC-calibrated, AC-compensated and de-skewed as specified in Chapter 2.5.1. In contrast to transformer-equipped topologies or the transformer-less half-bridge, in case of the transformer-less full-bridge, both connections of the DBD lamp float. This has negative influence on the voltage measurement accuracy since a differential voltage measurement with two voltage probes is required. In order to validate the electrical efficiency a thermal loss measurement at the heat-sink and the inductor was performed yielding the thermal loss $P_{loss,therm}$. Additionally, the losses in the semiconductors $P_{loss,\ calc}$ were calculated. The obtained waveforms of the experimental operation of the suggested HV-SPSP-topology are given in Figure 100. Solely two sinusoidal half-waves are generated that favourably lead to two ignitions per pulse. As indicated, PF of the first ignition is much higher compared to the second. The SiC MOSFETs switch with up to 43 kV/µs. After pulse termination, v_{LS} and v_{ERS} approach voltage levels determined by the different output capacitances of the respective bridge-leg. The brief turn-on of LS prior to the pulse permits a partial soft turn-on of ERS (see decrease of v_{ERS} in Figure 100B). The DBD voltage v_{DBD} is constant beyond the actual pulse but exhibits a significant negative offset as known from Figure 99B. A variation of pulse frequency is depicted in Figure 101A. The key parameters of the experiments are depicted in Figure 101B. As the pulse frequency rises, the power factor of the first voltage half-wave rises significantly due to a high power drained by the first ignition and a comparatively low RMS value of v_{DBD}. The outer ignition voltage $v_{DBD,ign}$ shrinks by several hundred volts. However, due to the higher peak current, the electrical efficiency of the circuit is slightly lessened. The significant measurement error of the differential voltage measurement at the elevated frequencies caused the discrepancy of the DBD power measurement during the actual pulse $P_{DBD,PULSE}$ and during the whole repetition period $P_{DBD,PER}$. Due to offset errors, $P_{DBD,PULSE}$ is significantly higher.

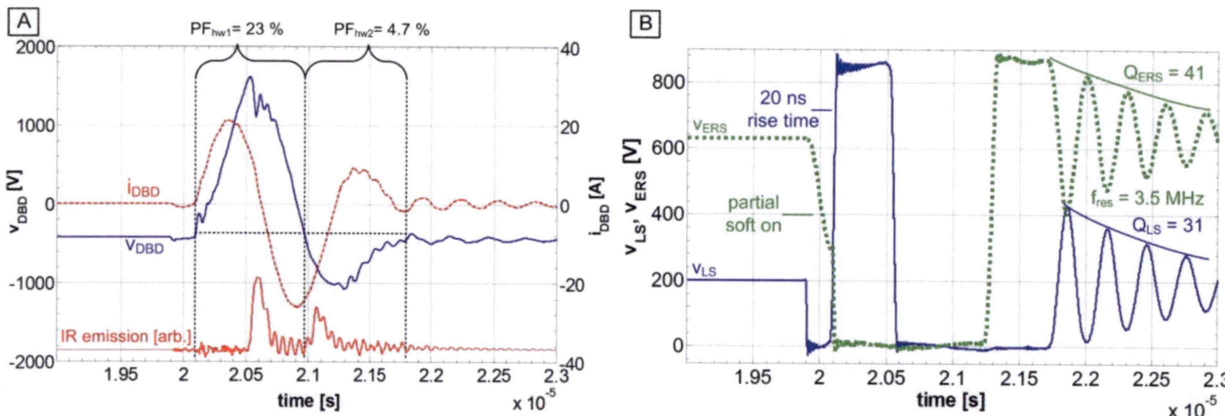

Figure 100: Experimental data of HV-SPSP-topology operation at V_s = 860 V, P_{in} = 107 W, f_{pulse} = 650 kHz, f_{res} = 25 kHz. A: DBD voltage v_{DBD}, DBD current i_{DBD} and NIR radiation. Note the decaying power factor PF from first to second ignition. B: Energy recovery switch voltage v_{ERS} and low-side switch voltage v_{LS}. Different quality factors of decaying parasitic series resonance are due to the different equivalent capacitance of the respective bridge-leg.

To conclude, the novel SPSP-topology and its variants presented here show significant advantages compared to the SP-topology described in Chapter 5.1. The prevention of a third lamp ignition enhances system efficiency and the possible high pulse frequencies enable for high power density. Power density is further enhanced by removal of C_{DC} and the HS diodes (page 138). As in case of the SP-topology, also here the removal of the transformer bears the possibility of achieving highest pulse frequencies.

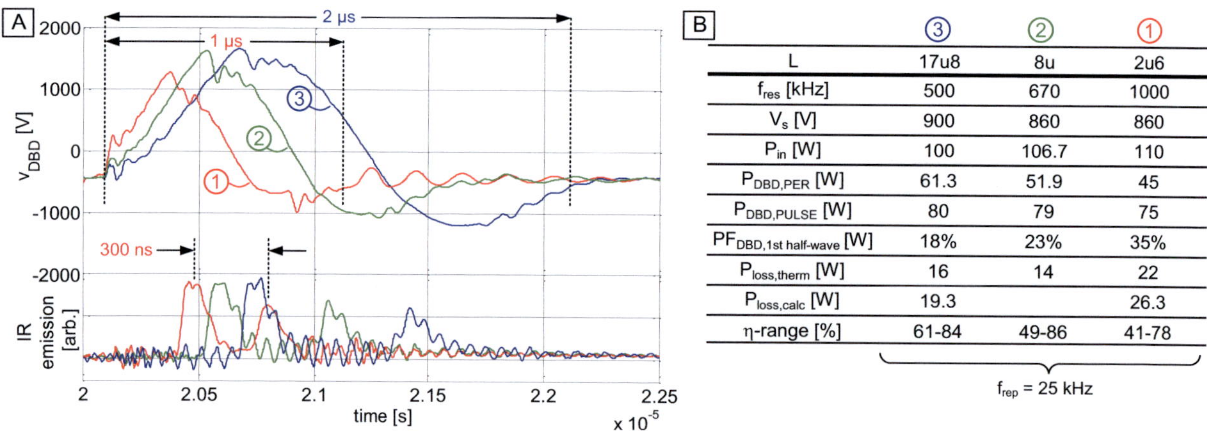

	③	②	①
L	17u8	8u	2u6
f_{res} [kHz]	500	670	1000
V_s [V]	900	860	860
P_{in} [W]	100	106.7	110
$P_{DBD,PER}$ [W]	61.3	51.9	45
$P_{DBD,PULSE}$ [W]	80	79	75
$PF_{DBD,1st\ half\text{-}wave}$ [W]	18%	23%	35%
$P_{loss,therm}$ [W]	16	14	22
$P_{loss,calc}$ [W]	19.3		26.3
η-range [%]	61-84	49-86	41-78

f_{rep} = 25 kHz

Figure 101: Comparison of different pulse widths achieved by different values for series inductance L. At pulse frequency of 1 MHz, delay between first and secondary ignition is as short as 300 ns. A: Waveforms obtained from experimental measurements. B: Table comparing the parameters of the experiments.

5.3 New Upswing Full-Bridge Topology

By adding a second path for energy feed-in, compared with the SPSP-topology, the full-bridge topology presented in this chapter is equipped with an additional degree of freedom. Although the full-bridge can be equipped with a transformer, it is especially suitable to generate the required high-voltage pulse out of a medium voltage source (Meisser 2010d). This allows for the use of medium-voltage power semiconductors that only have to handle the lamp current and not the by r multiplied lamp current as in the case of transformer-equipped topologies. Therefore, by principle, conduction losses could be reduced by a factor of r^2. However, multiple resonance periods that are necessary to gradually build up the required DBD voltage partly compensate the loss reductions. Without transformer, exceptional high resonance frequencies can be achieved. Because energy can be fed-in by both bridge-legs regardless of actual lamp current direction, compared to the upswing SP-topology operation mode, only the half amount of feed-in cycles is required which likewise cuts upswing time to half. By principle, the topology and operation mode would be suited to generate RF burst waves as suggested by Beleznai (2009).

In contrast to other full-bridge configurations for DBD drive (Kyrberg 2007b; Piquet 2007), the presented topology operates from medium voltage and does not store energy in the DBD lamp capacitance between the pulses. Instead, energy is recovered after the pulses and thereby offers the feature of low voltage during pulse pauses.

5.3.1 Schematic and Principal Mode of Operation

The schematic of the full-bridge upswing topology is depicted in Figure 102. Compared to the SPSP-topology depicted in Figure 91C, the right bridge-leg is fully-equipped with power semiconductor switches. Fast HS diode recovery is required during energy recovery cycles. Depending on the used power semiconductor switches and the intended switching scheme, fast external Schottky diodes are necessary. In contrast to all other novel topologies presented here so far, the full-bridge excites the resonant tank, which is set up by inductor L and the capacitance of the DBD C_{DBD}, for several times. This allows for a multiple voltage magnification at every time when energy is fed into the resonant circuit. The aim is to reach ignition voltage with a minimum amount of resonant swings prior to first ignition. This is because any AC voltage applied to the DBD before the first ignition generates charge carriers (pre-ionisation) that increase gas conductivity and consequently thermalise the gas. Referring to Chapter 3.4, this can impede homogeneity of the discharge and with that DBD efficiency.

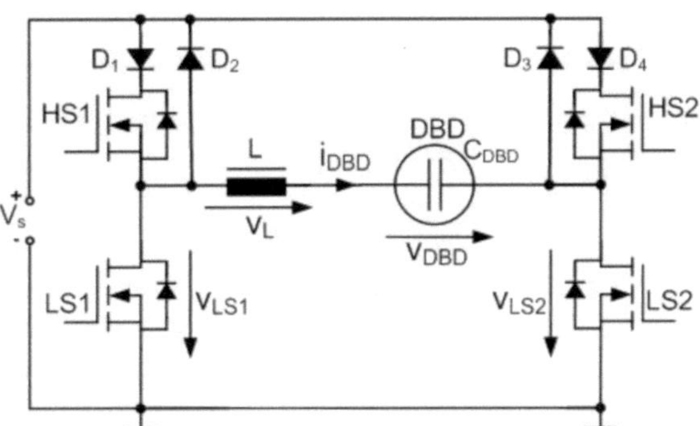

Figure 102: Schematic of the full-bridge topology that is favourably used for resonant upswing operation mode. Note that SiC Schottky diodes D1-D4 are necessary to eliminate energy recovery time losses that would otherwise be exhibited by the body diodes of Si MOSFET HS1 and HS2. Alternatively, the use of power semiconductor switches with very fast body diodes ($t_{rr} \ll 100$ ns) would be favourable.

In order to initiate resonant circuit upswing, one high-side switch and the low-side-switch of the opposite bridge-leg is turned on. Identical to the operation of the SP- and the SPSP-topology, as long as Q_S is sufficiently high, the DBD lamp current draws a sine half-wave and the DBD voltage reaches twice the input voltage level at maximum. Then, the upswing is proceeded by transiting to the opposite HS and LS under ZVS (or ZCS) condition. This further adds one V_S amplitude to v_{DBD}. According to Formula (5.31), after n feed-in cycles, the DBD voltage level is high enough to lead to ignition.

$$n = \frac{V_{DBD,ign}}{V_s} - 1 \tag{5.31}$$

After the voltage half-wave that leads to the first ignition, the following half-wave must be tuned to support an intensive second ignition resetting the internal charges. As possible with other series topologies, also here a current cut-off scheme can be performed or a recovery cycle can be initiated. After the second ignition, the major task is to prevent DBD tail current losses by swiftly reduce v_{DBD} to comparatively low voltage and minimise any oscillation if possible. However, this is the drawback of the upswing approach – also swinging down is necessary. The number of recovery cycles required to decrease v_{DBD} from the initial value $V_{DBD,init}$ down to zero volts is defined to:

$$n = \frac{V_{DBD,init}}{V_s} - 1 \tag{5.32}$$

This number of recovery cycles could be minimised whilst recovering energy into a voltage source whose level is much higher than that of V_S.

Because of the missing transformer, the DBD lamp capacitance can lay within the order of magnitude of the output capacitances C_{OSS} of the switches. In this case zero-voltage switching (ZVS) commutations are no longer effective since a significant percentage of circuit energy is required to charge the output capacitances. Therefore, for applications with low DBD lamp capacitance and low pulse frequency (implying low current), zero current switching (ZCS) is preferable. However, if lamp capacitance is large and pulse frequency is high, ZVS should be applied.

5.3.2 Experimental Results

Figure 103 shows experimental results obtained by driving a 3.18 nF flat DBD lamp with an additionally attached C_P = 1.87 nF capacitor at an electrical efficiency of more than 92 %. The corresponding NIR radiation, given in Figure 103A, indicates the first lamp ignition during the time of the second voltage half-wave. The ignition voltage obtained here lies several hundred volts below the level required to ignite the lamp by means of a basic sinusoidal pulse. The reason for this is the enhanced charge carrier density generated by the resonant upswing during the first voltage half-wave. To achieve a third ignition, further energy is fed in by closing HS1 and LS2 for a second time. Energy recovery is initiated by keeping all switches off. Delayed by the reverse recovery time of the LS body diodes, the current leaves the circuit by passing alternately D_2 and D_3. This is indicated in Figure 103B by alternating peaks of v_{LS1} and v_{LS2}, respectively. As can be derived from Figure 103A, a power of 15 W is consumed in the first half-wave by a loss process inside the DBD. This is almost 6.7 % of the total lamp power which restricts DBD lamp efficiency. The Power Factor (PF) of the second voltage half-wave, in which the first ignition occurs, has almost double the value compared to the following voltage half-waves. In contrast, the highest level of apparent power has to be distributed within the third half-wave. As is obvious, the third half-wave would not need such a high amplitude to lead to the second ignition. Fine-tuning the pulse shape by lowering energy input into the third half-wave could further increase PF and efficiency.

In Figure 104, results are given for a 1.15 MHz pulse generated to drive the same DBD lamp with attached C_P = 260 pF parallel capacitor. The electrical efficiency exceeded 80 %. Due to lower V_S, three resonant swings were necessary to reach ignition voltage. To achieve a strong second ignition, energy recovery is deferred by closing LS1. The rising edge of v_{LS1} at t = 2μs in Figure 104B indicates that energy recovery then occurs through turned-off LS2.

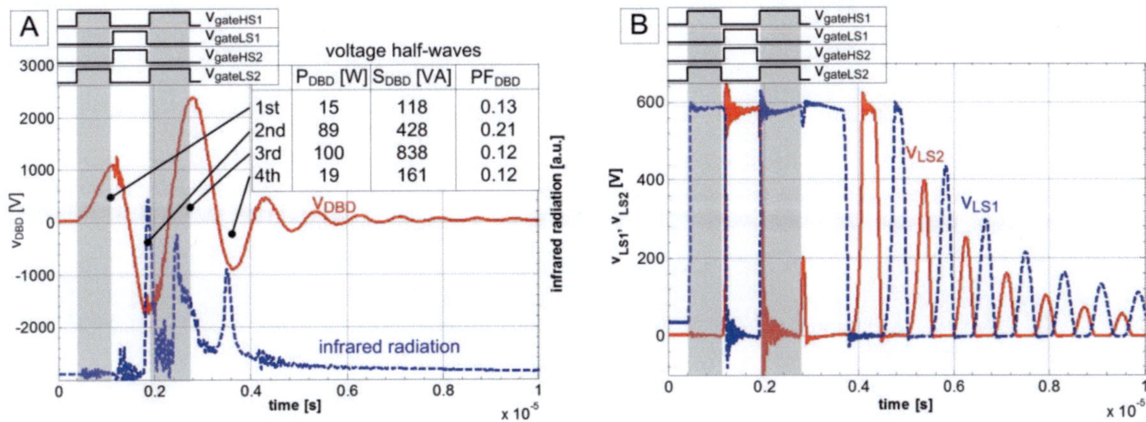

	P_{DBD} [W]	S_{DBD} [VA]	PF_{DBD}
1st	15	118	0.13
2nd	89	428	0.21
3rd	100	838	0.12
4th	19	161	0.12

Figure 103: Experimental waveforms obtained at V_s = 600 V, T_{rep} = 40 µs, f_{pulse} = 500 kHz, P_{DBD} = 228 W, η_{ECG} > 92 %. A: DBD voltage v_{DBD} and NIR radiation indicating three ignitions. B: Voltages across low-side switches LS1 and LS2. Note the spike of v_{LS1} around 5 µs. Here, LS2 and HS1 should be kept in on-state to further reduce losses.

Because measurements of the DBD voltage had to be performed differentially, efficiency calculations from mathematically offset-corrected waveforms have a rather high uncertainty. However, thermal loss measurements performed at the heat-sink to which all power semiconductors were attached proved in both cases an electrical efficiency well above 80 %. The efficiency is remarkably high, which is mainly due to low conduction losses and missing transformer losses.

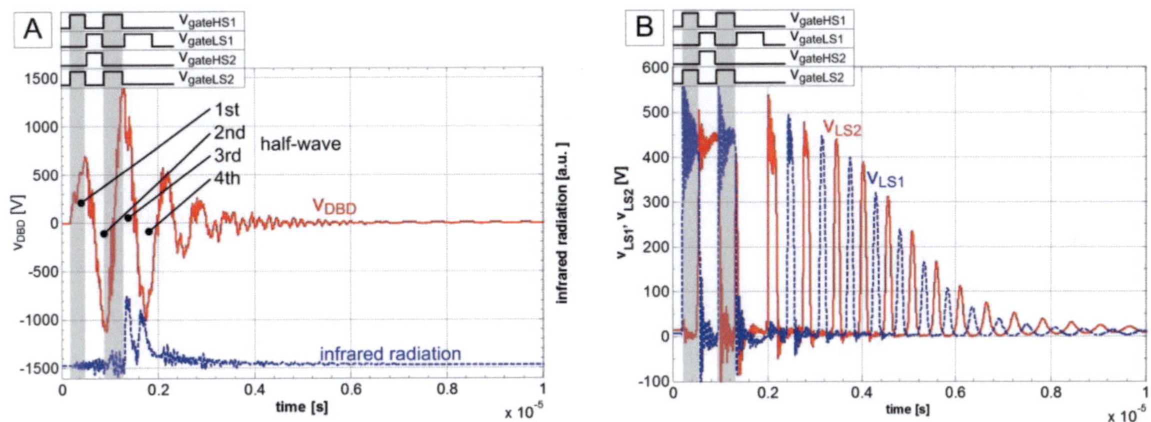

Figure 104: Experimental waveforms obtained at V_s = 450 V, T_{rep} = 40 µs, f_{pulse} = 1.15 MHz, P_{DBD} = 195 W, η_{ECG} > 80 %. A: DBD voltage v_{DBD} and NIR radiation indicating two ignitions. B: Voltages across low-side switches LS1 and LS2. Note that here, resonance without further energy fed-in or recovery is performed from 1.3 to 2 µs where LS1 is kept turned on.

Without doubt this operation mode is very promising when DBDs have to be operated at high pulse frequency from a low supply voltage level. Nevertheless, to obtain high DBD efficiency, the pulse repetition frequency needs to be lower compared to a SPSP-topology operation since the pulse duration is significantly longer. The number of energy recovery cycles could be reduced by recovering energy into a voltage source with higher amplitude than V_S. For instance, this could be a high-voltage capacitor. As mentioned before, this topology can be used to generate pulses with even more voltage half-waves supporting energy fed-in into the DBD. This bears the possibility of enhanced power density (Beleznai 2009) without deteriorated DBD efficiency.

6 Operation Modes and Resonance Behaviour of Parallel Topologies

Parallel topologies allow the pre-charge of an inductor in advance of applying the stored energy to the excimer DBD. That is, the inductor is virtually not "transparent to energy" as it first buffers energy instead of transferring it directly to the capacitive load. Then, triggered by an opening switch, the inductor freewheels into the capacitive part of the resonant circuit which is the DBD lamp. Compared to series topologies, higher initial DBD voltage slopes are possible since the inductor is pre-charged and can directly transfer its energy to the DBD. Parallel topologies provide huge and variable voltage amplification which also implies the ability to adapt to different load capacitances and therewith permit obviation of the transformer. However, most of these topologies can by principle generate unipolar pulses, only. Bipolar pulses are merely possible by utilising parasitic circuit components at the cost of reduced efficiency.

The very popular resonant flyback topology, also referred to as class-E converter, is found in commercial products such as the Planon® or the Linex® (Sowa 2004) and is patented for the use with DBD lamps (Okamoto 1999; Lecheler 2007). Due to its wide-spread use, the resonant flyback is shortly introduced in Chapter 6.1.

The novel patented modification named adaptive unipolar pulse topology (Meisser 2010a) which allows for active decoupling of DBD lamp and pre-charge circuit is presented in Chapter 6.2.

An alternative approach permits to obviate the transformer and instead uses low voltage switches combined in a matrix configuration (Meisser 2012d). This novel transformer-less resonant flyback is presented in Chapter 6.3.

6.1 Review of the Resonant Flyback Topology

The resonant flyback, also referred to as class-E converter, has already been used in DBD research (Carman 2004a) and commercial products by OSRAM (Sowa 2004). Flyback in essence means a non-continuous energy transfer form input to output. This energy is always intermediately stored in a magnetic storage element before being transferred to the output. In other words, the energy flow is non-continuous from input to output and vice versa which goes in hand with non-continuous currents. The magnetic storage used, an inductor with two separated windings also referred to as flyback transformer, is crucial for the circuit performance. It is pre-charged in advance of the pulse to be generated. Freewheeling of this transformer then applies a voltage slope to the DBD lamp. A resonant backswing entails both energy recovery of the lamp and magnetic reset of the coupled inductor. Although nothing but unipolar pulses can be generated by a single flyback, the popularity of this topology seems to be everlasting because it allows for a product that is simple and cost-effective at first glance. The following references report about resonant flyback topologies used in science and technology to drive DBDs:

A very low power resonant flyback design with a measured electrical efficiency of 50 % that drives a DBD at a power of 1 W is presented in Alonso (2004). The system is intended to be used for ozone generation. The authors use an operation mode in which a high-frequency, high-amplitude resonance (L_{stray} & C_{DBD}) is superimposed by a parasitic low-frequency, low-amplitude resonance (L_m & C_{DBD}). Although NIR radiation is not measured, it is assumed that unipolar excitation is performed with this circuit. A series resonant flyback incorporating a soft-switching unipolar switch is presented by Sugimura (2006). Park (2007a) uses two serially connected flybacks to operate a flat DBD lamp with distributed bipolar pulses (see Figure 21B) achieving a system efficiency of 36 lm/W. Each flyback transformer can be clamped by an additional switch necessary to transfer the pulse generated by the second flyback. Olivares (2009) presents a flyback that generates a pulse width of 450 ns at 100 kHz repetition frequency to drive a 14 W linear fluorescent lamp at a peak voltage of up to 2 kV. A lamp efficiency of 49 lm/W and a system efficiency of 22 lm/W is claimed. Liu (2011) presents a double switch resonant flyback inverter for a flat DBD lamp that operates in CCM permitting ZVS turn-on of the switches. A system efficiency of up to 22 lm/W is claimed and input power of up to 65 W is reported.

The flyback transformer does not necessarily require two electrically insulated windings. Heming (2009) suggests a centre-tap transformer to drive small DBD loads in applications such as analytical chemistry. Centre-tap transformers especially make sense where low transformer ratios are required.

Single flyback stages may be cascaded in order to enhance the output voltage level (Lecheler 2007). Alternatively, they may be connected in series with opposite polarity of the respectively generated voltage pulse (Park 2007a) or in parallel combined with decoupling diodes (El-Deib 2010a).

However, the resonant flyback topology has also some significant drawbacks. At least one switch has to face both the maximum primary-reflected lamp voltage as well as the high current level during pre-charge and energy recovery. The inevitable stray inductances of the coupled inductor and load connection cables restrict the maximum possible voltage slope, generate voltage overshoots that stress the opening switch and cause parasitic series resonance phenomena. The design of the coupled inductor is shortly addressed in Chapter 4.13.2.

The mentioned drawbacks fundamentally restrict the possible efficiency of flybacks. Hence, resonant flybacks are merely applicable for DBD lamps with low electrical peak energy such as small medium-pressure lamps. However, if prices of 1.2 kV and 1.7 kV class SiC power switches fall, resonant flybacks for DBD drive could cover an even broader range of applications than their DC-DC counterparts (Springett 2012).

6.1.1 Operation in Discontinuous Conduction Mode

The resonant flyback circuit and its operation mode seem to be simple only at first glance. Figure 105B shows the respective waveforms. The power semiconductor switch SW is used to pre-charge the primary inductance of the flyback inductor Tr in time range 1. When the intended peak current is reached, SW is switched off. The particular feature of the flyback inductor is that then a commutation of the current flow from the primary to the secondary winding occurs. It is important to note that this commutation entails abrupt current deterioration at the primary as well as a sudden current increase in the secondary transformer winding. Consequently, the resistive AC losses in the windings are much higher compared to those experienced in transformers with continuous current flow (as implemented in series topologies of Chapter 5).

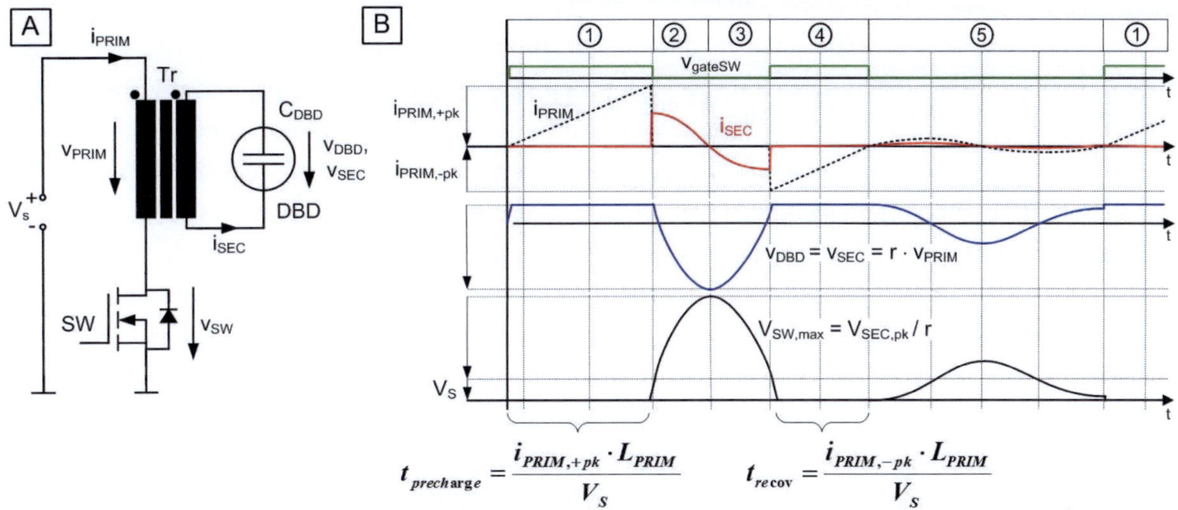

Figure 105: A: Schematic of the resonant flyback topology. B: Time-dependent voltages and currents for operation in discontinuous operation mode. Note the parasitic resonance that occurs during idle time in time range 5.

In time range 2, the primary current is determined by the output capacitance C_{OSS} of SW while the secondary current is defined by the characteristic capacitance of the DBD. Neglecting Tr's stray inductance, the resonance is mainly defined by the transformer's mutual inductance L_m and the primarily transferred C_{DBD}:

$$f_0 = \frac{\omega_0}{2\pi} = \frac{1}{2\pi \cdot r\sqrt{L_m \cdot C_{DBD}}} \tag{6.1}$$

L_m and C_{DBD} underlie restrictions. If these parameters are assumed to be constant, the higher the transformer ratio r, the lower the maximum pulse frequency can be. The wave-shape of DBD voltage and current is then found by modifying Formulae (5.11) and (5.12). For reasons of simplification, initial DBD voltage and supply voltage V_S are neglected.

$$v_{DBD}(t) = \left[\left[-\frac{I_{DBD,i}}{C_{DBD}}\right] \cdot \frac{\sin(\omega_d t)}{\omega_d}\right] e^{-\alpha t} \tag{6.2}$$

$$i_{DBD}(t) = \left[\left[\alpha I_{DBD,i}\right]\frac{\sin(\omega_d t)}{\omega_d} - I_{DBD,i}\cos(\omega_d t)\right] e^{-\alpha t} \tag{6.3}$$

This is valid as long as V_S is low compared to the peak value of V_{SW} which is usual in flyback topologies. The DBD voltage ramps down having an initial slope given in Formula (6.4) following a sine wave-shape according to Formula (6.2) and finally reaches its absolute minimum after about a quarter of the resonance period. Inductor energy is then zero since all resonant energy is stored in the DBD capacitance.

$$\frac{dv_{DBD,i}}{dt} = \frac{I_{DBD,i}}{C_{DBD}} \tag{6.4}$$

Ignition of the DBD occurs in advance of this point in time with temporary change of the equivalent circuit series resistance. This damps the resonance as energy is drawn and transformed into radiation. The resonant backswing leads to a sinusoidal increase of the DBD voltage until the secondary voltage equals the reflected supply voltage amplitude. In time range 4, the current migrates back to the primary winding while the body diode of SW conducts. This can be understood both as energy recovery and freewheeling of the primary inductance in order to compensate for the generated volt-seconds. Freewheeling is finished when the current crosses zero amplitude. At this time, v_{PRIM} is almost equal to the amplitude of the supply voltage. The reflected voltage v_{SEC} lies across the DBD capacitance. Hence, a significant amount of energy is still stored in the DBD. This energy initiates a parasitic resonance together with the flyback inductance. In contrast to serial topologies where the resonance could be tamed by including the C_{DC} capacitor or providing an alternative freewheeling path, this resonance is solely damped by resistive losses. Depending on the chosen repetition frequency, the switching on of SW may add significant loss. This is the case if the voltage across the switch is different from zero. In this case, additionally to C_{OSS} the reflected DBD capacitance is shorted. To prevent the related losses, the repetition frequency has to be chosen accordingly.

6.1.2 Operation in Continuous Conduction Mode

The CCM switching scheme leads to a seamless transition from one pulse to the next. This prevents the parasitic resonance to occur and also provides ZVS turn-on of SW. In this case, the sum of pre-charge time and energy recovery time is equal to the idle time between the pulses.

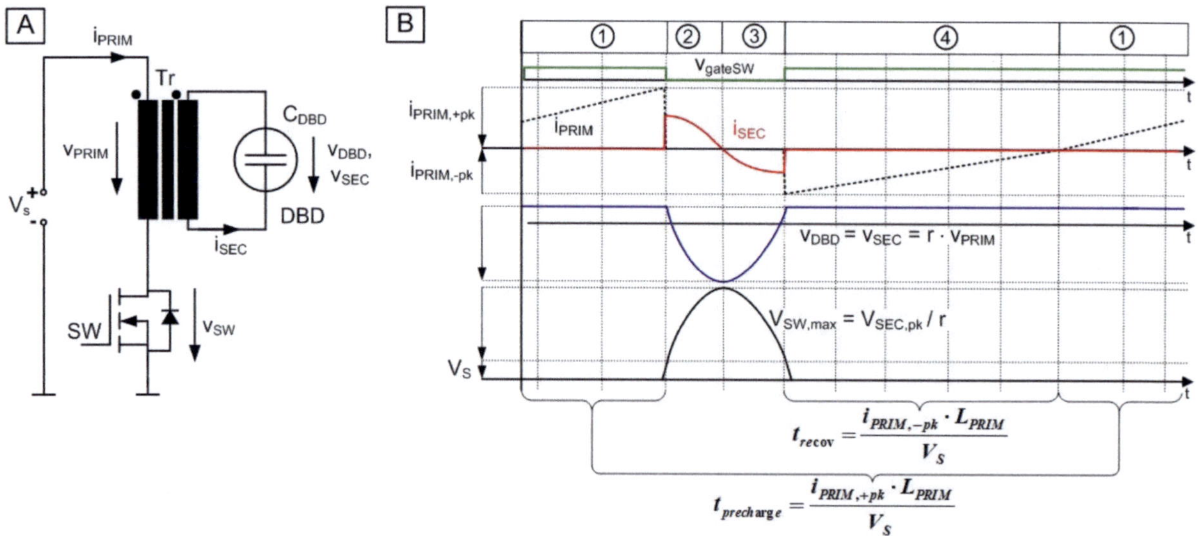

Figure 106: Special case of seamless current transition between pulses (CCM). Current is continuous and ZVS turn-on of SW is provided. To vary f_{rep} V_S must be adjusted accordingly. A: Simplified circuit. B: Operation wave-shapes.

To vary the idle time and hence to vary repetition frequency while keeping the pulse wave-shape constant, the supply voltage amplitude needs to be adjusted. The required voltage amplitude can be calculated as follows:

The real power consumed by the DBD depends on the difference of the energies stored in the inductor prior and after the pulse:

$$P_{DBD} = f_{rep} \cdot \left(E_{precharge} - E_{recov}\right) \tag{6.5}$$

Keeping in mind this equation, the peak inductor currents can be found. The pre-charge peak current $i_{precharge,pk}$ flowing through the primary has sufficient amplitude to support charging of the DBD capacitance up to peak voltage $V_{DBD,pk}$ while compensating for the real power that is consumed by the plasma during ignition. The energy balance between E_L and $E_{DBD,pk}$ yields:

$$i_{precharge,pk} = \sqrt{\frac{C_{DBD} V_{DBD,pk}^{\ 2}}{L}} \tag{6.6}$$

The remaining energy in the DBD after ignition and subsequent plasma discharge is reduced by the real power consumption of the DBD lamp. Consequently, the peak recovery current is defined to:

$$i_{recov,pk} = \sqrt{\frac{C_{DBD} \cdot V_{DBD,pk}^{\ 2} - 2\frac{P_{DBD}}{f_{rep}}}{L}} \tag{6.7}$$

The repetition frequency depends on the sum of pre-charge time, pulse time and recovery time.

$$f_{rep} = \frac{1}{t_{precharge} + t_{pulse} + t_{recov}} \tag{6.8}$$

The times $t_{precharge}$ and $t_{recovery}$ are given by the volt-seconds law. The currents $i_{precharge,pk}$ and $i_{recov,pk}$ as well as L correspond to the primary while C_{DBD} and V_{pk} correspond to the secondary transformer side. In case of low damping and low PF, the pulse frequency virtually equals the natural resonance frequency. Therefore:

$$t_{pulse} = \pi\sqrt{L \cdot r \cdot C_{DBD}} \tag{6.9}$$

Inserting all required times in Formula (6.8), one gets:

$$f_{rep} = \frac{1}{\frac{L \cdot i_{precharge,pk}}{V_S} + \pi\sqrt{L \cdot r \cdot C_{DBD}} + \frac{L \cdot i_{recov,pk}}{V_S}} \tag{6.10}$$

Rearranging for V_S then yields:

$$V_S = \frac{f_{rep} L \cdot \left(i_{precharge,pk} + i_{recov,pk}\right)}{1 - f_{rep}\pi\sqrt{L \cdot r \cdot C_{DBD}}} \tag{6.11}$$

The currents earlier determined can be inserted in Formula (6.11). The necessary supply voltage V_S for a given repetition frequency is then:

$$V_S = \frac{f_{rep} \cdot \left(\sqrt{L \cdot C_{DBD} V_{DBD,pk}^{2}} + \sqrt{L \cdot V_{DBD,pk}^{2} \cdot C_{DBD} - 2L \frac{P_{DBD}}{f_{rep}}} \right)}{1 - f_{rep} \pi \sqrt{L \cdot r \cdot C_{DBD}}} \qquad (6.12)$$

Formula (6.12) is defined for:

$$f_{rep} \overset{!}{<} 2 \cdot f_{pulse} \qquad (6.13)$$

V_S must be increased as the repetition frequency is raised or PF worsens. Therefore, it is suggested to similarly control both the supply voltage and the repetition frequency while using the peak pre-charge current as trigger.

6.1.3 Operation with Energy Freezing

With his energy freezing method, Sowa (2004) presents an alternative solution for a continuous resonant flyback operation. This approach overcomes the problem of a fixed repetition frequency in case of a continuous mode by providing a short circuit path where the magnetisation current can flow. This maintains the magnetisation of the inductor core by keeping the energy in the magnetic field. When the energy freezing switch SEF is then turned off, the current freewheels through the secondary winding into the DBD. The time range of this energy freezing is virtually unrestricted. However, according to ohm's law, the current will create a voltage drop that decreases the magnetic energy since the switch's on-resistance and the winding resistance are not zero. Therefore, excessive energy freezing durations generate significant loss.

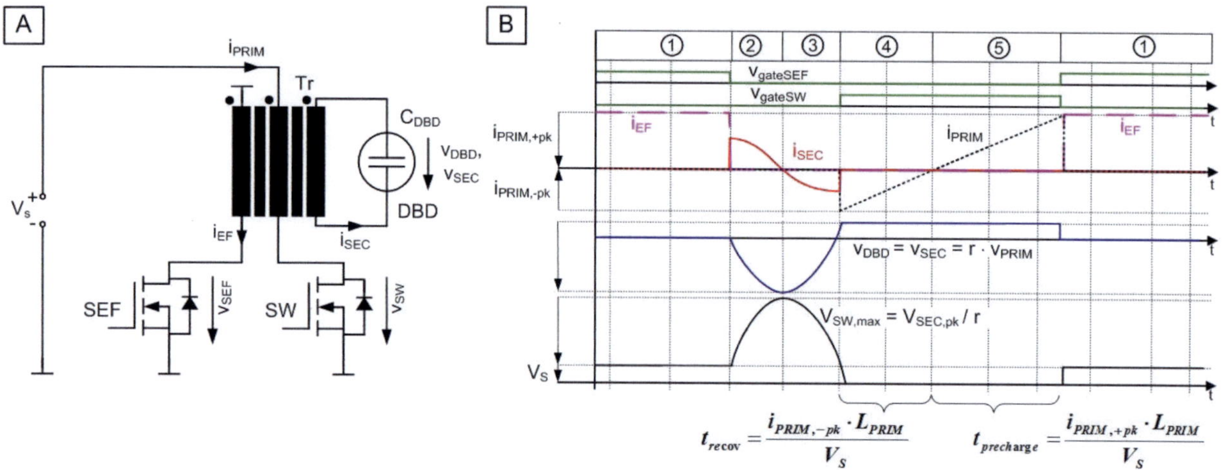

Figure 107: A and B: Energy Freezing circuit (A) and respective waveforms (B) according to Sowa (2004). As in CCM, current is continuous but inductive energy storage for long time-domains turns out to be inefficient.

It should be noted that the schematic covering the energy freezing circuitry (Sowa 2004) misses, in a strict sense, that the energy freezing switch must be able to **block voltage in both directions**. This further adds to the complexity of the circuit and provokes additional loss. The underlying principles of the energy freezing operation mode were also tested using the USPSP-topology of Chapter 7.3.

6.1.4 Experimental Results

In order to prepare investigations towards transformer-less resonant flyback derived topologies such as presented in Chapters 6.2 and 6.3, the traditional resonant flyback was shortly benchmarked. The advantages of the flyback such as high voltage amplification and adaptability to different load capacitances were evaluated driving a Planon® and a Planilum® DBD lamp. Figure 108A presents data obtained by driving a Planon® DBD lamp with a resonant flyback in CCM at 40 kHz repetition frequency. Although operated from 66.6 V supply voltage, due to the transformer ratio of $r \approx 7$, a significant voltage level of approximately 400 V was measured across the lamp during the idle time increasing the apparent power to S = 650 VA without contributing to the pulse itself. Further, a very sharp voltage spike is visible at t = 10 µs at the transformer's primary and also in the DBD current waveform.

This overshoot is generated by stray inductances mainly within the transformer and the primary circuit mesh due to the current transition from primary to secondary. The DBD ignites two times which is due to the 1 kV voltage overshoot after the primary ignition. The pulse duration is very long which can be explained by the huge transformer ratio of $r \approx 7$ that cuts down the achievable resonance frequency. The same circuit was also used to operate a Planilum® DBD lamp, which explains the choice and the paralleling of power semiconductors (Figure 108B). The Planilum® was driven with a pulse of 10 µs pulse duration from a supply voltage of 100 V. Only a power factor of PF = 8.5 % was achieved. The electrical efficiency was 70.2 %.

Neither the circuit configuration (Figure 108B) nor the generated pulse had been optimised. This would have included reduction of r and L_m in order to decrease pulse duration. Nevertheless, the results provide a benchmark to evaluate the topologies presented in the next chapters.

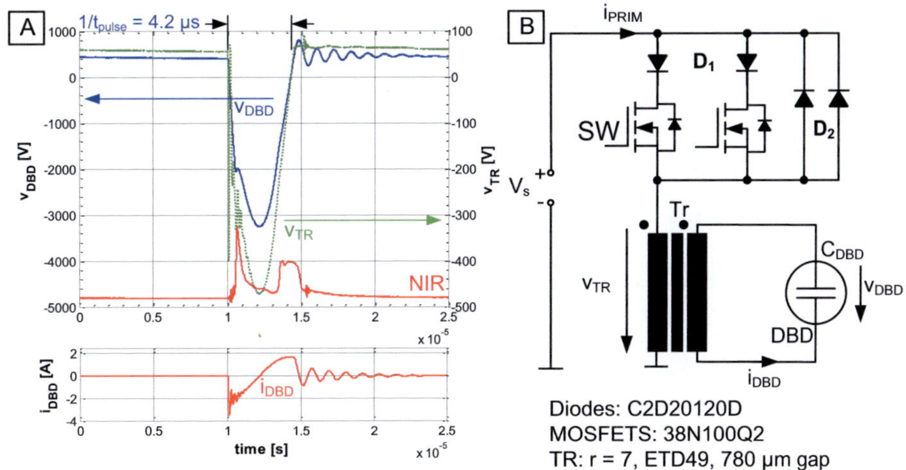

Figure 108: A: Experimental data obtained from the operation of a Planon® DBD lamp with resonant flyback depicted in B at V_s = 66.6 V, η = 69 %, P_{DBD} = 89.5 W, PF_{DBD} = 13.6 %, f_{rep} = 40 kHz, in CCM.

6.2 New Adaptive Unipolar Pulse Topology

Driven by the need for a transformer-less topology providing interim zero DBD voltage, the adaptive unipolar pulse (AUP-) topology had been developed and patented (Meisser 2010a). Like other parallel topologies, it is especially suited for driving DBDs from low supply voltage levels, e.g. 100 - 130 V. By variation of pre-charge time it is especially flexibly adaptable to changing load capacitances and ignition voltage levels. Additionally, the topology offers zero DBD voltage out of the pulses and allows for shorter pulses since no transformer ratio reduces f_{res}. Since the single inductor replaced the coupled inductors used in the resonant flyback topology, no current chopping between primary and secondary occurs. Therefore, higher efficiency of the magnetic component and much lower liability to harmful parasitic resonances exist.

6.2.1 Schematic and Principal Mode of Operation

The schematic of a basic topology variant is depicted in Figure 109. The schematic reminds of that of a two-switch inverted buck-boost converter assuming the DBD lamp to be a DC load. Also the inverted buck-boost allows for bi-directional energy transfer which is essential for the energy recovery feature required for efficient DBD drive. Beside the basic adaptive unipolar pulse topology variant, the topology can be combined with a high-voltage matrix switch (see Chapter 6.3) or with an additional (flyback-) transformer (Meisser 2010a). HS can also be laid out as super-cascode switch (Biela 2008).

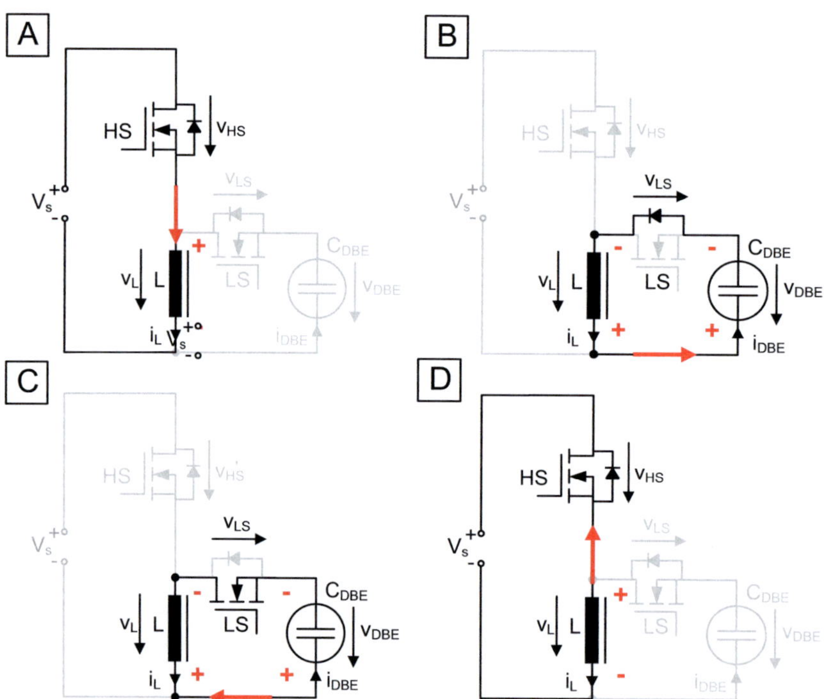

Figure 109: Schematic of the adaptive unipolar pulse topology in a basic variant depicting an operation sequence. A: Inductor pre-charge. B: Inductor freewheels through the DBD while body diode of LS conducts. C: Current backswing dicharging DBD capacitance. D: LS is turned off and current is recovered to the DC-link via naturally conducting the body diode of HS.

As long as switch LS is kept turned on, the operation principle resembles that of the resonant flyback. Instead of a flyback transformer, a flyback inductor is charged through HS which is a high-voltage rated fast power semiconductor switch equipped with a fast reverse diode. Since the inductor has only one winding, insulation issues are less severe than for the flyback transformer. Additionally, more space is available for multi-stranded litz-wire with a large cumulative cross-sectional area and large distances from turn to turn in order to minimise DC and high-frequency ohmic losses. The current through the winding is linearly rising and continuously transfers into a cosine wave-shape as soon as HS is turned off. The freewheeling inductor current brings about a soft turn-on of LS that is kept on during the pulse. The DBD connected in parallel to the flyback inductor is then charged with an initial maximum slew rate. Depending on the maximum voltage level the sinusoidal lamp voltage wave-shape leads to one or multiple lamp ignitions. As soon as the lamp voltage reaches zero volts again, LS is turned off and thereby prevents any significant voltage level across the lamp during energy recovery and the inductor pre-charge for the following pulse. Following this basic idea, topology variations were derived which are depicted in Figure 110. Figure 110B shows a mirrored variant with a V_S referenced DBD lamp. Figure 110C and Figure 110D demonstrate variants where the switches are supply rail referenced which dramatically reduces the requirements on gate drive insulation. From that perspective, variant D is the most favourable. In any of the transformer-less circuit variants depicted in Figure 110, LS needs to withstand the supply voltage level, only. Consequently, a cheap Si MOSFET can be chosen. Differently, HS experiences a maximum voltage level defined in Formula (6.14). Additionally, HS conducts the pre-charge current as well as the recovery current. Due to the unavailability of fast switches able to withstand several kilo-volts, the series connection of power semiconductors, preferably MOSFETs, is an option. Emerging technologies such as silicon carbide enable voltage ratings of 1.2 kV and more reducing the number of required series connected switches. If higher voltage and current rating is required, switches may be arranged to form a matrix switch such as presented in Chapter 6.3. Neglecting the transformer ratio r, the definitions of the pre-charge time and the current peak-level are identical to those of transformer-equipped pre-charge topologies.

$$V_{DS,HS} = V_S + V_{DBD,max} \tag{6.14}$$

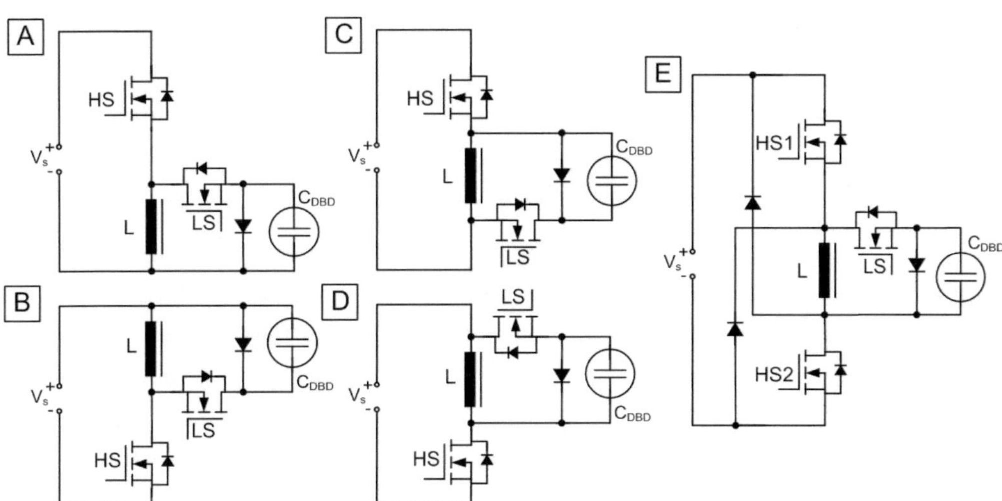

Figure 110: Circuit variations of the AUP-topology according to patent (Meisser 2010a). A: Basic topology with ground referenced DBD lamp, B: basic topology with V_S referenced DBD lamp, C: topology variant with switches supply rail referenced and lamp close to ground potential, D: topology variant with switches supply rail referenced and lamp close to V_S potential, E: three switch topology limiting all switch voltages to V_S.

6.2.2 Experimental Results

For a first evaluation of the topology, a pure capacitive load had been driven by the AUP-topology as depicted in Figure 110A. A 1 kV_{pk} pulse had been applied to a connected 4.7 nF capacitor. The circuit had been operated in DCM and 17.5 A peak current were transferred to the capacitor. LS was turned off shortly before v_{Cload} approached zero volts. At time 1.3 µs, energy recovery is finished and a parasitic parallel resonance between L and all parallel connected capacitive elements occurs. A comparatively high electrical efficiency of 86 % is obtained using two paralleled 1 kV 38N100Q2 Si MOSFETs as HS.

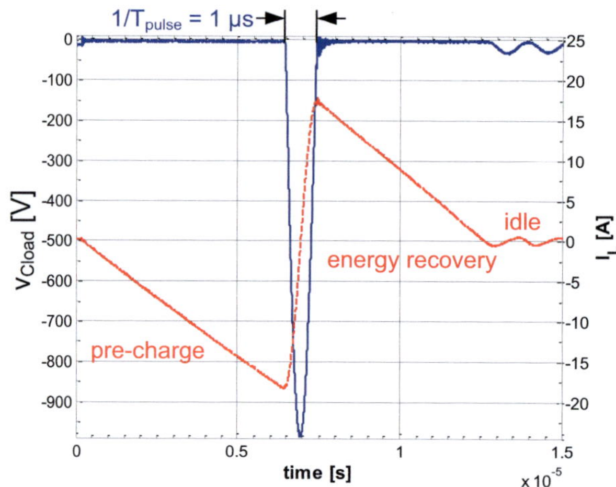

Figure 111: Adaptive unipolar pulse topology driving a pure capacitive load. L = 16 µH, C_{load} = 4.7 nF, P_{ECG} = 30 W, η_{ECG} = 86 %, T_{rep} = 25µs, V_S = 56 V. The load capacitor was made of series connected MKP capacitors. Peak energy stored $E_{Cload,pk}$ = 2.35 mJ – for comparison: 60x60 DBD lamp $V_{DBD,pk}$ = 2 kV, C_{DBD} = 3.18 nF E_{DBD} = 6.36 mJ.

The presented topology concept allows for DBD lamp operation with extremely short pulses since the pulse lengthening effects of a flyback transformer are prevented. Future work will cover the further investigation of this topology for DBD pulsed operation. The expected availability of SiC power semiconductor switches with breakdown voltages exceeding 2 kV will support further development of a simple and reliable pulse inverter basing on this concept. Alternatively, the matrix switch presented in the next chapter can be implemented. This would be feasible if extremely large lamps or lamp arrays are to be operated.

6.3 New Transformer-Less Resonant Flyback Using Matrix Switches

For decades, the transformer-equipped resonant flyback topology had been utilised to drive DBDs mainly with unipolar pulses. Here, an alternative is described that works without a transformer. Chapter 6.2 already showed a possible transformer-less topology equipped with a high-voltage switch. Excimer DBD lamps may require peak voltages exceeding 2 kV which bear the question how power semiconductor switches of such voltage rating and fast switching capability can be comprised of. As of this writing, single SiC switches of more than 1.7 kV rating were not commercially available. Consequently, at least two 1.2 kV rated switches would need to be connected in series as exemplarily done in Chapters 7.2 and 7.3. Using SiC JFETs a super-cascode incorporating several 1.2 kV devices can be created that is able to switch fast and operate reliably (Biela 2008). Alternatively, cost-effective Si MOSFETs can be combined to form a matrix switch. Tastekin (2011b) pursued an approach of purely series connected MOSFETs forming 10 KV switches assembled to a full-bridge to operate DBDs in pulsed mode.

Contrarily, in this chapter, a matrix switch design is presented that is implemented into a transformer-less resonant flyback topology (Meisser 2012d). Obviously, the matrix switch approach is universal and may be used for various kinds of pulsed power applications.

6.3.1 Matrix Switch Circuit Design

Matrix switches consist of multiple serially and parallel connected single switches that are simultaneously operated. In contrast to single high-voltage switches, matrix switches are universal, especially in case of a modular set-up, and bear possible cost reductions if a large amount of low-cost, low-performance components is used. Additionally, a certain level of redundancy can be implemented ensuring safe operation also in case that one switch in the matrix fails. Matrix switch configurations compared to single switches can also have higher switching and/or conducting performance if individually controlled fast MOSFETS and/or IGBTs are used in the modules, for instance.

The physical setup of the matrix switch presented in this paper follows a modular concept and incorporates supervisory circuits. The developed 5x2 matrix switch (MS) uses Coolmos™ MOSFETs with V_{DS} = 650 V and I_{DS} = 35 A rating each. The matrix switch has an R_{DSon} of 0.25 Ω, a peak current of 70 A and 2.7 kV voltage rating (at de-rating of 83 %) and contains five switch modules of which one is shown in Figure 112A. The five modules are connected to a common MS main board as sketched in Figure 113A. The main board layout ensures equal transient current sharing between the paralleled MOSFETs. As Figure 112B depicts, each module combines a dual MOSFET-driver, capacitive signal transmission (50 kV/μs immunity) and magnetic power transmission (5 KV_{pk}) with high-voltage insulation and over-temperature protection on a single, four-layer PCB. The PCB likewise acts as a press-plate to press down two paralleled MOSFETs onto a common heat-sink. Thereby, PCB, MOSFETs and heat-sink form a sandwich.

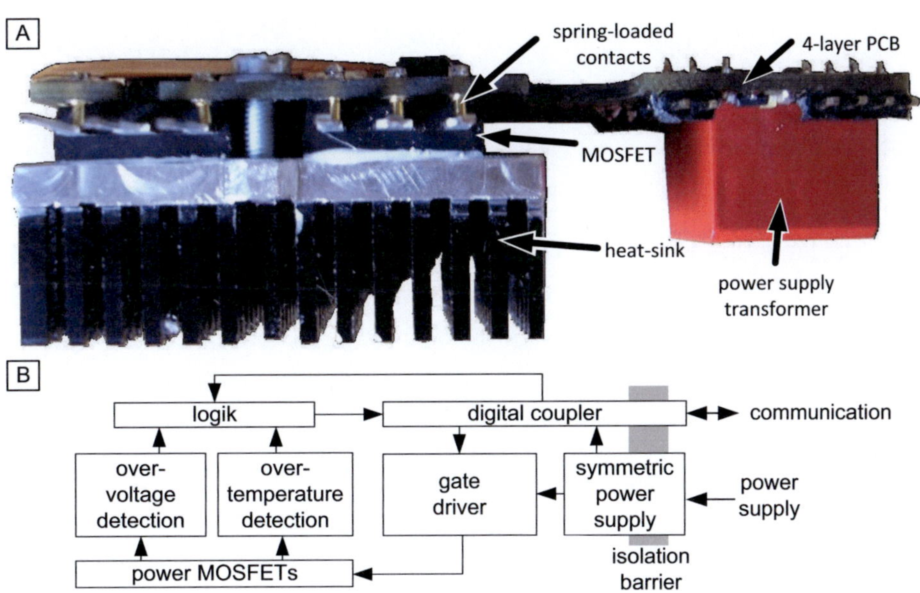

Figure 112: A: Picture of an individual matrix switch module indicating the main components. B: Functional block diagram of the modules.

The electrical contact to the MOSFETs is provided by spring loaded contacts (open arrows in Figure 113B). The MOSFET contact leads fit into sockets on the MS main board simplifying the solder-free replacement of individual modules. The heat-sinks of the individual switch modules are electrically isolated from the surroundings. An enhanced thermal cooling path is achieved since the MOSFETs are directly pressed onto the heat-sink. Common matrix switch designs use an individual transformer for both signal and power transfer to each module (Redondo 2011). In contrast, the insulation of each switch module is here performed by a compact power transformer and a digital coupler IC. This solution ensures maximum flexibility regarding switching duty cycle and provides enough power for high repetition frequencies. A schematic of the gate drive section of the modules is depicted in Figure 113B. The MOSFETs are driven by a high-current driver connected to bipolar power supply rails.

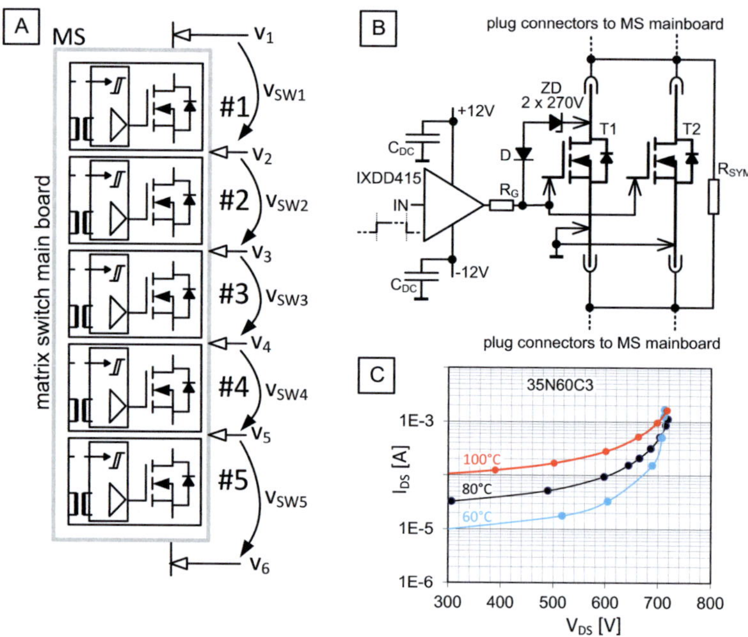

Figure 113: A: Block diagram of the 5x2 matrix switch main board. B: Schematic including the main parts of the gate drive section of each module. C: Avalanche behaviour of the used MOSFETs – temperature and voltage dependency of I_{DS} in off-condition (related to R_{DSoff}) obtained by measurements using the characterisation rig presented in Chapter 4.11.

As the individual modules of the matrix switch are connected in series, voltage sharing in static and dynamic case must be ensured. Static balancing is provided by a resistor branch connected in parallel to each module. For optimum sharing the current flowing through the balancing resistors should be higher than I_{DS} in off-state of the MOSFET in order to compensate for temperature influences on R_{DSoff} as shown in Figure 113C. Dynamic balancing was managed in three ways:

The used MOSFETs were taken from one batch and it is assumed that their characteristics are therefore very similar. As long as the dynamic output capacitances C_{OSS} similarily behave from MOSFET to MOSFET in the different modules, dynamic voltage sharing in off-state is achieved. In case of transient over-voltages intrinsic clamping occurs up to the limits of the MOSFET's avalanche capability.

In order to increase the clamping capability, each module contains a Zener-diode branch as shown in Figure 113B. A fast Schottky diode D prevents any current to flow in forward direction of the Zener-diodes. In case of v_{DS} overvoltage the Zener-diodes conduct. v_{DS} is then passively limited by the Zener-diodes' power dissipation. Depending on the used gate resistor R_G transient v_{DS} is also more or less actively limited by the increased conductivity of the MOSFET's channel due to the influence of the Zener current of the gate.

The signal inputs of each module are passed through a variable RC-filter that allows for an individual delay of the respective gate drive signal. By varying the delays, the transient voltage distribution between the modules can be optimised. Practically, this adjustment is done by adding delay to those modules who's Zener-diodes run hot indicating a transient over-voltage. This rather simple approach is justified since mainly turn-off of the switch may lead to significant losses and therefore needs to be optimised. This is also the case in the resonant flyback application that will be discussed in the following chapter.

6.3.2 MATRIX SWITCH IN APPLICATION – DBD DRIVE

The topology presented here is a two-component, transformer-less, resonant flyback. The matrix switch (MS) is embedded in this circuit as depicted in Figure 114A. The switch is connected in series to an inductor L and thereby controls the power flow into the inductor. The DBD lamp as load is connected in parallel to L. As long as MS is in on-state, current flows through L with constant positive slope determined by ratio V_S/L. In CCM as depicted in Figure 114B MS is only in off-state to permit the pulse during time range t_{pulse}. The turn-off of MS is most critical regarding the voltage distribution between the individual modules.

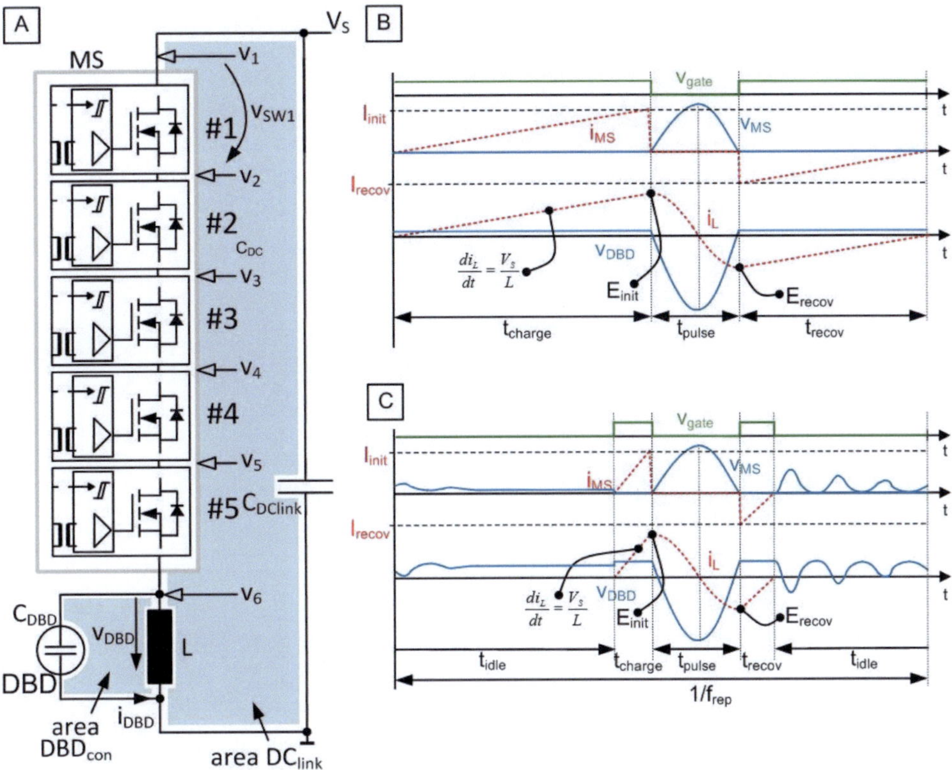

Figure 114: A: Schematic of the transformer-less resonant flyback with matrix switch. Blue: Commutation loops. B: Principal waveforms for CCM. C: Principal waveforms for DCM operation.

As mentioned in the previous chapter optimisation is possible by monitoring the temperatures of the Zener-diodes of the individual modules, for instance by means of an infra-red camera, and adjusting the R-C delay lines accordingly. The time delay of the module with the hottest Zener-diode needs to be reduced and vice versa. During off-state of MS L freewheels while ramping up DBD voltage v_{DBD} to ignition voltage level. At the time of maximum v_{DBD} amplitude also the voltage stress of MS is at maximum. Afterwards, C_{DBD} is discharged by the negative quarter-wave of i_L. As soon as v_{DBD} equals V_S, MS can turn on under ZVS condition in order to relieve its body diodes from conducting high current. MS stays turned on while the current transits through zero and again rises to charge L in advance of the next pulse.

In contrast, MS is virtually lossless turned off in DCM after recovery time t_{recov} as depicted in Figure 114B. Schottky diodes implemented in other topologies to enhance energy recovery are not necessary here since the comparatively low supply voltage level engenders a rather long recovery time with continuously decaying current. This effectively burkes reverse recovery-related problems. Reverse recovery times of the MOSFET body diodes are further reduced by keeping MS turned-on during t_{recov}. Pre-charging L for initiating the next pulse requires another turn-on of MS under ZCS condition and solely V_S level.

The power consumed by the DBD is defined by the difference of the energy levels stored in the inductor prior to the pulse and after pulse termination:

$$P_{DBD} = f_{rep} \cdot (E_{init} - E_{recov}) \tag{6.15}$$

The pre-charge peak current i_{init} defines the magnetic energy $E_{L,init}$ stored in inductor L. Assuming no losses between pulse start and DBD ignition at pulse centre, $E_{L,init}$ must be equal to the maximum energy stored in the DBD. Consequently, the inductor peak energy is determined by the energy level that needs to be transferred to the DBD which is roughly:

$$I_{init} = \sqrt{\frac{C_{DBD} \cdot V_{DBD,pk}^{2}}{L}} \tag{6.16}$$

The inductor pre-charge time is then given by:

$$t_{precharge} = \frac{V_{DBD,pk}\sqrt{L \cdot C_{DBD}}}{V_S} \tag{6.17}$$

The pre-charge time is set to store a determined energy in the inductor, which needs to be equal to the sum of peak DBD capacitor energy and real energy consumed by the DBD:

$$E_L = \frac{1}{2} L \cdot I_{L,pk}^{2} \stackrel{!}{=} \frac{1}{2} C_{DBD} \cdot V_{DBD,pk}^{2} + \frac{P_{DBD}}{f_{rep}} \tag{6.18}$$

This maximum energy is the key parameter for choosing a sufficient ferrite core. The energy handling capability of the pure core is enhanced by the introduction of an air gap. Inductor design is further dealt with in Chapter 4.13.1.

As the DBD has a low power factor in the range of 10 % (Sowa 2004), the low-damped circuit resonates approximately with natural resonance frequency determining the pulse duration to:

$$t_{pulse} = \pi\sqrt{L \cdot C_{DBD}} \tag{6.19}$$

The energy contained in the magnetic field of L at pulse termination is the difference of the initial energy and the power dissipated in the circuit during the pulse, mainly by the plasma discharge within the DBD. Likewise this difference is "reactive" energy that corresponds to reactive power transferred to charge and discharge the DBD capacitance. In first order, the ratio of power consumed by the DBD and reactive power is found to be (Sowa 2004):

$$PF_{DBD} = \frac{P_{DBD}}{f_{rep} \cdot E_{recov}} \approx 0.1 \tag{6.20}$$

The time required for inductor discharge (energy recovery) after pulse termination amounts to:

$$t_{discharge} = t_{recov} = \frac{V_{DBD,pk}\sqrt{\dfrac{L \cdot C_{DBD}}{PF_{DBD} + 1}}}{V_S} \tag{6.21}$$

Dividing Formula (6.17) by Formula (6.21) yields the simple ratio expressed in Formula (6.22). The conduction losses in the matrix switch can be calculated using Formula (6.23) and inserting Formulae (6.17) and (6.21) as well as the currents I_{init} and I_{recov}.

$$\frac{t_{precharge}}{t_{discharge}} = \sqrt{PF_{DBD} + 1} \tag{6.22}$$

$$P_{MS,cond} = \frac{1}{3} f_{rep} \cdot R_{DS,on} \left[t_{precharge} \cdot I_{init}^{2} + t_{recov} \cdot I_{recov}^{2} \right] \tag{6.23}$$

As the conduction losses $P_{MS,cond}$ directly correspond to the inductor charging and discharging times, which linearly depend on V_S, DCM operation with high V_S level outperforms CCM significantly.

The application circuit (Figure 114A) consists of two main commutation meshes – the DC_{link} commutation mesh and the DBD-load mesh. As high-voltage and current slopes occur in both meshes, the respective stray inductance needs to be minimised in order to damp parasitic series resonances.

Depending on the switch status of MS, the equivalent circuit supporting the parasitic resonance is depicted in Figure 115A and B respectively. Discarding dissipative circuit elements (ohmic losses, energy dissipation in the DBD plasma) and assuming the large-value DC-link capacitor to be a HF short, the parasitic series resonances seen in the measurement results are characterised by their natural resonance frequency.

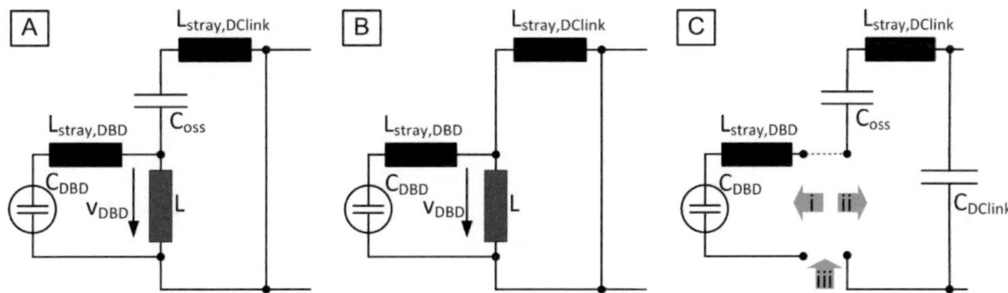

Figure 115: Equivalent circuit of resonant transformer-less flyback. A: With open matrix switch; B: with closed matrix switch; C: impedance measurement connection scheme.

Due to its high value and consequential high impedance at elevated frequencies, L is virtually transparent to these HF resonances. In case of turned off MS (Figure 115A) the parasitic series resonance frequency is given as:

$$f_{para,MSoff} = \frac{1}{2\pi\sqrt{\left(L_{stray,DBD} + L_{stray,DClink}\right) \cdot \frac{C_{DBD} \cdot C_{oss}}{C_{DBD} + C_{oss}}}} \tag{6.24}$$

In case of turned on MS this changes to:

$$f_{para,MSon} = \frac{1}{2\pi\sqrt{\left(L_{stray,DBD} + L_{stray,DClink}\right) \cdot C_{DBD}}} \tag{6.25}$$

As $C_{oss} \ll C_{DBD}$, this frequency is much higher than the resonance frequency in case MS is turned on or is naturally reverse conducting (L charge and discharge, Figure 115B).

In order to characterise the system, impedance measurements without DC-bias were performed as depicted in Figure 115C. This yielded the self-resonance of the DBD lamp (i) at 17 MHz ($L_{stray,DBD}$ = 318 nH) and the self-resonance of the commutation loop (ii) at 5.2 MHz ($L_{stray,DClink}$ = 120 nH). The open loop circuit with lamp attached (iii) showed a resonance at 15 MHz.

Note that C_{OSS} is highly voltage-dependent and therefore small signal measurements can only give a rough estimation of its value at elevated voltages.

6.3.3 Experimental Results

To gain time-domain measurement results of an application circuit, the matrix switch had been implemented into a resonant transformer-less flyback topology as shown in Figure 114A with measurement set-up depicted in Figure 116B. The voltages at each node were measured with wide-bandwidth, high-voltage probes. The output current i_{DBD} was measured with a wide-bandwidth current transformer. The NIR radiation of the DBD was detected by a PMT with attached RG780 filter. Two INFINIIUM-Series oscilloscopes were used to acquire the eight time-dependent signals.

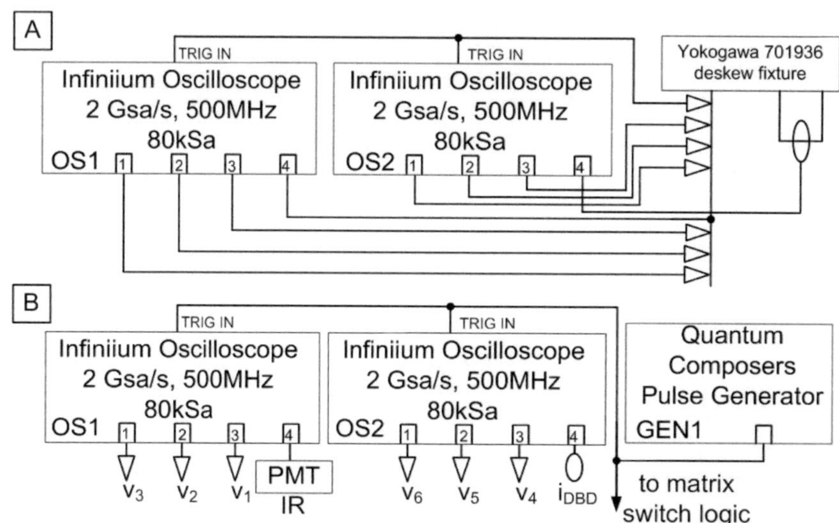

Figure 116: Measurement connection schemes. A: Measurement probe de-skewing. B: Connection to application circuit.

The trigger for both oscilloscopes was provided by a Quantum Composers pulse generator that simultaneously provided the gate signal for the matrix switch. The synchronisation of the trigger and the signal acquisition paths required de-skewing with a dedicated fixture (Figure 116A). Thus, after probe compensation, the equipment was connected as indicated in Figure 116B. A 1:1 voltage probe was used to connect the auxiliary trigger inputs of the oscilloscopes to the voltage signal output of the de-skew fixture. All other signals were then related to the point in time of the trigger event. The individual skew was set to a standard delay of 180 ns related to the trigger signal.

A first measurement was performed for CCM operation of the circuit. At the low supply voltage level of V_S = 25 V, a repetition frequency of 25 kHz and an input power of 27.2 W an inverter efficiency of only 68.5 % had been achieved. As shown in Figure 117, pulse duration was 641 ns and v_{DBD} slope was 13 kV/µs. Clearly to see are the points in time when the ignition of the lamp took place. Moreover, parasitic resonances are found in the current waveform with approximately 25 MHz and 10 MHz resonance frequency, respectively. These frequencies differ from those obtained by the impedance measurements discussed in the previous chapter. This is due to the strong voltage dependency of C_{OSS}. Nevertheless, the order of magnitude fairly conforms.

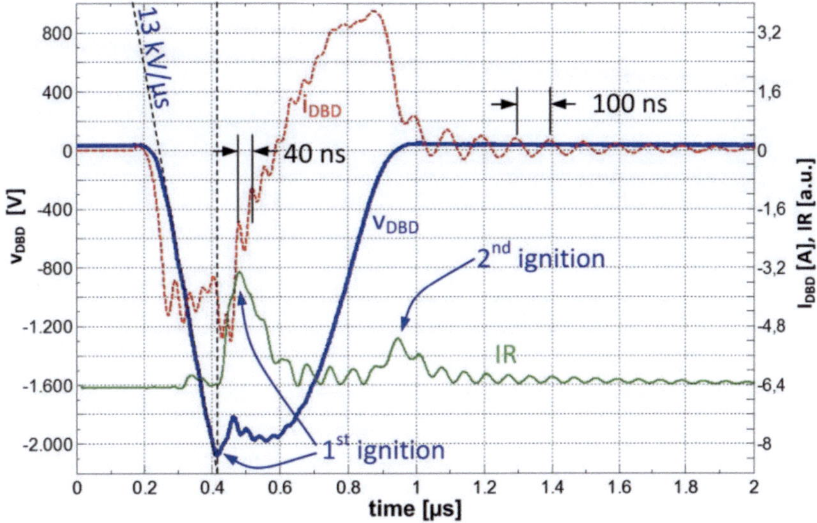

Figure 117: Measured waveforms of circuit in CCM operation. Round shape of voltage due to charging of C_{OSS} of matrix switches. Inverter drives a flat 0.04 m² DBD lamp with C_{DBD} = 310 pF and V_{ign} = 2 kV.

Figure 118 shows a comparison of the lamp voltage v_{DBD} for CCM and DCM. The measurement of the voltages of the individual modules of the matrix switch in CCM is shown in Figure 119. As can be seen, the Zener-diode breakdown voltage is not exceeded by any of the modules SW1 – SW5.

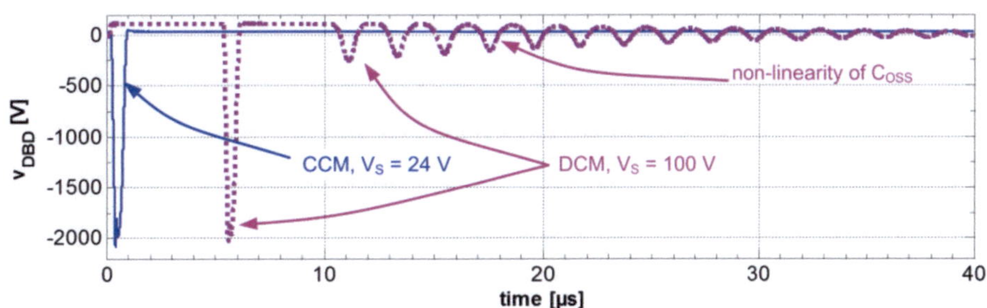

Figure 118: Experimental comparison of DCM and CCM operation. Note the resonance during idle time in DCM operation mode.

The switch with fastest turn-off is stressed with the highest voltage (Perret 2009). As the switches carry the same current, the voltage slopes are similar (300 - 400 ns in Figure 119A). A pulse frequency variation was performed by changing the inductance of L to gain the graphs shown in Figure 119B. The inverter efficiency drops from 86.5 % at t_{pulse} = 589 ns to 63.4 % at t_{pulse} = 160 ns. The high v_{DBD} slope of 33 kV/µs visibly excites the parasitic series resonance during the pulse. The frequency is higher than yielded by the impedance measurement due to the significantly smaller C_{OSS} caused by the elevated voltages. The change of C_{OSS} depending on v_{DS} can also be obtained during t_{idle} in Figure 119B. The presented matrix switch proved efficient and reliable operation in pulse mode up to 2.2 kV_{pk} peak voltage and 33 kV/µs slope. However, an automatic adjustment of the gate signal delay for each module by Zener-diode temperature analysis would further enhance the matrix switch performance.

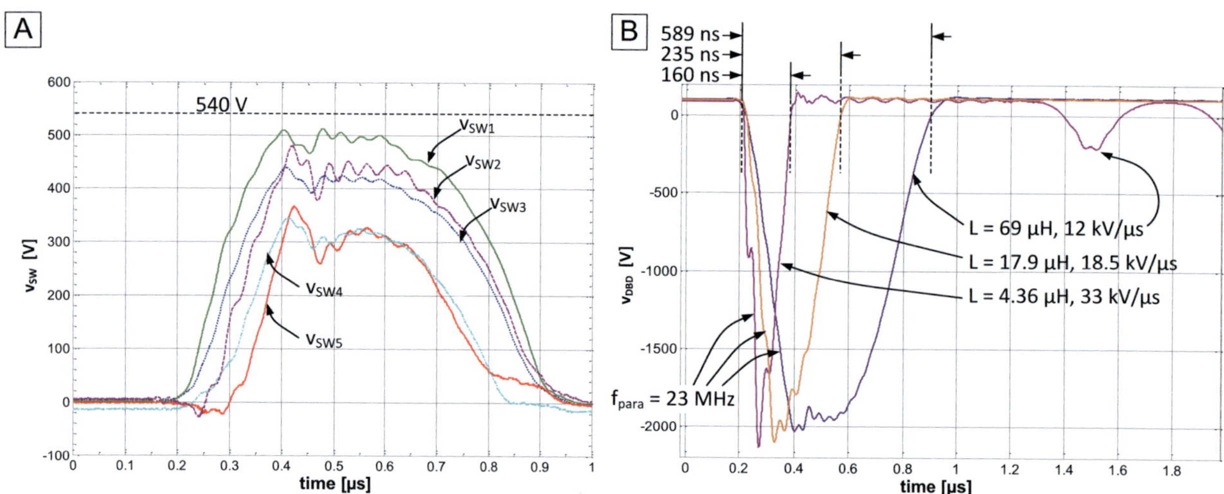

Figure 119: A: Measured individual switch waveforms in CCM operation. All switch voltages stay well below 540 V. B: Measured DBD voltage waveform. Pulse frequency variation by change of inductance of L. As the pulse frequency changes, the slope is naturally enhanced. Pulses as short as 160 ns equalling a pulse frequency of 3.13 MHz had been achieved.

The presented matrix switch equipped transformer-less flyback can be seen as a design study visualising the possibilities of using relatively cost-efficient Coolmos™ MOSFETs to equip transformer-less pulse inverters. The presented structure was designed for maximum flexibility. Especially the FOC-signal transmission and the individual gate driver for each voltage level may be exchanged by direct gate-drive transformers in order to create a cost-efficient design.

7 Operation Modes and Resonance Behaviour of Mixed Mode Topologies

This type of topologies allows both modes of operation: resonant energy recovery and pre-charge of the inductor. This is achieved by allowing the inductor current path to be independent of the current path through the DBD. The major difference to the family of parallel topologies is that here, the inductor can be connected in parallel and in series to the DBD. One of the first attempts to follow the idea of independent inductor pre-charge and subsequent pulse shaping by the inductor was the use of the auxiliary resonant commutated pole inverter (ARCPI) with the lamp connected to one leg of the full-bridge configuration (Winkelmann 2003). The same topology was later named adaptive pulse gear (Trampert 2004). By varying the stored energy, the slope of the lamp voltage could be set to the recommended value. The topology was simplified towards the patented adaptive pulse gear topology and operation method (Heering 2004; Daub 2010) which allowed for a certain degree of recovery of energy stored in the lamp. However, it was found that by using this energy a second ignition can be supported (Paravia 2007). The stray inductance of the transformer determined the shape of the negative half-wave. Hence the excessive current flow and voltage overshoots restricted the reliability and efficiency of this approach. This topology is shortly reviewed in Chapter 7.1.

Within the scope of this dissertation, two novel mixed mode topologies were developed. Although, the schematic is similar to the patented schematic of the adaptive pulse topology, the operation method of the novel universal sinusoidal pulse (USP-) topology is different and will be presented in Chapter 7.2. The USP-topology as well as a further development, the combination of the USP-topology with the SPSP-topology of Chapter 5.2 had already been published (Meisser 2011b) and are presented in Chapter 7.3.

7.1 Brief Review of the Adaptive Pulse Topology

The topology and the operation method were patented (Heering 2004) while the further development and investigation of this topology was main part of the dissertation of Daub (2010). In this dissertation, Daub claims up to 72 % topology efficiency driving a 15" Planon® lamp. Then again, driving one to four coaxial DBD lamps with 2 kV ignition voltage and C_{DBD} = 48 pF each, the reported efficiency is only around 60 % while the individual lamp consumed a power of 10 W. The adaptive pulse topology is originally intended to generate unipolar pulses. Sure enough, bipolar pulses can be composed by utilising parasitic circuit elements.

The adaptive pulse (AP-) topology including main parasitics is depicted in Figure 120A.

The inductor L is pre-charged through the turned on switches S_1 and S_2. The pre-charge time directly sets the energy level stored inside L which is later transferred to the DBD lamp. With this, the slew rate of the positive voltage rise for a given C_{DBD} can be determined. Therewith, the flexible adjustment to changing DBD properties is possible. However, this feature is also naturally provided by other topologies, especially those intending an inductor pre-charge. If S_2 is turned off at the defined current level, this allows the inductor to freewheel into the DBD lamp through D1 leading to a first voltage rise. In the exemplary waveform depicted in Figure 120B, its slew rate is 6.6 kV/µs. Beneficially, S1 is kept turned on as long as i_L is positive since in this case, further energy can be drawn from V_s. The DBD is charged until the ignition voltage is reached and then, in a resonant manner, the current direction changes and energy is partly fed back into the supply V_S. Recovery occurs naturally and cannot be actively controlled as possible by LS in series topologies.

Up to here, the waveform mimics those generated by resonant flybacks (see Chapter 6.1). But differently, in order to generate a small negative voltage backswing, S_2 can be turned on during or after the energy recovery phase. The parasitic stray inductances of the transformer and the attached lamp generate a negative lamp voltage. Due to its restricted maximum voltage level, the negative voltage half-wave possibly does not support the second ignition of the DBD sufficiently. The generated wave-shape is comparable to those presented by Somekawa (2005). In both cases, heavy ringing occurs subsequent to the bipolar pulse. This is since no further energy recovery is possible through which the ringing could be damped.

Despite of the intrinsic restrictions on efficiency set by the inductor pre-charge, the commercial availability of high-voltage fast SiC power switches could allow to achieve better circuit performance than reported by Daub (2010). The drawbacks of insufficient energy recovery and restricted bipolar pulse properties lead to the development of the family of universal pulse topologies presented in the following chapters.

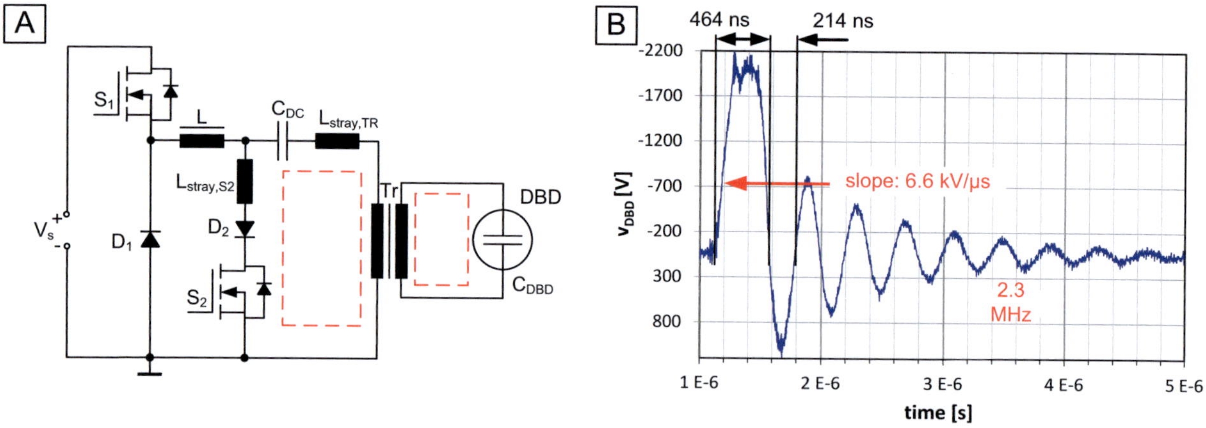

Figure 120: A: Schematic of adaptive pulse (AP-) topology. Beside the transformer stray inductance, the marked areas contribute to the effective stray inductance. This is utilised to generate negative backswing of lamp voltage as shown in B. Here, a 0.04 m² flat Xe excimer DBD lamp with V_{ign} = 2 kV and C_{DBD} = 310 pF is driven by an experimental setup of the AP-topology.

7.2 New Universal SP-Topology

The need to drive DBDs with higher pulse frequencies and nevertheless high electrical efficiency can be fulfilled by the proposed family of universal sinusoidal pulse (USP-) topologies. The basic scheme of a transformer-equipped half-bridge variant is shown in Figure 121A and B. It combines the adaptability of the adaptive pulse topology with a strong bipolar excitation of the lamp derived from the sinusoidal pulse topology. The implemented step-up capability provided by pre-charge switch S_{PC} allows the reduction of transformer ratio r in order to enhance the maximum possible resonance frequency. The SP-topology only supports a maximum voltage overshoot factor of two. In contrast, as the inductor pre-charge enhances the total voltage amplification, higher primary peak voltage is obtained from a lower supply voltage with the new USP-topology. Only the switch S_{PC} and the respective diode D_{PC} have to withstand the primary-reflected maximum lamp voltage. As S_{PC} and D_{PC} only conduct the pre-charge current, on-state losses are cut. Favourably, the half-bridge switches face the reduced supply voltage. This allows the use of low-voltage switches with very low R_{DSon} and fast switching behaviour. The series diode D_{PC} blocks the body diode of S_{PC} and thereby provides bipolar operation of the circuit.

Therefore, this topology is a promising attempt to generate bipolar high-frequency, high-voltage pulses out of low voltage power supplies.

Figure 121B shows the key waveforms. In time-domain 1 of Figure 121B, turned on switches HS and S_{PC} initiate the pre-charge of inductor L. The current linearly ramps up. As the inductor must be designed to store the DBD peak energy level, differently to serial topologies the peak current is reached in advance of charging the DBD capacitance. Even more important the peak current level during the first positive slope of DBD voltage is twice as large as compared to series topologies.

In time domain 2, the actual pulse is initiated by opening S_{PC}. The energy stored in L forces the current to ramp up the DBD voltage v_{DBD}. In contrast to the SP-topology, the slope of v_{DBD} is initially maximal. Before the lamp voltage reaches its peak value, commutation under ZVS condition from HS to LS is performed.

In time domain 3, the energy cycling between the lamp capacitance and the inductor leads to a subsequent negative and an additional positive peak of lamp voltage. If equipped with sufficient voltage amplitude, each sine half-wave of the pulse leads to an ignition of the DBD lamp. After the second positive voltage peak, energy recovery is initiated by turning off LS. Current is pushed by L through the intrinsic anti-parallel diode of HS back into the intermediate supply V_s.

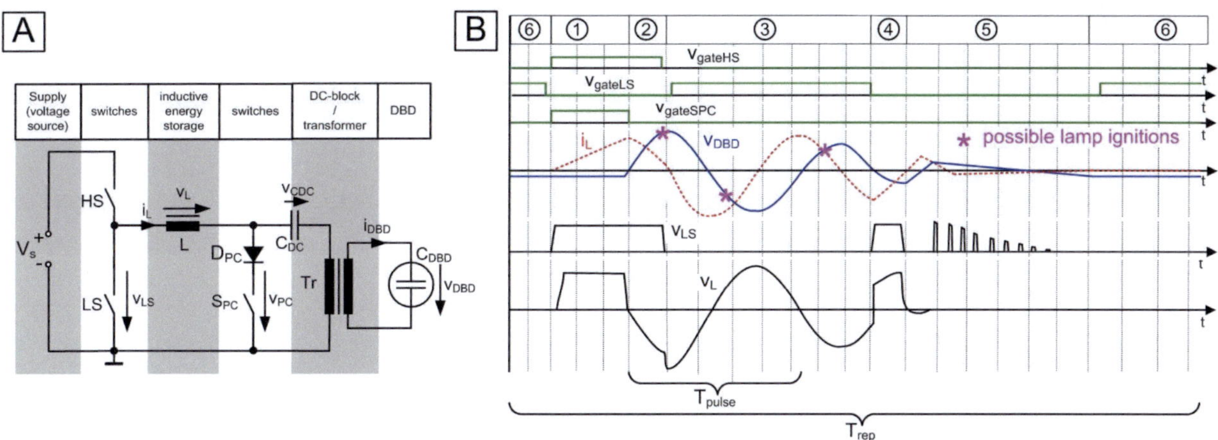

Figure 121: Universal Sinusoidal Pulse (USP-) topology. A: Main circuit blocks and simplified circuit diagram. B: Switch timing and characteristic waveforms of inductor current i_L, inductor voltage v_L, lamp voltage v_{DBD}.

The imbalance of the voltage time products of the bipolar pulse leaves the transformer core with a proportional energy offset after pulse termination. Due to the unsymmetrical pulse shape, the energy left in the circuit is higher than in the SPSP-topology (Meisser 2010c). In time domain 5 of Figure 121C, the small current that is forced by the mutual inductance L_m of the transformer discharges the DBD to negative voltage amplitude. Subsequently, in time domain 6, LS is turned on in order to allow the transformer to freewheel. This also effectively prevents the parasitic parallel resonance of C_{DBD} and L_m to occur (Meisser 2010b). The most critical element in the circuit is the high-voltage switch S_{PC} as it has to face the maximum primary voltage and is also stressed by peak voltages induced by stray inductances.

Possible topology variations are shown in Figure 122. The DBD lamp can be driven by a transformer or the transformer is omitted by the use of high-voltage switches. A further variation includes a full-bridge configuration. As one upper bridge-leg is only used for the purpose of energy recovery, diode D_{ERS} is used instead of a high-side switch.

As C_{DBD} and maximum lamp voltage $V_{DBD,pk}$ are given for a certain application, evidently, the maximum possible resonance frequency f_S is fundamentally restricted by the current carrying capability of the semiconductors and magnetic devices. Formula (7.1) gives the respective peak current. Taking into account the low power factor, the contribution of DBD real power can be neglected and formula (7.2) is obtained.

$$I_{DBD,pk} = \sqrt{4\pi^2 \cdot C_{DBD} \cdot f_{pulse}^{\ 2} (C_{DBD} \cdot V_{DBD,pk}^{\ \ 2} + P_{DBD} \cdot T_{rep})} \tag{7.1}$$

$$I_{DBD,pk} = 2\pi \cdot C_{DBD} \cdot f_{pulse} \cdot V_{DBD,pk} \tag{7.2}$$

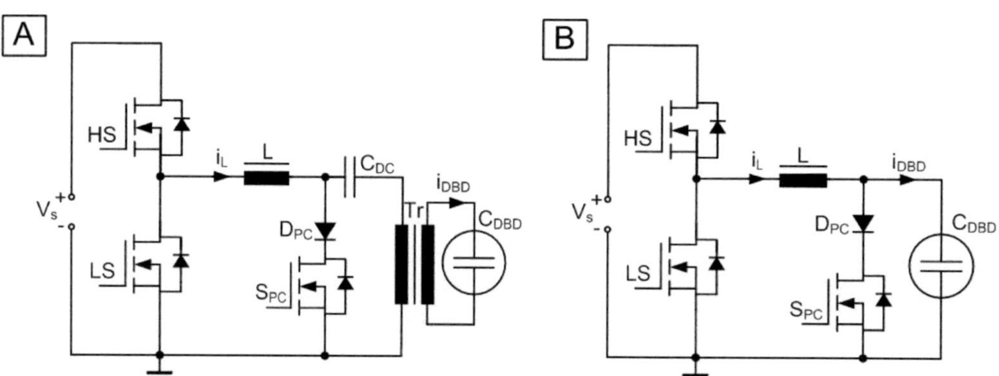

Figure 122: Family of universal sinusoidal pulse topologies. A: Transformer-equipped half-bridge variant. B: Transformer-less half-bridge variant presented in Chapter 7.2.4.

Therefore, switches with greatly enhanced performance can be used. Beside their lowest on-resistance, the low gate charge Q_g also reduces switching losses of the half-bridge. Thereby, switch utilisation is greatly enhanced and higher currents are possible.

However, pre-charge switch S_{PC} is the bottleneck of the circuit and therefore has to be chosen carefully. To minimise on-state losses during pre-charge, SJEP120R063 silicon carbide normally-off JFETs in combination with C2D20120D SiC Schottky diodes were used. To optimise their static and dynamic behaviour, a tailored driving circuit was developed which is presented in Chapter 4.12.3.

Although, a transformer is used in this circuit variation, the transformer ratio r is smaller than required for use in the SP-topology, depending on the maximum voltage of the pre-charge switch. This enables the operation with higher pulse frequency than possible with the SP-topology.

7.2.1 TIME DOMAIN BEHAVIOUR

Referring to Figure 121, the topology permits two principal operation modes regarding the energy transfer from the pre-charged inductor to the DBD. Either HS is turned off as soon as S_{PC} is turned off or, alternatively, the supply voltage source V_S is kept in the circuit via the still turned on HS. In the first case the DBD voltage follows a pure sinusoidal wave-shape as defined in Formula (6.2). In the second, more preferable case additional energy is drawn from the supply source reducing the required level of energy stored in the pre-charged inductor and changing the DBD voltage wave-shape towards:

$$v_{DBD}(t)=\left[\frac{C_{DBD}\cdot R_S\cdot V_{DBD,i}+2L\cdot i_{L,i}-V_S\cdot R\cdot C_{DBD}}{2\cdot L\cdot C_{DBD}\cdot\omega_d}\right]\sin(\omega_d t)\cdot e^{-\alpha\cdot t} + \left(V_{DBD,i}-V_S\right)\cos\omega_d e^{-\alpha\cdot t}+V_S \tag{7.3}$$

In Formula (7.3), the pure damped sinus is expressed by the first term. The second and third term represent the additional support by the supply voltage source V_S. Next, the power loss in the switches during inductor pre-charge is calculated.

The power lost in all series resistances R_S in the circuit rises quadratically with time. Integration across the pre-charge time reveals:

$$E_{loss,on}=\int_0^{t_{precharge}}\left(\frac{R_S\cdot I_{pk}^2}{t_{precharge}^2}\cdot t^2\right)dt=\frac{1}{3}R_S\cdot I_{pk}^2\cdot t_{precharge} \tag{7.4}$$

Assuming that the peak energy stored in the DBD must also be stored in L yields:

$$P_{loss,on}=\frac{2}{3}\frac{R_S\cdot E_{DBD}\cdot t_{precharge}\cdot f_{rep}}{L} \tag{7.5}$$

This simple formula visualises that in terms of power losses, a short-time pre-charge with high supply voltage level is more preferable than a longer lasting pre-charge at lower supply voltage level. Formula (7.5) can be rewritten to include the supply voltage level:

$$P_{loss,on}=\frac{2}{3}R_S\cdot f_{rep}\frac{E_{DBD}\cdot I_{pk}}{V_S} \tag{7.6}$$

7.2.2 EXPERIMENTAL RESULTS

First experimental results are shown in Figure 123. The DBD lamp was driven at a power of 87 W with an electrical efficiency of 67 % while thermal measurements according to Chapter 4.9 revealed a maximum efficiency of 77.8 %. The used 17.9 μH inductor stored energy of 4.4 mJ in advance of each pulse. The pulse frequency is 440 kHz. Mischievously, as visible in Figure 123B, the turn-on of LS occurred at 50 ns, which is too late as to benefit from the provided ZVS condition. Remarkably, Figure 123A indicates a parasitic oscillation which disturbs the first slope of the DBD voltage by adding a significant 11 MHz HF component.

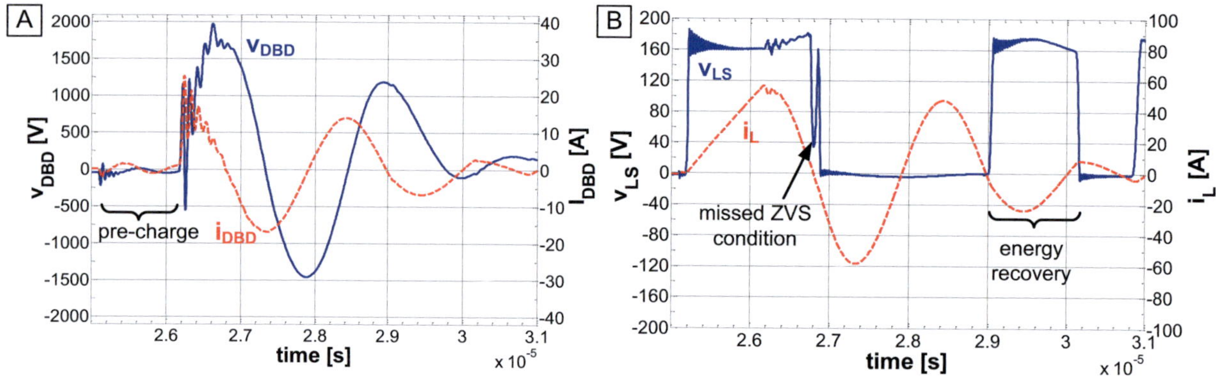

Figure 123: Characteristic waveforms of transformer-equipped half-bridge USP-topology gained by experiment. $V_S = 175$ V, $t_{charge} = 1$ µs, $T_{rep} = 50$ µs, $f_{pulse} = 440$ kHz, $\eta = 67.5$ %, $\eta_{th} = 77.8$ %. A: DBD voltage and corresponding current. B: Voltage across LS and current through inductor L. Note the linear current growth during pre-charge.

7.2.3 Parasitic Resonances and Their Damping

The reason for the 11 MHz parasitic series resonance visible in Figure 123A will be explained by the simplified circuit given in Figure 124. For pulse initialisation, the charged inductor L generates a steep voltage step across S_{PC}. This excites the connected high-frequency series resonance circuit consisting of the stray inductances of the DBD, transformer and connection cables and the DBD capacitance C_{DBD}. The HF current path passes the S_{PC} output capacitance $C_{oss,eff}$, while i_L stays almost unaffected since $L/L_{stray} > 10$.

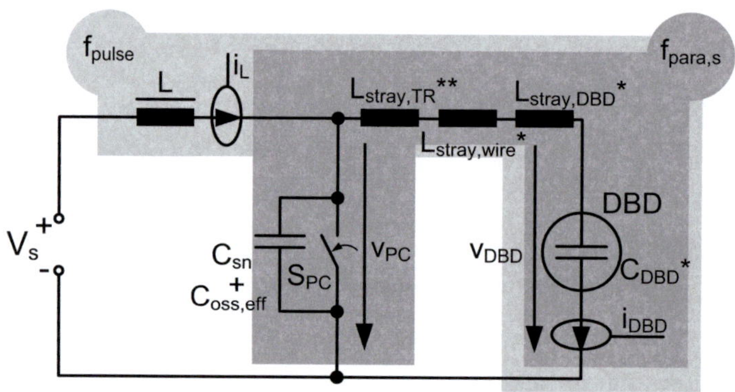

Figure 124: Electric equivalent circuit in time-domain 2 – first ramp-up of DBD voltage. Two series resonant circuits interact. * primary reflected values in case of transformer-equipped design, ** appears only in transformer-equipped design.

Therefore, the amplitudes of HF current are split between the inductive paths with the reciprocal ratio which is proved by comparison of the currents depicted in Figure 123A and B. It is of highest importance to reduce the stray inductance of the transformer by an interleaved, full breadth winding design since wiring and intrinsic DBD inductance can hardly be lowered. Therewith, a stray inductance of only 0.1‰ of the mutual inductance has been achieved. The total output capacitance $C_{OSS,SPC}$ of S_{PC} plays a major role concerning the frequency $f_{para,s}$ according to Formula (7.7) and the amplitude of the parasitic resonance. According to Formula (7.8), $C_{OSS,SPC}$ limits the initial voltage slope and hence limits the maximum overshoot of v_{DBD} due to the parasitic resonance phenomenon.

$$f_{para,s} = \frac{1}{2\pi\sqrt{\left(\sum L_{stray}\right) * \left(\frac{C_{DBD} \cdot \sum C_{PC}}{C_{DBD} + \sum C_{PC}}\right)}} \quad \text{with} \quad \begin{aligned} L_{stray} &= L_{stray,TR} + \frac{L_{stray,wire} + L_{stray,DBD}}{r^2} \\ C_{PC} &= C_{OSS,SPC} + C_{sn} \end{aligned} \tag{7.7}$$

$$v_{SPC}'(t=0) \propto \left(\frac{2L \cdot i_{L,initial} - V_S \cdot R \cdot C_{OSS,SPC}}{2L \cdot C_{OSS,SPC}}\right) \tag{7.8}$$

Concerning the main pulse resonance, $C_{OSS,SPC}$ is parallel to C_{DBD} but has minor influence on the actual pulse frequency defined in Formula (7.9) as $C_{DBD,prim}/C_{OSS,SPC} > 10$.

$$f_{pulse} = \frac{1}{2\pi\sqrt{(L + L_{stray}) \cdot (C_{DBD} + C_{PC})}} \tag{7.9}$$

In order to reduce the voltage overshoot due to the parasitic series resonance, the initial voltage slope of the pulse must be restricted. This restriction is achieved by an additional snubber capacitor C_{sn} in parallel to S_{PC}. This also helps to reduce turn-off loss of S_{PC}. Furthermore, it reduces $f_{para,s}$ and also decreases the electromagnetic radiation of the circuit.

However, C_{sn} is not an ordinary snubber capacitor with the aim to damp resonances in the switch leg. Instead, it reduces the voltage slope in order to minimise excitation of the parasitic resonance. The results gained with an appropriately chosen capacitance are presented in Figure 125. As the turn-off loss is significantly reduced, the electrical efficiency rises by up to 2 %. Although, the voltage overshoot amplitude is significantly reduced, the quality factor of the parasitic resonance stays high while oscillation continues. It is assumed that this is due to the reduced damping by the first ignition. The peak voltage is reduced for higher values of C_{sn}.

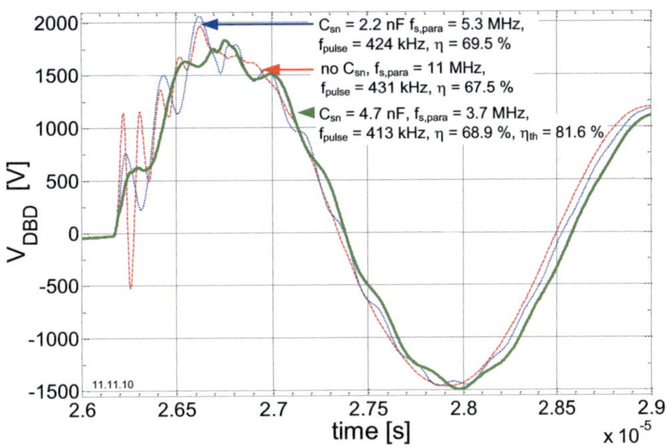

Figure 125: Wave-shape enhancement with added capacitor C_{sn}. Parasitic resonance is significantly damped leading to an efficiency rise while pulse frequency changes only marginally.

In conclusion, the transformer-equipped circuit variation benefits from the principal advantages of the topology while preventing the rather sophisticated series connection of multiple transistors and diodes in order to form a high-voltage switch required for the transformer-less variation. In contrast, the avoidance of a transformer not only reduces volume and weight of the circuit. Furthermore, there is potential for higher pulse frequency and lower power loss, for a given DBD lamp.

7.2.4 Transformer-less Version

Using serially connected high-voltage switches and diodes, the principle of inductor pre-charge can be used to drive a DBD lamp without a transformer. Referring to the basic circuit depicted in Figure 122D, only one switch must be capable of withstanding the lamp voltage amplitude. The two half-bridge switches must only be rated for supply voltage amplitude. As the peak voltage of the sample lamp is, depending on pulse shape and pulse repetition frequency, in the range of 1.7 to 1.9 kV, the high-voltage switch has at least to withstand 2 kV. Experimentally, this had been achieved by the series connection of two SiC 1.2 kV JFETs of type SJEP120R063. In order to maximise switching performance, a tailored driving circuit had been designed. It provides a high current in times of switch transitions while restricting current through the intrinsic gate-source diode in conducting state. The exact time at which lower and upper switch of this high-voltage combination change state can be adjusted independently. Thereby, the turn-off time can be adjusted to stress both parts with the same (dynamic) voltage. To protect the switches from transient overvoltage conditions, Zener diode arrays from drain to gate limit the maximum V_{DS}, especially at turn-off, for both switch parts independently. Compared to the transformer-equipped topology variation, for the same load and approximately the same pulse resonance frequency, the quality factors of both pulse resonance and parasitic series resonance lie one order of magnitude higher. As a result, multiple energy recovery cycles occur and the voltage overshoot, induced by the parasitic series resonance, is higher. In order to reduce the voltage overshoot, an extra capacitor C_{sn} was connected in parallel to the total S_{PC} switch combination. As Figure 126 indicates, this reduces the overshoot and enhances circuit efficiency.

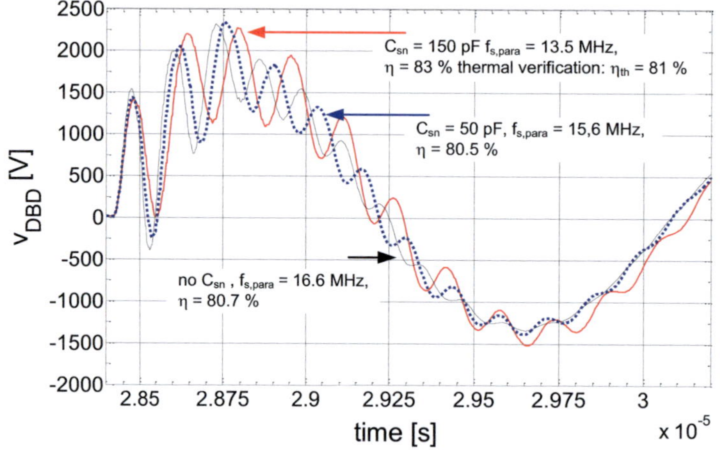

Figure 126: DBD voltage for different extra capacitors, connected in parallel to S_{PC}. Frequency and initial overshoot of parasitic resonance decrease while $Q_{para,s}$ rises with higher capacitance values.

However, for higher C_{sn} values, the parasitic oscillation occurs for a longer time. With a C_{sn} of 150 pF, an electrical efficiency of 83 % was obtained while thermal verification supports 81 % efficiency.

Both the USP and the USPSP-topology presented in the next chapter are well suited to drive DBDs from low supply voltage levels. The inbuilt capability of high voltage amplification also means that higher pulse frequencies can be achieved as the transformer ratio can be reduced or the transformer can be removed completely.

7.3 NEW TRANSFORMER-LESS UNIVERSAL SPSP-TOPOLOGY

Bidirectional energy recovery, independent of actual inductor current direction, is possible with a full-bridge configuration showed in Figure 127A. In contrast to the SPSP-topology of Chapter 5.2, an inductor pre-charge is possible by the additional high-voltage switch S_{PC} changed towards a unipolar device by means of the series connected high-voltage diode D_{PC}. Basing on the principles of inductor pre-charge and the possibility for bidirectional energy recovery, this topology is referred to as universal single period sinusoidal pulse (USPSP-) topology. The components of the right bridge-leg, namely low voltage energy recovery switch ERS and diode D_{ER} contribute only minor to overall losses. Advantageously, they reduce count and total duration of the necessary energy recovery cycles as they provide an alternative recovery path. In case of the transformer-less circuit variation as depicted in Figure 127B, it may be unfavourable that none of the DBD supply cables is permanently connected to a reference potential. However, because the flying potential is only of V_S amplitude, this has little impact. In order to highlight the strengths of this topology, the transformer-less variant will be discussed more closely. Figure 128A shows this circuit variant as a design tailored for a maximum output voltage of 2.4 kV. S_{PC} consists of two series connected 1.2 kV SiC JFETs with attached anti-parallel fast small-signal diodes. D_{PC} consists of a 2x2 matrix of 1.2 kV SiC Schottky diodes. The left bridge-leg and ERS are fast 200 V 25 mΩ Si MOSFETs and D_{ERS} is a SiC Schottky diode.

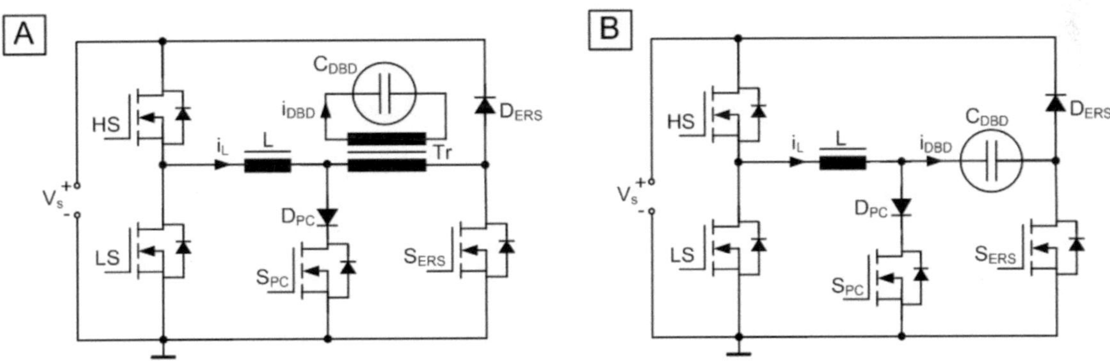

Figure 127: Schematic variations of the USPSP-topology. A: Transformer-equipped variant. B: Transformer-less variant.

In normal operation mode e.g. DCM, the circuit works as follows. According to Figure 128B, L is charged via HS and D_{PC}/S_{PC} up to a current level defined by the energy required to charge the DBD up to ignition voltage level. During that time, also ERS is turned on in order to ensure zero DBD voltage. Then, S_{PC} is turned off and L freewheels and charges C_{DBD} with positive current. As long as HS and ERS are kept turned on, supply source V_S contributes energy to that charge-up. As soon as the intended peak DBD voltage is attained, ZVS commutation from HS to LS is initiated. Subsequently, the current becomes negative and discharges C_{DBD} down to a negative voltage peak. Shortly before the current becomes positive again, ERS could be turned off under ZVS condition. Alternatively, ERS is turned off later during the positive current half-wave with significant switch loss. The latter could be feasible for reasons of ignition support. Then, as long as significant current cycles, all switches are kept in off-state. By multiple cycles, energy is recovered into the supply voltage source. The energy level that is left after the last recovery cycle is the reason for the non-zero voltages V_{LS} and V_{ERS} during idle time. They have no negative impact. At best, a part of this energy can be used to reduce the switching losses at initiation of the next pulse. For this purpose, S_{PC} is activated first leading to a ZVS turn-on of ERS. Energy is transferred from $C_{OSS,LS}$ into L and back resulting in a turn-on of HS at reduced voltage drop. Alternatively to that operation mode, instead of energy recovery, energy can be stored in L during idle time (time domain 6). This scheme is depicted in Figure 128C. Contrarily to the DCM operation mode energy is frozen in L as soon as v_{DBD} passed through a full sinusoidal period. The energy freezing is achieved by turning on all switches except for HS. In contrast to energy freezing in resonant flybacks as presented in Chapter 6.1.3, current levels lowered by a factor of r occur here. Main losses are generated in D_{PC}.

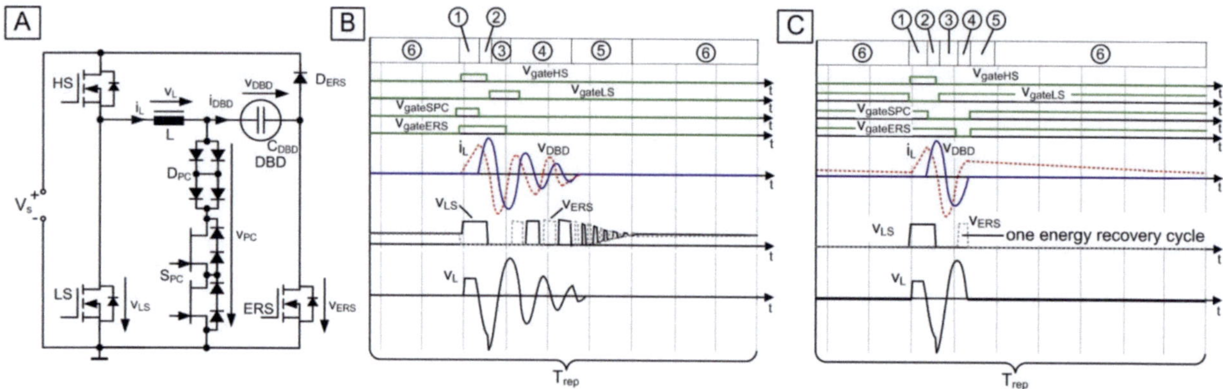

Figure 128: Transformer-less, bidirectional energy recovery variation of universal pulse topology. A: Extended schematic. B: Standard switching scheme. C: "Energy freezing" switching scheme.

Further changes of the principal USPSP-topology may include the design of S_{PC} as a bidirectional switch which would allow for a non-resonant, pure inductive energy recovery (see Figure 38B on page 77). This would prevent multiple recovery cycles and would be well suited to terminate a burst wave which is suggested by Beleznai (2008c) (see Figure 19 on page 53) with the scope to increase the DBD's power density. The higher conduction losses that would be generated by the bipolar switch would be compensated by the low-loss operation of the bridge switches that maintain the burst.

7.3.1 Experimental Results

Experimental results in form of measured waveforms are shown in Figure 129. A pulse frequency of 588 kHz is obtained while the pulse terminates shortly after solely four energy recovery cycles.

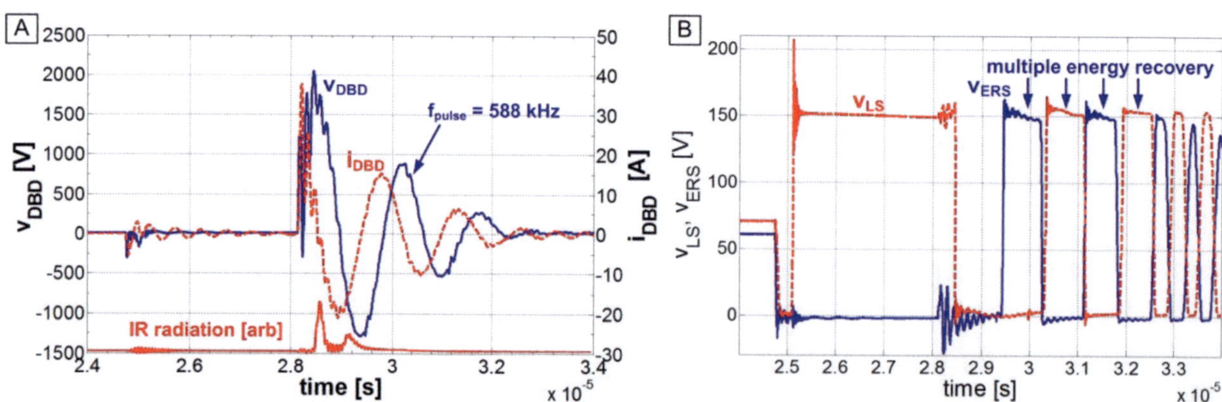

Figure 129: Full-bridge with standard switching pattern and multiple energy recovery cycles. V_S = 155 V, T_{rep} = 50 µs, η = 71.5 %, $η_{th}$ = 79 %. A: DBD lamp voltage corresponding current and measured NIR emission indicating two lamp ignitions. B: Voltages across LS and ERS.

Figure 129B shows that four energy recovery cycles are required to transfer the energy stored in the DBD back to the voltage supply source. In the configuration of Figure 128A, the circuit attains an electrical efficiency of 71.5 % being supplied by a supply voltage level of V_S = 155 V. This value had been verified by thermal measurements at the power semiconductor heat-sink and the inductor.

7.3.2 Energy Freezing Operation Mode

The energy freezing switching scheme according to Figure 128C which is comparable to that proposed by Sowa (2004) for resonant flyback DBD drivers has been experimentally investigated. The latter has the advantage of driving the lamp with an ideal bipolar pulse and true zero voltage during idle time, which could further improve DBD efficiency. However, as magnetic storage for rather long time-domains is inefficient, the electrical efficiency is poor. As a compromise, several recovery cycles may be initiated before the DBD and L are shorted.

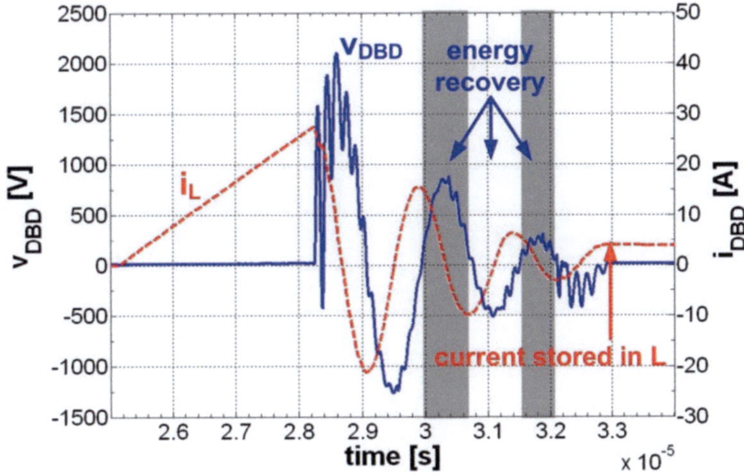

Figure 130: Full-bridge with energy freezing scheme and 3 recovery cycles. Energy stored in inductor (≙ 2.8 W) is completely dissipated. Operation at V_S = 160 V and T_{rep} = 50 µs whith an achieved efficiency of η = 69 %.

By doing that, efficiency rises from 47 % with one recovery cycle to 69 % with three recovery cycles. The waveforms belonging to the latter approach are shown in Figure 130. At a repetition period of 50 µs, the current of 4 A that is left after the last recovery cycle is completely dissipated during the idle time when the inductor freewheels.

However, energy freezing may make sense in applications where the idle time is short and in the range of pulse duration. Additionally to losses during energy storage in the inductor, the turn-on of HS for pre-charge initiation may no longer occur under ZCS condition and is therefore lossy.

In conclusion, transformer-less topology variants such as the one presented here are by principle suited for a very high frequency pulsed operation of DBDs as the requirements on the circuit quality factor are lower and no transformer reduces the possible resonance frequency. However, for all presented topologies but especially for the members of the parallel and mixed mode topology families, the reduction of the DBD's stray inductance is mandatory for a high performance operation at high pulse frequency.

8 Evaluation of Topology Performance

The scope of this final chapter is to compare and qualitatively assess the performance of the topologies presented in the previous chapters. DBD lamps are operable under an extensive variety of operation modes, which dramatically change their electrical behaviour. For this reason, no absolute quantitative rating of topologies can be provided. Instead, indications are presented here which support the qualitative results. An overview of the DBD lamps used in this work and the achieved efficiencies is provided in Figure 131. In order to allow for a comparison, the different DBD lamps are arranged according to the maximum energy stored within their capacitance. Note that depending on the applied peak voltage, the maximum energy of the respective DBD lamp may vary between operation points.

The 60x60 Planilum® DBD lamp having an ignition voltage of around 2 kV and a total capacitance of 3.18 nF had been chosen to be the benchmark load and was operated by most of the topologies. This lamp was the most demanding load due to its high energy requirement and significant parasitic elements (see Chapter 2.5.3). As a matter of fact, virtually every topology under investigation shows advantages for a certain application:

As can be derived from Figure 131, the upswing full-bridge topology achieves an outstanding electrical efficiency of 92 % at a pulse frequency of 500 kHz and still 80 % at 1.15 MHz. It is believed that the burst mode operation Beleznai (2008c) based on this topology and operation mode will see further improvement and is one way to achieve high power densities without the cost of lowered efficiency. The high efficiency of the upswing full-bridge topology is explained by the comparatively low conduction and switching losses experienced by the power semiconductor switches and the missing transformer losses. The principle of the resonant upswing allows for high voltage amplification and in turn for the use of cheap, fast and well-conducting medium-voltage Si MOSFET switches.

A comparison of the SP- and SPSP-topology revealed the improvements regarding DBD voltage pulse wave-shape and inverter efficiency possible with the SPSP-topology. System efficiency rose by more than 10 % while the apparent power absorbed by the DBD lamp shrunk by more than 25 %. This topology is thus suited for enhanced operation of mid-size DBDs by the use of mature Si power semiconductor switches.

Generally, transformer-less topologies have proven to generate pulses at higher efficiency and at higher pulse frequencies than their transformer-equipped counterparts. The reason is that the transformer not only transforms voltage but less-favourably also leads to enhanced primary current and above all reduces the maximal achievable pulse frequency by its ratio r. As presented in this work, the transformer is also linked to parasitic resonance effects that may lower efficiency and reliability of the inverter.

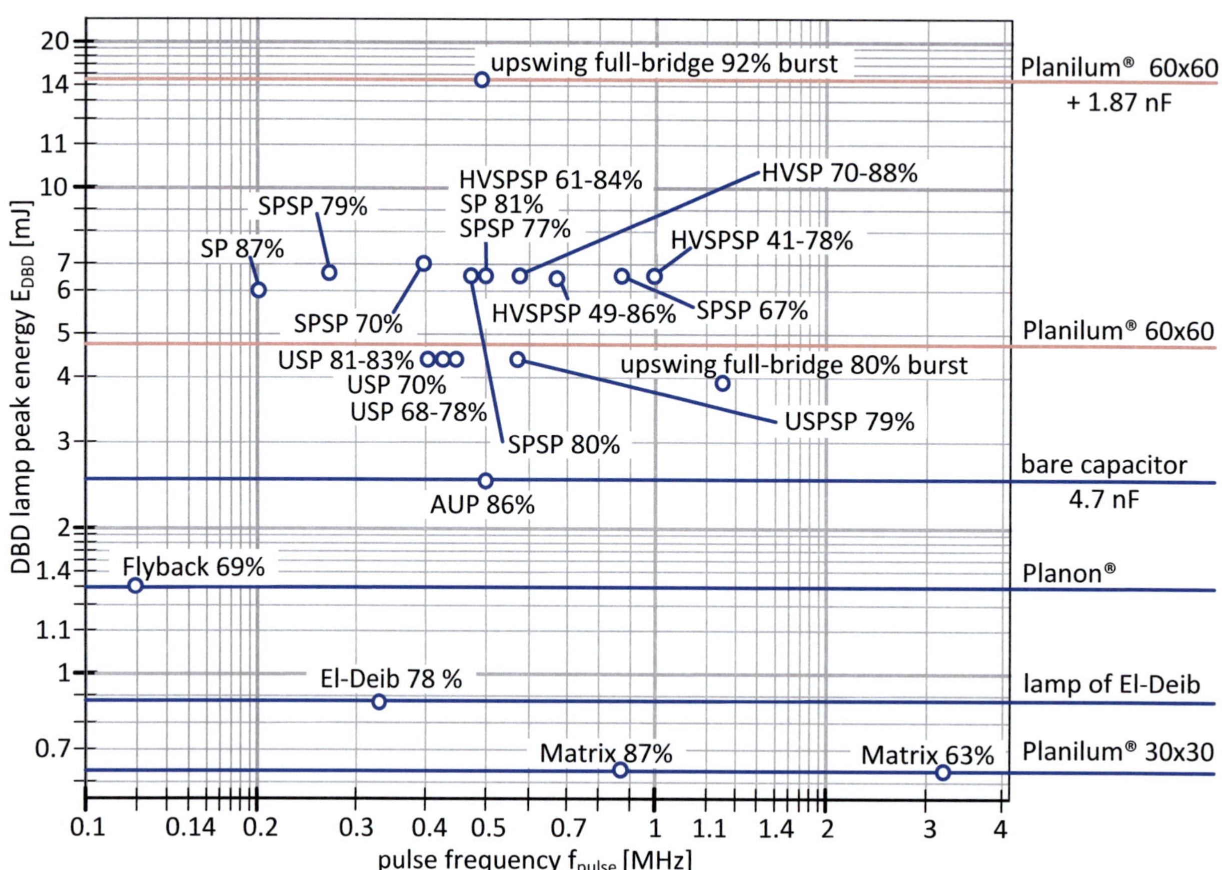

Figure 131: Comparison of the efficiencies of topologies presented in this work. Different DBD lamps are classified regarding their peak energy E_{DBD} (logarithmic, y-axis, see Formula (2.27) on page 47) including the used Planilum® and Planon® lamps and the lamp mentioned in a publication of El-Deib (El-Deib 2010a). Topologies and operation modes are classified by their pulse frequency (logarithmic, x-axis). Electrical efficiencies of the respective operating point are provided.

The single switch topologies derived from the resonant flyback topology have the advantage of very low component count and the ability to generate unipolar pulses of very high frequency. They are therefore predestined for applications where the DBD lamp exhibits comparatively low maximum energy and cost efficiency is paramount. However, those topologies may become uneconomic for large DBD lamps with high energy demand considering that the extensive losses generated will cause a huge cooling effort.

In contrast, transformer-less SP- and SPSP-topologies offer the ability to drive large DBD lamps with extensive energy demands but are based on costly high-voltage SiC power semiconductors. Their future will lay in UV / VUV industrial installations where reliable high temperature operation is of higher importance than lowest cost.

The transformer-less variants of the innovative mixed mode topologies (USP- and USPSP-topology) are best suited for operating medium-sized lamps especially from low-voltage inputs in the range of 100 V. This allows for the use of cost-efficient Si MOSFETs in the bridge-legs leaving one high-voltage SiC power semiconductor switch as the one and only significant cost factor. In contrast to the aforementioned transformer-less topologies, both the USP- and the USPSP-topology provide a great initial DBD voltage slope without any preceding voltage excitation.

The variability of the pulse shape that is provided by series and mixed mode topologies should be utilised to further tailor and optimise the operation of DBD lamps regarding their discharge behaviour and efficiency.

9 Summary

This work extends the variety of driving circuits for the efficient high-frequency pulsed operation of excimer dielectric barrier discharge (DBD) lamps. It covers the discussion of innovative circuit topologies which are presented here for the first time. Various experimentally obtained data support the feasibility of the presented topology concepts. Moreover, various novel operation modes of these topologies are discussed. The focus was set on the prevention of parasitic resonances that can negatively affect the reliability and efficiency of pulse generators. The conducted research unveiled a simple and efficient way of shifting the frequency of parasitic resonances by applying dedicated switching schemes to the power semiconductor switches. On top of that, topology variations that allow abandonment of the transformer were developed and experimentally validated.

The thesis outlines the history and gives an overview of contemporary DBD applications, state-of-the-art DBD radiation sources, DBD operation modes and inverter concepts. Selected properties of DBD excimer light sources are discussed leading to a suggested optimised wave-shape for pulsed operation of DBDs. The influence of the different DBD capacitances on the energy support of the transient DBD plasma and on the energy that needs to be provided by the inverter is analysed. Being the fundamental base of this work resonant circuits are discussed regarding their time-dependent behaviour and their properties in the frequency domain.

For the first time, the impedance characterisation technique is utilised to comprehensively quantify parasitic electrical elements of DBD lamps, the inverter power stage, power magnetics and the gate drive circuits. To the author's knowledge, impedance measurement has not yet been used to that extent in order to describe critical real-world circuit components. The investigations disclosed critical component properties supporting the adaption of power stages and gate drive circuits to the different topologies under investigation. The power stage inductance was reduced down to three nH achieved with a specifically developed SiC power module. The module design and manufacture was conducted by the author at the PEATER lab at the University of Warwick, UK.

The static-on characterisation of power switches including the internal thermal path of the devices is a further speciality presented here. Contrarily to pulsed measurement techniques, the devices were mounted on a temperature-controlled heat-sink and continuously operated at the desired current level. Cutting-edge silicon carbide (SiC) power semiconductors that were implemented in several transformer-less topology variants were characterised with the help of this characterisation rig. Measurements demonstrate that datasheet parameters provided for low temperature operation are not suited for capturing the ohmic losses and thermal runaway phenomena of devices implemented in real-world circuits.

The novel topologies developed and validated in this work can operate DBDs with a variety of pulse wave-shapes, principally with unipolar and bipolar pulses. While driving a 3.18 nF DBD lamp having a vast peak energy of 14 mJ with a 500 kHz pulse generated by the novel resonant upswing topology, an electrical efficiency of 92 % was obtained. This high electrical efficiency is caused by the minimised losses in the full-bridge due to low voltage operation and simultaneously reduced current due to the missing transformer current amplification. A smaller DBD lamp with 0.63 mJ was operated by a resonant transformer-less flyback with a unipolar pulse that exceeded 3 MHz. The inverter equipped with a five-stage matrix switch achieved an electrical efficiency of 63 %. This work proves that an enhancement of the pulse frequency from 385 kHz to 3.1 MHz not only improves the DBD lamp efficiency but also more than doubles the lamp's radiation output.

To date, a classification scheme allocating pulse inverter topologies to families sharing key properties was required. As a consequence, a classification for pulse inverter topologies driving capacitive loads such as DBDs was developed and applied for known and novel topologies.

The benefit of the presented approaches to pulse generator structures and components is not limited to inverters for DBD operation. The underlying concepts can be transferred to other applications where short pulses are required for driving mainly capacitive loads. To name but a few, pumping of Lasers, driving of Laser-diodes, operating of piezo-actuators and driving of DBD reactors require efficient inverter solutions which may be based on the presented concepts.

This dissertation builds, amongst others, upon eight published first-author conference papers and two patent applications. Further publications on the properties of the SiC half-bridge modules will follow.

10 Outlook

In retrospect, the DBD light source has seen many positives, such as the commercialisation of the Planon®, Linex® and Xeradex® light sources, and negatives, like the discontinuation of the Planon® and the Planilum® lamps. Notwithstanding this development, the future of the DBD VUV light source appears to be bright.

Due to the fact that LEDs make only little progress to achieve reasonable efficiencies at wavelengths of 365 nm and lower (Kneissl 2010), they are highly unlikely to represent a competing technology to DBDs in the VUV in the near future. Apart from that, higher efficiency and higher power density, which go hand in hand with increased lifetime are the keys to broaden the fields of application and develop potential markets for DBDs. By way of example, the global UV wastewater treatment market is expected to grow by 18 % annually. The global UV light market will cross the USD 2 billion margin by 2018 (Kneissl 2010). It is the author's believe that this substantial continuous growth will again raise the interest in DBD technology.

Concerning the DBD VUV lamp, further innovations are required. As suggested by Beleznai (2008c), the power density of DBD lamps could be further increased by operating the DBD with high-frequency bursts. This has the advantage of being generated based on the resonant upswing and energy recovery technique presented in Chapter 5.3 (Meisser 2010d). Also based on that chapter, modifications in the geometrical structure of the DBD can be implemented.

The presented topologies still do not cover all possible variations. It seems worthwhile to pay special attention to the further investigation of the adaptive unipolar pulse topology (Chapter 6.2) and the universal single period sinusoidal pulse topology (Chapter 7.3). For instance, the latter could be modified in order to generate bursts but terminate them immediately without the need for multiple energy recovery cycles.

Despite of many decades of research, the DBD light source still conceals a myriad of secrets. Based on the current state of knowledge, the impact of the pulse wave-shape should be further investigated. Building on the suggestions regarding an optimum pulse wave-shape provided in Chapter 3.4, analysis should be based on interpreting wavelength and time-resolved measurements in the VUV and NIR. This will broaden the understanding of the time constant dependences of the excimer generation processes.

Based on the results of these investigations the emphasis should also be put on the development and verification of DBD pulse inverter control techniques. Indeed, this topic has been intentionally omitted from this work, even though it is of importance to the system performance and reliability. Related topics are pure electrical ignition detection, DBD power measurement, ensuring ZVS condition under changing load conditions and self-optimising energy recovery.

Concerning power electronics for high frequency operation, the further steps towards a "system in a module" are of high importance. These steps include the implementation of the gate driver, its buffer capacitors as well as the DC-link capacitors in the power electronics module in order to further reduce parasitic inductance and enhance the heat dissipation capability of the gate driver.

V. REFERENCES

Abuishmais, I. and T. Undeland (2011): "Dynamic characterization of 63 mOhm, 1.2 kV, normally-off SiC VJFET". IEEE 8th International Conference onPower Electronics and ECCE Asia, pages: 1206-1210.

Agarwal, A. (2011): "Advances in SiC MOSFET Performance". SiC & GaN User Forum - Potential of Wide Bandgap Semiconductors in Power Electronic Applications, Birmingham, UK.

Aiba, A., H. Himeji, et al. (2008): "Excimer-Lampe". Patent, DE102009022823A1.

Alonso, J. M., J. Cardesin, et al. (2004): "Low-power high-voltage high-frequency power supply for ozone generation". IEEE Transactions on Industry Applications **40** (2), pages: 414-421.

Barnard, J. E. and H. D. R. Morgan (1903): "Upon the Bactericidal Action of Some Ultra-Violet Radiations as Produced by the Continuous-Current Arc". Proceedings of the Royal Society of London **72**, pages: 126-128.

Beleznai, S. (2009): "Development of mercury-free dielectric barrier dicharge light sources". earned degree: PhD, Budapest University of Technology and Economics.

Beleznai, S., G. Mihajlik, et al. (2006): "High-efficiency dielectric barrier Xe discharge lamp: theoretical and experimental investigations". Journal of Physics D: Applied Physics **39** (17), pages: 3777 - 3787.

Beleznai, S., G. Mihajlik, et al. (2010): "High frequency excitation waveform for efficient operation of a xenon excimer dielectric barrier discharge lamp". Journal of Physics D: Applied Physics **43** (1), pages: 1 - 9.

Beleznai, S., G. Mihajlik, et al. (2008a): "Improving the efficiency of a fluorescent Xe dielectric barrier light source using short pulse excitation". Journal of Physics D: Applied Physics **41** (11), pages: 1 - 6.

Beleznai, S., G. Mihajlik, et al. (2008b): "Improving the efficiency of a fluorescent Xe dielectric barrier light source using short pulse excitation". Journal of Physics D: Applied Physics **41** (11), pages: 115202.

Beleznai, S., P. Richter, et al. (2008c): "New modulated driving signal for efficient excitation of DBD discharges". HAKONE XI, Oleron Island.

Bente, M., G. Avramidis, et al. (2004): "Wood surface modification in dielectric barrier discharges at atmospheric pressure for creating water repellent characteristics". Holz als Roh- und Werkstoff **62** (3), pages: 157-163.

Berzelius, J. J. (1824): "Untersuchungen über die Flussspathsäure und deren merkwürdigsten Verbindungen". Annalen der Physik **77** (5), pages: 1-48.

Bhosle, S., G. Zissis, et al. (2006): "Implementation of an efficiency indicator in an electrical modeling of a Dielectric Barrier Discharge Lamp". Conference Record of the 2006 IEEE Industry Applications Conference, 41st IAS Annual Meeting, pages: 1784-1790.

Biela, J., D. Aggeler, et al. (2008): "5kV/200ns Pulsed Power Switch based on a SiC-JFET Super Cascode". Proceedings of the IEEE International Power Modulators and High Voltage Conference, pages: 358-361.

Biela, J., M. Schweizer, et al. (2011): "SiC versus Si - Evaluation of Potentials for Performance Improvement of Inverter and DC-DC Converter Systems by SiC Power Semiconductors". IEEE Transactions on Industrial Electronics **58** (7), pages: 2872-2882.

Bilodeau, T. M., W. G. Dunbar, et al. (1989): "High-voltage and partial discharge testing techniques for space power systems". Electrical Insulation Magazine, IEEE **5** (2), pages: 12-21.

Boeuf, J. P., C. Punset, et al. (1997): "Physics and Modelling of Plasma Display Panels". J. Phys IV France, pages: C4-2 - C4-14.

Bolton, J. R. and R. Cater (1994): "Homogeneous photodegradation of pollutants in contaminated water: an introduction". From: "Surface and Aquatic Environmental Photochemistry". G. Melz, R. Zepp and D. Crosby, CRC Press.

Bolton, J. R. and C. A. Cotton (2008): "The Ultraviolet Desinfection Handbook". American Water Works Association.

Borisov, V. M., V. A. Vodchits, et al. (1998): "Powerful highly efficient KrF lamps excited by surface and barrier discharges". QUANTUM ELECTRONICS **28** (4), pages: 297-303.

Borris, J. (2010): "Oberflächenstrukturierung mittels Plasma-Printing bei Atmosphärendruck". "Plasma+Oberfläche". **1** pages: 8-10.

Boyd, I. W. and I. I. Liaw (2010): "Excimer ultraviolet sources for thin film deposition: a 15 year perspective". Laser Applications in Microelectronic and Optoelectronic Manufacturing, Proc. SPIE 7584, San Francisco, California, USA, pages: 75840C 75841 - 75814.

Boyd, I. W. and J.-Y. Zhang (1997a): "Low Temperature Si Oxidation with Excimer Lamp Sources". Proceedings on the MRS spring meeting, pages: 343 - 354.

Boyd, I. W. and J. Y. Zhang (1997b): "New large area ultraviolet lamp sources and their applications ". Nuclear Instruments and Methods in Physics Research Section B: Beam Interactions with Materials and Atoms **121** (1-4), pages: 349-356.

Brehaut, S. and F. Costa (2006): "Gate driving of high power IGBT through a Double Galvanic Insulation Transformer". IEEE 32nd Annual Conference on Industrial Electronics, pages: 2505-2510.

Brockmeyer, A. (1997): "Dimensionierungswerkzeug für magnetische Bauelemente in Stromrichteranwendungen". Dissertation, earned degree: PhD, Rheinisch-Westfälische Technische Hochschule Aachen.

Carman, R. J. and R. P. Mildren (2003): "Computer modelling of a short-pulse excited dielectric barrier discharge xenon excimer lamp (lambda ~172 nm)". J. Phys. D. **36** (1), pages: 19-33.

Carman, R. J., R. P. Mildren, et al. (2004a): "High-pressure (> 1 bar) dielectric barrier discharge lamps generating short pulses of high-peak power vacuum ultraviolet radiation". Journal of Physics D-Applied Physics **37** (17), pages: 2399-2407.

Carman, R. J., B. K. Ward, et al. (2004b): "Double discharges in bi-polar pulsed dielectric barrier discharge lamps in Xenon and their influence on the VUV (lambda=172nm) production efficiency". 10th International Symposium on the Science and Technology of Light Sources, Toulouse, FRANCE, IOP Publishing Ltd, pages: 651-652.

Chabert, P. and N. Braithwaite (2011): "Physics of radio-frequency plasmas". Cambridge University Press.

Chen, W., J. Zhang, et al. (2004): "Surface modification of polyimide with excimer UV radiation at wavelength of 126 nm". Thin Solid Films **453-454**, pages: 3-6.

Chung-Wook, R., K. Hye-Jeong, et al. (2003): "Multilevel voltage wave-shaping display driver for AC plasma display panel application". IEEE Journal on Solid-State Circuits **38** (6), pages: 935-947.

Coppa, B. (2012): "SiC semiconductor market growth & competitive landscape". Accessed: 28.01.2013 from: http://www.examiner.com/article/sic-semiconductor-market-growth-competitive-landscape.

Crystan "Magnesium Fluoride (MgF 2)". Accessed: 21.06.2013 from: http://www.crystran.co.uk/userfiles/files/magnesium-fluoride-mgf2-data-sheet.pdf.

Daub, H. (2010): "Ein adaptives Impuls-EVG zum Betrieb von dielektrisch behinderten Gasentladungslampen". Dissertation, earned degree: Dr.-Ing., KIT Scientific Publishing.

Dichtl, J. (1998): "Grundlegende experimentelle Untersuchungen zur Realisierung Hg-freier Leuchtstofflampen mittels Excimer - Entladungen in Xenon". Diploma Thesis, Universität Karlsruhe (TH).

Diez, R., H. Piquet, et al. (2012): "Current-Mode Power Converter for Radiation Control in DBD Excimer Lamps". IEEE Transactions on Industrial Electronics **59** (4), pages: 1912-1919.

Diez, R., J. P. Salanne, et al. (2007): "Predictive model of a DBD lamp for power supply design and method for the automatic identification of its parameters". European Physical Journal-Applied Physics **37** (3), pages: 307-313.

DIN5031-10 (2000): "Optical radiation physics and illuminating engineering; photobiologically effective radiation, quantities, symbols and actions".

EDC_magazine (2003): "Silvana Osram Booth". Accessed: 26.02.2013 from: http://www.edcmag.com/ext/resources/legacy/EDC/2003/11/Files/Images/93435.jpg.

Eden, J. G. and S. J. Park (2010): "Microcavity plasma lighting: thin sheet lamps for general illumination". LS12WLED3, Eindhoven, Netherlands, pages: 271-272.

Eden, R. (2012): "Market Forecasts for Silicon Carbide & Gallium Nitride Power Semiconductors". APEC 2012, Orlando, Florida.

El-Deib, A. (2010a): "Modeling of and Driver Design for a Dielectric Barrier Discharge Lamp". Dissertation, earned degree: PhD, University of Toronto.

El-Deib, A., F. Dawson, et al. (2010b): "A reduced order model for the Dielectric Barrier Discharge (DBD) lamp using the Proper Orthogonal Decomposition (POD) technique". LS12WLED3, Eindhoven, Netherlands, pages: 261-262.

El-Deib, A., F. Dawson, et al. (2010c): "Comparison between current controlled and voltage controlled pulsed drivers for a Dielectric Barrier Discharge (DBD) lamp". LS12WLED3, Eindhoven, Netherlands, pages: 287-288.

Eliasson, B., M. Hirth, et al. (1987): "Ozone synthesis from oxygen in dielectric barrier discharges". J. Phys. D: Appl. Phys. (11), pages: 1421-1437.

Eliasson, B. and U. Kogelschatz (2000): "Discharge reactor and uses thereof". Patent, US6136278.

Falkenstein, Z. (1998): "Applications of dielectric barrier discharges". Proceedings of the 12th International Conference on High-Power Particle Beams, pages: 117-120.

Falkenstein, Z. (2001a): "Development of an excimer UV light source system for water treatment". IUVA, pages: 1-6.

Falkenstein, Z. (2001b): "Surface cleaning mechanisms utilizing VUV radiation in oxygen-containing gaseous environments". Lithographic and Micromachining Techniques for Optical Component Fabrication, pages: 1-10.

Falkenstein, Z. and J. J. Coogan (1997): "Microdischarge behaviour in the silent discharge of nitrogen - oxygen and water - air mixtures". Journal of Physics D: Applied Physics **30** (5), pages: 817-825.

Fan, W., L. Dong, et al. (2007): "Measurement of spatio-temporal dynamics of patterns in dielectric barrier discharge by using a special optical design". 3rd International Symposium on Advanced Optical Manufacturing and Testing Technologies: Advanced Optical Manufacturing Technologies, pages: 67220B-67220B.

Finantu-Dinu, E. (2007): "Planare dielektrisch behinderte Entladungen". Dissertation, earned degree: PhD, University of Wuppertal.

Fließ, R. (1989): "Entwicklung und Aufbau eines Hochspannungs- Mittelfrequenz- Generators für den Betrieb stiller (dielektrisch behinderter) Entladungen". earned degree: Diploma, University of Karlsruhe.

Flores-Fuentes, A. A., R. Pena-Eguiluz, et al. (2006): "A multicell converter model of DBD plasma discharges". 11th Latin American Workshop on Plasma Physics, Mexico City, MEXICO, pages: 188-191.

Florez, D., R. Diez, et al. (2012): "DBD excimer lamp power supply with fully controlled operating conditions". 13th International Conference on Optimization of Electrical and Electronic Equipment, pages: 1346-1352.

Galai, L., F. Costa, et al. (2012): "EMI disturbance in Double Galvanic Insulation Transformer for high voltage insulation- Application on multilevel converter for 10kV SiC Mosfet drivers". 6th Asia-Pacific Conference onEnvironmental Electromagnetics, pages: 210-213.

Gellert, B. and U. Kogelschatz (1991): "Generation of excimer emission in dielectric barrier discharges". Applied Physics B: Lasers and Optics **52** (1), pages: 14-21.

Grebennikov, A. and N. O. Sokal, Eds. (2007): "Switchmode RF Power Amplifiers". Elsevier.

Große-Ophoff, M. (2007): "Saubere Luft im Autoinnenraum". from.

Guivan, M. M., A. A. Malinina, et al. (2011): "Experimental and theoretical characterization of a multi-wavelength DBD-driven exciplex lamp operated with mercury bromide/rare gas mixtures". Journal of Physics D: Applied Physics **44** (22), pages: 1-11.

Haaks, D. (1980): "Kinetik der Xe_2^* und Kr_2^* Excimere". Dissertation, earned degree: PhD, Universität Gesamthochschule

Haehre, K., M. Meisser, et al. (2012a): "Characterization and comparision of commercially vailable Silicon Carbide (SiC) power switches". PEMD 2012, Bristol, GB, pages: 1-6.

Haehre, K., M. Meisser, et al. (2012b): "Switching Speed-Control of an Optimized Capacitor-Clamped Normally-On Silicon Carbide JFET Cascode". EPE-PEMC, Novi Sad, Serbia, pages: DS1a.11-11 - DS11a.11-15.

Harry, J. E. (2010): "Introduction to Plasma Technology". Wiley-VCH.

Härter, H. (2010): "LEDs verbrauchen laut einer Studie während der Produktion mehr Energie als Kaltkathodenlampen". In: Elektronikpraxis, Accessed: 08.05.2013 from: http://www.elektronikpraxis.vogel.de/opto/articles/266157/.

Hartmann, M., A. Musing, et al. (2007): "Switching transient shaping of RF power MOSFETs for a 2.5 MHz, three-phase PFC". 7th Internatonal Conference on Power Electronics (ICPE), pages: 1160-1166.

Hashino, S. and S. Toshihisa (2010): "Separation measurement of parasitic impedance on a power electronics circuit board using TDR". Energy Conversion Congress and Exposition (ECCE), pages: 2700-2705.

Hassall, G. and P. W. Smith (1994): "High Repetition Rate Generator for pulsed capacitively coupled Discharges". IEE, pages: 7/1-7/3.

Heering, W., H.-P. Daub, et al. (2004): "Vorrichtung zur Erzeugung von elektrischen Spannungsimpulsfolgen, insbesondere zum Betrieb von kapazitiven Entladungslampen und ihre Verwendung". Patent, DE102004021243B3.

Heming, R., A. Michels, et al. (2009): "Electrical generators driving microhollow and dielectric barrier discharges applied for analytical chemistry". Analytical and Bioanalytical Chemistry **395** (3), pages: 611-618.

Heraeus (2013): "Quarzglas für die Optik - Daten und Eigenschaften". HQS-SO V1.3_D.

Herring, C., J. Bulson, et al. (2012): "Microplasma Planar Lighting". Eden Park Illumination.

Hitzschke, L., S. Jerebic, et al. (1998): "Leuchtstofflampe". Patent, WO0001998049712A1.

Hombach, A. and R. Oliver (2008): "Dielectric barrier discharge lamp configured as a coaxial double tube having a getter". Patent, US000008174191B2.

Hombach, A., O. Rosier, et al. (2010): "Dielectric barrier discharge lamp configured as a double tube ". Patent, US000008237364B2.

Honda, K. and Y. Naito (1955): "On the Nature of Silent Electric Discharge". J. Phys. Soc. Japan. **10** (1007-1011).

Hopf, E. (2012): "SiC-JFET-Pionier Semisouth am Abgrund". In: Markt & Technik, Accessed: 12.11.2012 from: http://www.elektroniknet.de/bauelemente/news/article/92792/?type=99.

Hosselet, L. M. L. F. (1971): "Ozonbildung mittels elektrischer Entladungen". Retrieved: TH-Report 71-E-19 from. TH-Report 71-E-19

Huang, Q., Q. Meng, et al. (2007): "Dielectric barrier discharge lamp system and driving method thereof having reletively better performance in startup and re-startup of dimming". Patent, US020080303448A1.

Huber, A., A. Veser, et al. (1996): "Schaltungsanordnung zur Erzeugung von Impulsfolgen, insbesondere für den Betrieb von dielektrisch behinderten Entladungen". Patent, EP0781078B1.

Ilmer, M., R. Lecheler, et al. (2000): "Hg-free flat panel light source PLANON - a promising candidate for future LCD backlights". SID International Symposium, San Jose, Californien, USA, Society for Information Display.

IOT_GmbH. (2012): "IOT GmbH web page" in. Retrieved: 04.01.2013, from: http://www.iot-gmbh.de/sites/uv technik/excirad172 technische daten.htm.

IUPAC (2007): "Excimer Lamp". "IUPAC Compendium of Chemical Terminology".

IXYS (2009): "datasheet DE475102N21A ". IXYS.

Jinno, M., H. Motomura, et al. (2003): "Fundamental Research on Xenon and Xenon-Rare Gas Pulsed Dielectric Barrier Discharge Fluorescent Lamps". XXVI ICPIG 2003, Greifswald, Germany, pages: 1-2.

Jou, S. Y., C. T. Hung, et al. (2011): "Enhancement of VUV Emission from a Coaxial Xenon Excimer Ultraviolet Lamp Driven by Distorted Bipolar Square Voltages". Contributions to Plasma Physics, pages: 907-919.

Jüstel, T. (2007): "Fluorescent Lamp Phosphors - Is there still News ?". PGS, Soul, South Korea, pages: 1-25.

Jüstel, T. (2011): "VUV Phosphors for Plasma Display Panels VUV Phosphors for Plasma Display Panels and Excimer Discharge Lamps". Materials Valley Workshop.

Jüstel, T., H. Nikol, et al. (2002): "Device for Disinfecting Water Comprising a UV-C gas discharge lamp". Patent, US6398970B1.

Kaiser, K. L. (2004): "Electromagnetic Compatibility Handbook". CRC PRESS.

Kaminski, N. and O. Hilt (2011): "GaN Device Physics for Electrical Engineers". ECPE SiC GaN user forum, Birmingham, UK, pages: 1-33.

Kapels, H., R. Rupp, et al. (2001): "SiC Schottky diodes: A milestone in hard switching applications". PCIM, Nuremberg, pages: 1-6.

Kawanaka, J., T. Shirai, et al. (2001): "1.5 kW high-peak-power vacuum ultraviolet flash lamp using a pulsed silent discharge of krypton gas". Applied Physics Letters **79** (23), pages: 3752-3754.

Kazimierczuk, M. K. (2008): "RF Power Amplifiers". Wiley&Sons Ltd.

Kazimierczuk, M. K. (2009): "High-Frequency Magnetic Components". John Wiley & Sons.

Khachen, W. and J. R. Laghari (1990): "Polypropylene for high voltage high frequency airborne applications". Conference Record of the 1990 IEEE International Symposium on Electrical Insulation, pages: 80-83.

Kim, G. Y., J. B. Park, et al. (2012): "Study of atmospheric pressure chemical vapor deposition by using a double discharge system for SiOx thin-film deposition with a HMDS/Ar/He/O-2 gas mixture". Journal of the Korean Physical Society **61** (3), pages: 397-401.

Kindel, E. and C. Schimke (1996): "Measurement of Excited States Density and the VUV-Radiation in the Pulsed Xenon Medium Pressure Discharge". Contributions to Plasma Physics **36** (6), pages: 711-721.

Kling, R. (1997): "Untersuchung an hocheffizienten Excimerentladungslampen". Dissertation, earned degree: Dr.-Ing., Universität Karlsruhe (TH).

Kneissl, M., T. Kolbe, et al. (2010): "Report within WP2.5: Compact Units for Decentralised Water Supply. ". Retrieved: D 2.5.13 from. D 2.5.13

Kogelschatz, U. (1991): "Silent-Discharge Driven Excimer Uv Sources and Their Applications". Symposium of the Spring Conference of the European Materials Research Society: Laser Surface Processing and Characterization, Strasbourg, France, Elsevier Science, pages: 410-423.

Kogelschatz, U. (1993): "UV production in dielectric barrier discharges for pollution control". Retrieved: G34 Part B, from. G34 Part B.

Kogelschatz, U. (2000a): "Fundamentals and applications of dielectric-barrier discharges.". HAKONE VII: Int. Symp. On High Pressure Low Temperature Plasma Chemistry, Greifswald, Germany, pages: 1-7.

Kogelschatz, U. (2002): "Filamentary, patterned, and diffuse barrier discharges". IEEE Trans. on Plasma Science **30** (4), pages: 1400-1408.

Kogelschatz, U. (2003a): "Dielectric-barrier discharges: Their history, discharge physics, and industrial applications". Plasma Chemistry and Plasma Processing **23** (1), pages: 1-46.

Kogelschatz, U. (2003b): "Excimer lamps: History, discharge physics, and industrial applications". 6th International Conference on Atomic and Molecular Pulsed Lasers, Tomsk, RUSSIA, SPIE - International Society of Optical Engineering, pages: 272-286.

Kogelschatz, U., B. Eliasson, et al. (1999): "From ozone generators to flat television screens: history and future potential of dielectric-barrier discharges". 14th International Symposium on Plasma

Chemistry, Prague, Czech Republic, International Union Pure Applied Chemistry, pages: 1819-1828.

Kogelschatz, U., H. Esrom, et al. (2000b): "High-intensity sources of incoherent UV and VUV excimer radiation for low-temperature materials processing". Applied Surface Science **168** (1–4), pages: 29-36.

Kogelschatz, U. and J. Salge (2008): "High Pressure Plasmas: Dielectric-Barrier and Corona Discharges - Properties and technical Applications". From: "Low Temperature Plasmas". R. Hippler, H. Kersten, M. Schmidt and K. H. Schoenbach. Weinheim, Wiley-VCH. **2**.

Köhler, T. (1989): "Offener UV/VUV-Excimerstrahler und Verfahren zur Oberflächenmodifizierung von Polymeren". Patent, DE19920693C1.

Koudriavtsev, O., S. Moiseev, et al. (2003): "Estimation of load matching condition for dielectric barrier discharge load". IEICE Transactions on Fundamentals of Electronics Communications and Computer Sciences **E86A** (1), pages: 244-247.

Koudriavtsev, O., W. Shengpei, et al. (2000): "Power supply for silent discharge type load". Industry Applications Conference, pages: 581-587.

Kudryavtsev, O., T. Ahmed, et al. (2004): "Efficient Simulation Method for Dielectric Barrier Discharge Load ". Journal of Power Electronics **4** (3), pages: 188-196.

Kundu, S. K., E. M. Kennedy, et al. (2012): "Experimental investigation of alumina and quartz as dielectrics for a cylindrical double dielectric barrier discharge reactor in argon diluted methane plasma". Chemical Engineering Journal **180** (0), pages: 178-189.

Kunze, K. (2001): "Ein Plasma für die Miniaturisierung von Analysesystemen - die dielektrisch behinderte Entladung". Diploma Thesis, earned degree: Diploma, University of Dortmund.

Kyrberg, K. (2005): "Circuit arrangement having a converter without a transformer but with an inductor for the pulsed operation of dielectric barrier discharge lamps". Patent, US7355351B2.

Kyrberg, K., H. Guldner, et al. (2007a): "Half-Bridge and Full-Bridge Choke Converter Concepts for the Pulsed Operation of Large Dielectric Barrier Discharge Lamps". Power Electronics, IEEE Transactions on **22** (3), pages: 926-933.

Kyrberg, K., H. Güldner, et al. (2007b): "Half-bridge and full-bridge choke converter concepts for the pulsed operation of large dielectric barrier discharge lamps". IEEE Transactions on Power Electronics **22** (3), pages: 926-933.

Lambrecht. (1998): "Untersuchungen an Quecksilberhochdrucklampen zur effizienten Erzeugung ultravioletter Strahlung". from.

Lawal, O., B. Dussert, et al. (2008): "Proposed Method For Measurement Of The Output Of Monochromatic (254 nm) Low Pressure UV-Lamps". "IUVA News". No.: 1. **10**.

Lecheler, R. and W. Sowa (2007): "Converter circuit having class E converter modules". Patent, US7274578B2.

Lee, J. K. and J. P. Verboncoeur (2008a): "Plasma Display Panel". From: "Low Temperature Plasmas", Wiley-VCH.

Lee, S. Y., J. S. Gho, et al. (2008b): "Analysis of Pulse Power Converter for Plasma Application". 34th Annual Conference of the IEEE-Industrial-Electronics-Society, Orlando, FL, IEEE, pages: 505-509.

Legrini, O., E. Oliveros, et al. (1993): "Photochemical Processes for Water Threatment". Chemical Review **93**, pages: 671-698.

Liaw, I. I. and I. W. Boyd (2008): "The Development and Application of UV Excimer Lamps in Nanofabrication". From: "Functionalized Nanoscale Materials, Devices and Systems". A. Vaseashta and I. N. Mihailescu, Springer Netherlands. pages: 61-76.

Liu, S. (2002): "Electrical modeling and unipolar-pulsed energization of dielectric barrier discharges". Dissertation, earned degree: Dr.-Ing., Universität Karlsruhe (TH).

Liu, S. and M. Neiger (2001): "Excitation of dielectric barrier discharges by unipolar submicrosecond square pulses". J. Phys. D. **34** (11), pages: 1632-1638.

Liu, S. and M. Neiger (2003a): "Double discharges in unipolar-pulsed dielectric barrier discharge xenon excimer lamps". J. Phys. D. **36** (13), pages: 1565-1572.

Liu, S. and M. Neiger (2003b): "Electrical modelling of homogeneous dielectric barrier discharges under an arbitrary excitation voltage". J. Phys. D: Appl. Phys. **36**, pages: 3144-3150.

Liu, Z., W. J. Wang, et al. (2011): "A Novel ZVS Double Switch Flyback Inverter and Pulse Controlled Dimming Methods for Flat DBD Lamp". IEEE Transactions on Consumer Electronics **57** (3), pages: 995-1002.

Lomaev, M. I., D. V. Shitz, et al. (2002): "Influence of excitation pulse form on barrier discharge excilamp efficiency". pages: 38-45.

Lomaev, M. I., E. A. Sosnin, et al. (2012): "Excilamps and their applications". Progress in Quantum Electronics **36** (1), pages: 51-97.

Lorch, W., Ed. (1987): "Handbook of water purification". Ellis Horwood Ltd.

Lorentz, D. C., O. J. Eckstrom, et al. (1973): "Excimer formation and decay processes in rare gasses". Accessed: 28.08.1973 from.

Lunitronix (2010): "Ultraslim LED Panels". Accessed: 08.05.2013 from: http://www.leds.de/LED-Lampen-und-Leuchten/Ultraslim-LED-Panels/.

Maniktala, S. (2004a): "Switching Power Supply Design & Optimization". McGraw-Hill Professional.

Maniktala, S., Ed. (2004b): "Switching Power Supply Design & Optimization". McGraw-Hill Professional.

Manley, T. C. (1943): "The Electric Characteristics of the Ozonator Discharge". Journal of the Electrochemical Society **84** (1), pages: 83-96

Marinos, J. (2010): "Planar Magnetics Advancements for Power Electronics Applications". APEC.

Mehnert, R. (1999): "Excimer UV Curing in Printing". from: http://www.iaea.org/inis/collection/NCLCollectionStore/_Public/31/016/31016304.pdf.

Meisser, M. (2010a): "Schaltungsanordnung und Verfahren für den Betrieb einer kapazitiven Last". Patent, EP000002389047A1.

Meisser, M. and K. Haehre (2012a): "Impedance Characterization of High-Frequency Gate Drive Circuits for Silicon RF MOSFET and Silicon-Carbide Field-Effect Transistors". PCIM, Nuremberg, Germany, pages: 1588-1594.

Meisser, M., K. Haehre, et al. (2012b): "Impedance characterization of high frequency power electronic circuits". PEMD 2012, Bristol, GB, pages: 1-6.

Meisser, M. and R. Kling (2012c): "Dielektrisch behinderte Entladungs-Lampe". Patent, DE102012017779.8.

Meisser, M., R. Kling, et al. (2011a): "Transformerless high voltage pulse generators for bipolar drive of Dielectric Barrier Discharges". PCIM 2011, Nuremberg, Germany, pages: 259-264.

Meisser, M., R. Kling, et al. (2011b): "Universal Resonant Topology for High Frequency Pulsed Operation of Dielectric Barrier Discharge Light Sources ". APEC 2011, Fort Worth, TX, USA, pages: 1-8.

Meisser, M., K. Messerschmidt, et al. (2012d): "MOSFET Matrix-Switch for Pulsed Operation of Plasma Optical Radiation Sources ". EAPPC-BEAMS 2012, KIT campus, Karlsruhe, Germany, pages: 1-4.

Meisser, M., M. Paravia, et al. (2010b): "Resonance Behaviour of a Pulsed Electronic Control Gear for Dielectric Barrier Discharges". PEMD, Brighton, UK, EIT, pages: 1-6.

Meisser, M., M. Paravia, et al. (2010c): "Single period sinusoidal pulse generator for efficient drive of dielectric barrier discharges". PCIM, Nuremberg, Germany, pages: 808-813.

Meisser, M., M. Paravia, et al. (2010d): "Driving of dielectric barrier discharge lamps with a pulsed transformer-less full-bridge topology". LS12, Eidhoven, Netherlands, pages: 39-40.

Meunier, J., P. Belenguer, et al. (1995): "Numerical model of an ac plasma display panel cell in neon-xenon mixtures". Journal of Applied Physics **78** (2), pages: 731-745.

Mildren, R. P. and R. J. Carman (2000): "Enhanced efficiency from a Xe excimer barrier discharge lamp employing short-pulsed excitation". International Conference on Atomic and Molecular Pulsed Lasers III, USA, SPIE-Int. Soc. Opt. Eng. Proceedings of Spie, pages: 283-290.

Mildren, R. P. and R. J. Carman (2001a): "Enhanced performance of a dielectric barrier discharge lamp using short-pulsed excitation". J. Phys D: Appl. Phys **34** (1), pages: L1-L6(1).

Mildren, R. P. and R. J. Carman (2001b): "Enhanced performance of a dielectric barrier discharge lamp using short-pulsed excitation". Journal of Physics D-Applied Physics **34** (1), pages: L1-L6.

Mildren, R. P., R. J. Carman, et al. (2001c): "Visible and VUV images of dielectric barrier discharges in Xe". Journal of Physics D - Applied Physics **34** (23), pages: 3378-3382.

Morimoto, Y. and S. Nozawa (1999): "Effect of Xe2* light .7.2 eV. on the infrared and vacuum ultraviolet absorption properties of hydroxyl groups in silica glass". Physical Review B **59**, pages: 4066-4073.

Müller, S. and R. J. Zahn (1996): "On various kinds of dielectric barrier discharges". Contr. to Plasma Physics **36** (6), pages: 697-709.

Murnick, D. E., N. M. Masoud, et al. (2009): "Fluorescent Excimer Lamps". Patent, US2011/0254449A1.

Neff, W., B. Kelmis, et al. (1997): "Excitation of gas discharge with short voltage pulses ". Patent, DE000019717127A1.

Neiger, M. (1992): "Dielectric barrier discharges: an unusual new light source". 6th International Symposium on the Science and Technology of Light Sources (LS6), Budapest, Hungary, pages: 75-82.

Newman, D. S. and M. J. Brennan (1995): "The Dielectric Barrier Discharge: A Bright Spark for Australia's Future". Australian Journal of Physics **48**, pages: 543-556.

Nozaki, K. O. a. T. (2002): "Ultrashort pulsed barrier discharges and applications". Pure and Applied Chemistry **74** (3), pages: 447-452.

Okamoto, M. (1999): "Dielectric Barrier Discharge Lamp Light Source". Patent, EP1176853A1.

Olivares, V. H., M. Ponce-Silva, et al. (2007): "DBD Modeling as a function of waveforms slope". 38th IEEE Power Electronic Specialists Conference, Orlando, FL, IEEE, pages: 1417-1422.

Olivares, V. H., M. Ponce, et al. (2009): "Short Pulsed Voltage Dielectric Barrier Discharge on Tubular Fluorescent Lamps". Electronics, Robotics and Automotive Mechanics Conference, 2009. CERMA '09., pages: 417-422.

Oppenländer, T. (2003): "Photochemical Purification of Water and Air". Wiley-VCN.

Oppenländer, T. and J. Waizmann (1997): "Photochemische Wasserbehandlung mit einem KrCl*-Excimer-UV-Rundrohrstrahler in einem Flachbettreaktor". Chemie Ingenieur Technik **69** (10), pages: 1470-1474.

Osram (2002): "Osram stellt neues Linex-Lampensystem vor: Turbolicht für Scanner und Kopierer". Accessed: 11.02.2010 from: http://www.osram.de/osram_de/Presse/Fachpresse/Speziallampen/2002/Press614094.jsp.

OSRAM (2012): "Xeradex Excimer ultraviolet lamps, single-ended".

Pal, U. N., P. Gulati, et al. (2011): "Analysis of Discharge Parameters in Xenon-Filled Coaxial DBD Tube". IEEE Transactions on Plasma Science **39** (6), pages: 1475-1481.

Panoisis, E., N. Merbahi, et al. (2008): "Comparative Power Balance of Dielectric Barrier Discharges under Unipolar and Bipolar Pulsed Excitation". SFE 2008, Paris, pages: 1-6.

Paravia, M. (2006): "Optimierter Pulsbetrieb zur Effizienzsteigerung von Xe-Excimer-Flachlampen". earned degree: Diploma, Universität Karlsruhe, Lichttechnisches Institut.

Paravia, M. (2009a): "internal report". from.

Paravia, M. (2010): "Effizienter Betrieb von Xenon-Excimer-Entladungen bei hoher Leistungsdichte". earned degree: Dr.-Ing., KIT.

Paravia, M., M. Meisser, et al. (2009b): "Plasma Efficiency and Losses for pulsed Xe Excimer DBDs at high Power Densities". GEC, Sagatoga Springs, NY, USA.

Paravia, M., M. Meisser, et al. (2008a): "Arguments for increased efficiency of a Xe excimer DBD by pulsed instead of sinusoidal excitation". 61st Annual Gaseous Electronics Conference, GEC, Dallas, TX, USA.

Paravia, M., K. Trampert, et al. (2008b): "Threshold current density for homogeneous excitation of pulsed xenon excimer DBD.". IEEE International Conference on Plasma Science, ICOPS, Karlsruhe, Germany.

Paravia, M., K. E. Trampert, et al. (2007): "Homogenisation of a pulsed dielectric barrier Xe discharge using falling voltage edge for secondary ignition". LS 11, Shanghai, FAST-LS, pages: 499-500.

Park, J.-H., I.-K. Lee, et al. (2008): "Planar Light-Ssource Pulse-Type Driving Circuit Adopting a Transformer". Patent, WO002009022804A2.

Park, J. H., I. K. Lee, et al. (2007a): "High efficiency inverter systems for driving mercury-free flat fluorescent lamps". 4th Power Conversion Conference (PCC-Nagoya 2007), Nagoya, JAPAN, IEEE, pages: 688-691.

Park, S.-J., J. G. Eden, et al. (2004): "Microdischarge devices with 10 or 30μm square silicon cathode cavities: pd scaling and production of the XeO excimer ". Applied Physics Letters **85** (21), pages: 4869-4871.

Park, S. J., J. D. Readle, et al. (2007b): "Lighting from thin (< 1 mm) sheets of microcavity plasma arrays fabricated in Al/Al2O3/glass structures: planar, mercury-free lamps with radiating areas beyond 200 cm(2)". Journal of Physics D-Applied Physics **40** (13), pages: 3907-3913.

Paul, R. (1987): "Aufbau und Untersuchung einer neuartigen Nanosekunden-Blitzlampe in Form einer gepulsten XeCl-Excimer-Glimmentladung". earned degree: Diploma, Karlsruhe Institute of Technology.

Pemen, A., E. v. Heesch, et al. (2012): "Quasi Resonant Pulse Modulator For Surface-Dielectric-Barrier-Discharge Generation". EAPPC-BEAMS, Karlsruhe, Germany.

Perret, R., Ed. (2009): "Power Electronics Semiconductor Devices". Hoboken, John Wiley & Sons, Inc.

Pflumm, C. (2003): "Simulation homogener Barrierenentladungen inklusive der Elektrodenbereiche". Dissertation, earned degree: Dr.-Ing., Universität Karlsruhe (TH).

Pipa, A. V., M. Schmidt, et al. (2007): "Ultraviolet (UV) emissions from a unipolar submicrosecond pulsed dielectric barrier discharge (DBD) in He-Air mixtures". 5th EU-Japan Symposium on Plasma Processing, Belgrade, SERBIA, IOP Publishing Ltd., pages: 12014-12014.

Piquet, H., S. Bhosle, et al. (2012): "Control of the UV flux of a XeCl dielectric barrier discharge excilamp through its current variation". Quantum Electronics **42** (2), pages: 157-164.

Piquet, H., S. Bhosle, et al. (2007): "Innovative power supply concepts for DBD excilamps". 8th Int. Conf. on Atomic and Molecular Pulsed Lasers, Tomsk, RUS, SPIE - International Society of Optical Engineering, pages: 93810-93810.

Piquet, H., R. Diez, et al. (2010): "DBD lamp converter design using an electrical model of the load". Mathematics and Computers in Simulation **81** (2), pages: 420-432.

Qsil (2011): "Materialspezifikation Quarzglas ilmasil PS ". S-101-01-PS

quark-tec "Excimer VUV lamp". from: http://www.quark-tec.com/english/product/5.html.

Radium. "New Surface Treatment Solutions" in. Retrieved: 05.04.2007, from: http://www.xeradex.de/Prospekt.pdf.

Redondo, L. and J. F. Silva (2011): "Solid State Pulsed Power Electronics". "Power electronics handbook". M. H. Rashid, Elsevier.

Reich, L., S. Beleznai, et al. (2005): "Dielectric barrier discharge lamp". Patent, EP1615258B1.

Research, I. (2013): "Market for GaN and SiC Power Semiconductors Set to Rise By Factor of 18 from 2012 to 2022". Accessed: 26.04.2013 from: http://www.imsresearch.com/press-release/market for gan and sic power semiconductors set to rise by factor of 18 from 2012 t o 2022&cat id=35&from=?goback=%2Egmp 147061%2Egde 147061 member 235404026.

Riese, J. (1991): "Untersuchung an unipolar gepulsten dielektrisch behinderten Entladungen in XeCl-Excimer-Gasgemischen". earned degree: Diploma, Universität Karlsruhe.

Roth, M. (2001): "Experimentelle Untersuchungen zu Verlustmechanismen des Kathodenfalls in Barrierenentladungen". Dissertation, earned degree: Dr.-Ing., Universität Karlsruhe (TH).

Roth, M. and R. Wittkötter (2005): "Dielektrische Barriere- Doppelrohr Konfiguration". Patent, DE102005006656A1.

Salvermoser, M. and D. E. Murnick (2003): "Efficient, stable, corona discharge 172 nm xenon excimer light source". Journal of Applied Physics **94** (6), pages: 3722-3731.

Salvermoser, M. and D. E. Murnick (2004): "Corona Discharge Lamps". Patent, EP1794856B1.

Salvermoser, M. J., U. Kogelschatz, et al. (2009): "Influence of humidity on photochemical ozone generation with 172 nm xenon excimer lamps". Eur. Phys. J. Appl. Phys. **47** (2), pages: 22812-p22811 - 22812-p22816.

Schiene, W. (2004): "DBD lamp with integrated multifunction means". Patent, WO002006006139A1

Schorpp, V. (1988): "Untersuchung inkohärenter Excimeremission aus stillen Entladungen in XeCl". earned degree: Diploma, Universität Fridericiana Karlsruhe.

Schorpp, V. (1991): "Die dielektrisch behinderte Edelgas-Halogen-Excimer-Entladung: Eine neuartige UV-Strahlungsquelle". Dissertation, earned degree: Dr.-Ing., Universität Fridericiana Karlsruhe.

Schreiber, A., B. Kühn, et al. (2005): "Radiation resistance of quartz glass for VUV discharge lamps". Journal of Physics D: Applied Physics **38** (17), pages: 3242-3250.

Schülting, L. (1993): "Optimierte Auslegung induktiver Bauelemente für den Mittelfrequenzbereich". Dissertation, earned degree: Dr.-Ing, Rheinisch-Westfälische Technische Hochschule Aachen.

Schwarz-Kiene, P. (2001): "Betriebsgeräte und Verfahren zur effizienten Erzeugung ultravioletter Strahlung". Dissertation, earned degree: Dr.-Ing., Universität Karlsruhe.

Schwarz-Kiene, P. and W. Heering (1998): "Electronic ballast for pulsed operation of Dielectric Barrier Discharges". Light Source 8, Greifswald, Germany, pages: 278-279.

Sewraj, N., N. Merbahi, et al. (2009): "VUV spectroscopy and post-discharge kinetic analysis of a pure xenon mono-filamentary dielectric barrier discharge (MF-DBD)". Journal of Physics D: Applied Physics **42** (4), pages: 1-13.

Shao, T., Z. Niu, et al. (2011): "ICCD Observation of Homogeneous DBD Excitated by Unipolar Nanosecond Pulses in Open Air". IEEE Transactions on Plasma Science **39** (11), pages: 2062-2063.

Shao, T., D. Zhang, et al. (2010): "A Compact Repetitive Unipolar Nanosecond-Pulse Generator for Dielectric Barrier Discharge Application". Plasma Science, IEEE Transactions on **38** (7), pages: 1651-1655.

Shin, W. H., J. H. Choi, et al. (2006): "Bidirectional pulse plasma power supply for treatment of air pollution". 37th IEEE Power Electronics Specialist Conference (PESC 2006), Jeju Island, South Korea, IEEE, pages: 3143-3148.

Shoyama, T. and Y. Yoshioka (2006): "Theoretical Study of Methods for Improving the Energy Efficiency of NOx Removal from Diesel Exhaust Gases by Silent Discharge". IEEJ Transactions on Fundamentals and Materials **126** (7), pages: 695-702.

Shuhai, L. and M. Neiger (2003): "Double discharges in unipolar-pulsed dielectric barrier discharge xenon excimer lamps". J. Phys. D: Appl. Phys. **36** (13), pages: 1565-1572.

Siemens, W. v., Ed. (1857): "Ueber die elektrostatische Induction und die Verzögerung des Stroms in Flaschendrähten". in: "Poggendorffs Annalen der Physik und Chemie".

Sobottka, A., L. Prager, et al. (2008): "Argon Excimer Lamp". Accessed: 16.11.11 from: www.uni-leipzig.de/~iom/muehlleithen/.../sobottka argon lamp.pdf.

Sokal, N. O. (2001): "Class-E RF Power Amplifiers". "QEX". No.: 9.

Somekawa, T., T. Shirafuji, et al. (2005): "Effects of self-erasing discharges on the uniformity of the dielectric barrier discharge". J. Phys. D. (12), pages: 1910-1917.

Sonntag, C. v. (1988): "Disinfection with UV radiation". Tenth Brown Boveri symposium on process technologies for water treatment, pages: 150-179.

Sosnin, E. A., S. M. Avdeev, et al. (2010): "A new Hg free DBD lamps for applications in medicine". LS12WLED3, Eindhoven, Netherlands, pages: 515-516.

Sowa, W. and R. Lecheler (2004): "Lamp driver concepts for dielectric barrier discharge lamps and evaluation of a 110 W ballast". Industry Applications Conference, pages: 1379-1385 vol.1372.

Springett, N., R. Schrader, et al. (2012): "Depletion-mode SiC VJFET simplifies high voltage SMPS". PCIM, Nuremberg, Germany.

Stevanovic, L., K. Matocha, et al. (2010): "Realizing the full potential of silicon carbide power devices". IEEE 12th Workshop onControl and Modeling for Power Electronics (COMPEL), pages: 1-6.

Stevens, B. and E. Hutton (1960): "Radiative Life-time of the Pyrene Dimer and the Possible Role of Excited Dimers in Energy Transfer Processes". Nature **186**, pages: 1045-1046.

Stockwald, K. (1991): "Neuartige Xenon- und Xenon/Quecksilber-Lampen im UV-VUV Spektralbereich". Dissertation, earned degree: Dr.-Ing., Universität Karlsruhe (TH).

Strydom, J. T., M. A. de Rooij, et al. (2004): "A comparison of fundamental gate-driver topologies for high frequency applications". Nineteenth Annual IEEE Applied Power Electronics Conference and Exposition (APEC), pages: 1045-1052 vol.1042.

Sugimura, H., B. Saha, et al. (2006): "Single Reverse Blocking Switch Type Pulse Density Modulation Controlled ZVS Inverter with Boost Transformer for Dielectric Barrier Discharge Lamp Dimmer". CES/IEEE 5th International Power Electronics and Motion Control Conference (IPEMC) pages: 1-5.

Tarasenko, V., S. Avdeev, et al. (2009): "High Power UV and VUV Excilamps and Their Applications". Acta Physica Polonica A **116** (4).

Tastekin, D., F. Blank, et al. (2011a): "A novel approach for a Marx topology with bipolar output voltages for Dielectric Barrier Discharge". EPE, Birmingham, pages: 1-8.

Tastekin, D., Q. K. Nguyen, et al. (2011b): "Pulsed Voltage Converter with bipolar output voltages up to 10 kV for Dielectric Barrier Discharge". 8th International Conference on Power Electronics ICPE - ECCE Asia, Jeju Island, Korea.

Tiggelman, M. P. J., K. Reimann, et al. (2007): "Reducing AC impedance measurement errors caused by the DC voltage dependence of broadband high-voltage bias-tees". IEEE International Conference on Microelectronic Test Structures (ICMTS), pages: 200-205.

Trampert, K. (2009): "Ladungstransportmodell dielektrisch behinderter Entladungen". earned degree: Doktor-Ingenieur, University of Karlsruhe.

Trampert, K. E., H. P. Daub, et al. (2004): "Novel Hg-free, flat and large excimer lamps operated by an adaptive pulse gear". Light Sources 2004, Toulouse, France, Inst. of Phys., pages: 509-510.

Trampert, K. E., M. Paravia, et al. (2007): "Total spectral radiant flux measurements on Xe excimer lamps from 115 nm to 1000 nm". Proceedings of the SPIE, SPIE, pages: 661-647.

Trompeter, F.-J. (2001): "Barrierenentladungen zum Abbau von Schadstoffen in motorischen Verbrennungsabgasen". earned degree: PhD, TH Aachen.

Ushio (2000): "Excimer Irradiation Unit - Excimer general brochure". No.: 08-4-2000XD.

Ushio (2006): "CiMAX-200 Excimer lamp systems". Ushio. No.: TPD CiMAX (E) - 06/05.

Ushio (2009): "XeFL". Ushio. No.: S-XEFL-1109.

UV-Solutions (2011a): "UV Wavelength chart". Accessed: 01.02.2013 from: http://www.uvsns.com/images/Wavelength%20Chart.pdf.

UV-Solutions (2011b): "UVS 172-01® Specifications". Accessed: 16.11.11 from: www.uvsns.com.

Voeten, S. J., F. J. C. M. Beckers, et al. (2011): "Optical Characterization of Surface Dielectric Barrier Discharges". IEEE Transactions on Plasma Science **39** (11), pages: 2142-2143.

Volkova, G. A., N. N. Kirillova, et al. (1984): "Vacuum-Ultraviolet Lamps With a Barrier Discharge in Inert Gases ". Journal of Applied Spectroscopy **41**, pages: 1194-1197.

Vollkommer, F. and L. Hitzschke (1994): "Verfahren zum Betreiben einer inkohärent emittierenden Strahlungsquelle". Patent, WO94/23442.

Vollkommer, F. and L. Hitzschke (1998): "Dielectric Barrier Discharge". 8th International Symposium on the Science and Technology of Light Sources (LS-8), Greifswald, Germany, pages: 51-60.

Voss, S., H. Roßmanith, et al. (2001): "The influence of air gap position and winding position on the inductance in high frequency transformers". 9th European Conference on Power Electronics and Applications Graz, Graz, Austria.

Waffenschmidt, E. and B. Ackermann (2001): "Size advantage of coreless transformers in the MHz range". 9th European Conference on Power Electronics and Applications Graz, Graz, Austria.

Wagner, H. E., R. Brandenburg, et al. (2003): "The barrier discharge: basic properties and applications to surface treatment". Vacuum **71** (3), pages: 417-436.

Waite, M. (2008): "Glass and light - Planilum® - The exciting new lighting technology from Saint-Gobain". Intelligent Glass Solutions (1), pages: 44-45.

Wammes, K. (2002): "LICHTQUELLE". Patent, WO002004032179A2

Wammes, K. (2013): "Funktionsprinzip - e^3-Lampen gehören in die Familie der Niederdruckentladungs-Lampen. " in. Retrieved: 06.03.2013, from: http://www.wammes.eu/funktionsprinzip.

Wammes, K., L. Hitzschke, et al. (1999a): "Elektronisches Vorschaltgerät für Entladungslampe mit Dielektrisch Behinderten Entladungen". Patent, EP 1110 434 B1.

Wammes, K., F. Vollkommer, et al. (1999b): "Operational method and electronic ballast for a discharge lamp comprising dielectrically impeded discharges ". Patent, US000006353294B1.

Winkelmann, R. (2003): "Digital geregelter Pulsbetrieb von dielektrisch behinderten Gasentladungen". Diplomarbeit, earned degree: Diploma, Universität Karlsruhe.

Wolf, O. (2000): "Experimentelle Untersuchungen an dielektrisch behinderten Entladungen zur Schadstoffentfernung aus Abgasen". Dissertation, earned degree: Dr.-Ing., Universität Karlsruhe (TH).

Yokokawa, Y., M. Yoshioka, et al. (1998): "Device for operation of a discharge lamp". Patent, US006084360A.

Yu, J. J. (2004): "Formation of high-quality advanced high-k oxide layers at low temperature by excimer UV lamp-assisted photo-CVD and sol-gel processing". Chemical Research in Chinese Universities **20** (4), pages: 396-402.

Yukimura, K., H. Murakami, et al. (2007): "Parametric survey on NO removal in an intermittent dielectric barrier discharge by one-cycle sinusoidal power source". 16th IEEE International Pulsed Power Conference, pages: 1068-1073.

Zacher, S. (2004): "Modellierung und Messung des Abtrags einfacher Öle mittels atmosphärischer Barrierenentladung". dissertation, earned degree: PhD, Universität Fridericiana Karlsruhe.

Zen, S., Y. Teramoto, et al. (2012): "Surface treatment of dye-sensitized solar cell using dielectric barrier discharge". 2012 Joint Electrostatics Conference.

Zhan, H., C. Li, et al. (2007): "Homogeneous dielectric barrier discharge in air for surface treatment". Conference on Electrical Insulation and Dielectric Phenomena (CEIDP), pages: 683-686.

Zimmer. (2011): "LMG500 Präzisions-Leistungsmessgerät". Accessed: 31.08.2012 from: http://www.zes.com/download/produkte/zes_lmg500_datenblatt_d.pdf.

Zrilli, T. (2005): "Plasma und plasmakatalytische Verfahren zum NOx-Abbau im Dieselabgas". dissertation, earned degree: PhD, Universität Fridericiana Karlsruhe.

VI. ABBREVIATIONS, UNITS AND NOTATION

In the table below, all relevant abbreviations, notations and variables are listed and specified regarding their unit (in typical order of magnitude), a short description and the page of first introduction within this work.

Variable:	Unit:	Description:	Page:
α	[Ω/μH]	attenuation of a damped resonant circuit	71
A_{active}	[cm²]	total active area, mainly of flat DBD lamps	31
A_{CU}	[mm²]	cross-sectional area of a copper conductor	98
$A_{CU,prim}$	[mm²]	cross-sectional area of a copper conductor of primary winding	107
$A_{CU,sec}$	[mm²]	cross-sectional area of a copper conductor of secondary winding	107
A_e	[mm²]	cross-sectional area of a magnetic core	99
A_L	[nH]	inductivity constant, core dependent	99
AP	-	adaptive pulse (topology)	169
ARCPI	-	auxiliary resonant commutated pole inverter	169
AUP	-	adaptive universal pulse (topology)	157
B_{max}	[T]	maximum magnetic flux density	98
C_b	[nF]	barrier capacitance of DBD	29
C_{block}	[nF]	DC-blocking capacitor	87
C_{charge}	[pF]	(var.) capacitance of the charge cloud appearing after ignition	29
C_{DBD}	[nF]	total DBD capacitance	44
$C_{DBD,bare}$	[nF]	bare DBD capacitance (excluding optional C_p)	41, 51
C_{DBD+P}	[nF]	DBD capacitance including parallel capacitor C_P	42
$C_{DBD,prim}$	[nF]	DBD capacitance, primary-transferred (transformer)	104
$C_{DBD,sec}$	[nF]	DBD capacitance, secondary-transferred (transformer)	104
C_{DC}	[μF]	direct current blocking capacitor or bypass capacitor	95
C_{DS}	[pF]	drain-source capacitance	90
CF	-	characteristic factor qualifying the impact of $V_{g,ext}$	49
C_g	[pF]	gap capacitance of DBD	29
C_{iw}	[pF]	inter-winding capacitance of a transformer	104

C_{link}	[µF]	intermediate circuit capacitor, DC-link capacitor	69
CM	-	common mode (noise, signals)	105
$C_{OSS(,HS)}$	[pF]	power electronics switch output capacitance (of HS)	114
C_p	[nF]	parallel capacitance that may be attached to bare DBD	41
C_{pp}	[nF]	equivalent parallel capacitance	120
$C_{p,L}$	[nF]	parallel capacitance of an inductor	104
$C_{p,Tr}$	[nF]	parallel capacitance of a transformer	104
C_S	[pF]	equivalent series capacitance	137
C_{shift}	[nF]	level-shift capacitor, mainly in gate drives	95
C_{sn}	[pF]	snubber capacitor	175
C_{suprt}	[nF]	equivalent capacitance in parallel to the gap	50
$C_{th,device}$	[J/K]	thermal capacity of a power dissipating device	88
CW	-	continuous wave, opposite to pulsed operation	55
δ	[mm]	skin depth	98
D	-	diode in general	162
D_1, D_4	-	diode in series to HS	114, 147
D_2, D_3	-	HS Anti-parallel diode	114, 147
d_b	[mm]	total dielectric barrier width of a DBD lamp	29
DBD	-	dielectric barrier discharge (lamp or reactor)	19
D_{ER}	-	energy recovery diode	135
DFN	-	package style, dual flat no-lead	94
d_{gap}	[mm]	gap width of a DBD lamp	29
DIN	-	Deutsche Industrie-Norm (DE), German industrial standard	20
DIP	-	package style, dual in-line	94
D_{PC}	-	pre-charge diode making S_{PC} unidirectionally conducting	171
d_{ww}	[mm]	distance of two wires	98
ECG	-	electronic control gear, electronic ballast, inverter	19
$E_{DBD(,pk)}$	[mJ]	electrical energy stored within the DBD's capacitances	47, 100
$E_{DBD,rel}$	[mJ/kV^2	relative gap energy	49
EE	-	energy ratio: E accessible to plasma/ E stored in the DBD	48
$E_{g,ext}$	[mJ]	electrical energy provided to plasma assuming v_g = constant	47
$E_{g,ign}$	[mJ]	electrical energy provided to plasma assuming $V_{g,ext} = 0$ V	48
E_g	[mJ]	maximum energy accessible by the plasma	51
$E_{L,pk}$	[mJ]	maximum energy stored in inductor L	80, 100

EMC	-	electro-magnetic compatibility	55
EMR	-	electro-magnetic radiation	81
$E_{precharge}$	[mJ]	maximum inductor energy after inductor pre-charge	155
E_{recov}	[mJ]	maximum energy stored in inductor for energy recovery	155
ERS	-	energy recovery switch	135
ESL_{Cp}	[nH]	equivalent series inductance of parallel capacitor C_p	43
ESL_{Clink}	[nH]	equivalent series inductance of DC-link capacitor C_{link}	82
ESR_{Cp}	[Ω]	equivalent series resistance of parallel capacitor C_p	43
ESR_{Clink}	[nH]	equivalent series inductance of DC-link capacitor C_{link}	82
E_{tot}	[mJ]	total energy stored in a resonant circuit	72
ε_0	[Fm^{-1}]	absolute dielectric constant (permittivity) $8.854187817 \times 10^{-12}$	29
e_B	-	exponent of B in Steinmetz formula	98
e_f	-	exponent of f in Steinmetz formula	98
ε_{rb}	-	relative dielectric constant	29
f_0	[kHz-	(natural) resonance frequency of a circuit	69
f_c	[kHz]	frequency of control signals applied to switches	69
f_d	[kHz]	resonance frequency of a damped resonant circuit	71
FOC	-	fibre-optic cable	93
FR4	-	common PCB material based on glass fibre epoxy compound	85
f_{res}	[MHz]	resonance frequency, in general	104
$f_{res,DBD}$	[MHz]	natural resonance frequency of the DBD lamp	44
$f_{res,DBD+p}$	[MHz]	natural resonance frequency of the DBD lamp with attached C_p	44
$f_{res,max,I}$	[MHz]	maximum possible resonance frequency, current-limited	80
$f_{res,max,Q}$	[MHz]	maximum possible resonance frequency, Q_S-limited	80
F_{Pdc}	-	primary-secondary power loss distribution factor (transformer)	107
f_{pp}	[MHz]	parasitic parallel resonance frequency	119
f_{ps}	[MHz]	parasitic series resonance frequency	119
FWHM	-	full-width half-maximum, describes optical line width	24
HS	-	high side switch (of a bridge configuration)	69
i_{blind}	[A]	blind component flowing through C_g	29
i_{Cp}	[A]	current through parallel capacitor C_p	41
$I_{C,i}$	[A]	initial current through capacitor C	73
$i_{DBD(')}$	[A]	total current through DBD (primary transferred)	29
$i_{DBD,bare}$	[A]	current through bare DBD (excluding optional C_p)	41

$I_{DBD,i}$	[A]	initial DBD current	78
I_{DS}	[A]	drain-source current	90
i_{Lm}	[A]	current through mutual inductance of transformer	118
$I_{L,pk}$	[A]	inductor peak current	79
$I_{L,rms}$	[A]	inductor rms current	78
i_{MS}	[A]	current through matrix switch	163
INZV	-	interim non-zero voltage: voltage during idle time is different from	55
I_{pk}	[A]	peak current of L **or** C_{DBD} in a resonant circuit	71
i_{plasma}	[A]	plasma current	29
$i_{plasma,Cp}$	[A]	current provided by parallel capacitor C_p supporting plasma	42
$i_{precharge,pk}$	[A]	maximum pre-charge current	155
i_{prim}	[A]	primary current, related to transformer	104
IR	-	infrared	134
$i_{recov,pk}$	[A]	maximum recovery current	155
i_{sec}	[A]	secondary current, related to transformer	104
IZV	-	interim zero voltage: voltage during idle time is equal or almost	55
K	-	K-factor describing the influence of conductor proximity	98
L	[μH]	inductor, inductive part of the resonant circuit	69
l_e	[cm]	average length of the magnetic path	99
l_g	[mm]	total air gap length	100
L_m	[mH]	mutual inductance of the transformer	104
L_{min}	[μH]	minimum possible inductance	80
LS	-	low-side switch (of a bridge configuration)	69
$L_{S(P)}$	-	inductor, inductive part of the series (parallel) resonant circuit	86
$L_{stray(,Tr)}$	[μH]	stray inductance of the transformer	104
$L_{stray,DBD}$	[μH]	parasitic DBD stray inductance	43
l_w	[m]	length of a copper conductor	98
μ	[Hm^{-1}]	electromagnetic permeability	98
μ_0	[N/A^2]	magnetic field constant, value: $4\pi 10^{-7}$	99
μ_C	[N/A^2]	electromagnetic permeability of a magnetic core without air gap	100
μ_{eff}	[N/A^2]	effective permeability of a magnetic core with air gap	100
μ_r	-	relative permeability	99
MLCC	-	multi-layer ceramic capacitor	87
MS	-	matrix switch	162

MTTH	-	mean time to half brightness	27
η_{DBD}	[lm/W]	DBD lamp efficiency	139
$\eta_{(ECG)}$	[%]	electrical efficiency of the inverter	134
n	-	number of energy recovery cycles	148
N	-	number of turns of a winding, related to magnetics	106
NBOH	-	non-bridging oxygen hole centres, referred to as degradation effect	32
NIR	-	narrow infrared	35
osr	-	overshoot ratio of resonant peak voltage to supply voltage	73
P_0	-	factor of the Steinmetz-Formula, also referred to as core constant	98
P_{DBD}	[W]	real power consumed by the DBD	43
PDP	-	plasma display panel	26
PF	-	power Factor, also λ, ratio of absolute real power to apparent	30
P_{loss}	[W]	power loss, generally	88, 98
P_{MS}	[W]	matrix switch power loss	165
PMT	-	photo-multiplier tube	46
POF	-	polymer optical fibre	93
P_R	[W]	real power dissipated in resistor R	75
p_{Xe}	[mbar]	partial xenon pressure	31
Q_b	[var]	reactive power of a resonant circuit	71
Q_G	[As]	gate charge	90
Q_{GD}	[As]	gate-Drain charge, referred to Miller-effect	90
$Q_{S(P)}$	-	quality factor of a series (parallel) resonant circuit	71, 86
ρ_{CU}	[Ωmm^2	resistivity of copper, 17,8 × 10^{-3} Ωmm^{2}m^{-1}	98
r	-	transformer ratio	104
R_{AC}	[Ω]	AC resistance of a conductor	98
R_B	[Ω]	gate resistor, connected to source	95
R_{bound}	[Ω]	resistance determining boundary to aperiodic circuit behaviour	71
$R_{Cu,Tr}$	[Ω]	parasitic winding resistance of a transformer	104
$R_{Cu,L}$	[Ω]	parasitic winding resistance of an inductor	104
$R_{(Cu,)DBD}$	[Ω]	parasitic resistance of the DBD	43, 104
R_{DC}	[Ω]	DC resistance of a conductor	98
$R_{DS,on}$	[Ω]	drain-source resistance	165
RG	-	reverse geometry (related to SMD chip resitors)	97
R_{HC}	[Ω]	gate resistor, connected to gate, high current path	96

r_{inner}	[mm]	inner radius of inner winding (defined by bobbin)	107
R_{LC}	[Ω]	gate resistor, connected to gate, low current path	96
$R_{L,s}$ / R_S	[Ω]	series resistor as part of a damped resonant circuit	69, 86
$R_{L,p}$ / R_P	[Ω]	parallel resistor as part of a damped resonant circuit	72, 86
R_m	[Ω]	magnetising resistance	104
rms	-	root mean square	139
r_{outer}	[mm]	outer radius of outer winding (defined by bobbin)	107
R_p	[Ω]	equivalent parallel resistor	120
R_{plasma}	[Ω]	resistance of the plasma ignited within the DBD's gap	38
r_w	[mm]	radius of a wire	98
σ	[$\Omega^{-1}m^{-1}$]	electrical conductivity	98
SBD	-	Schottky barrier diode	145
SDBD	-	surface dielectric barrier discharge	33
S_{DBD}	[kVA]	apparent power merely required to charge and discharge C_{DBD}	43
SI	-	Système International d'Unités (FR)	20
SOA	-	safe operating area	117
SP	-	sinusoidal pulse (topology)	113
S_{PC}	-	pre-charge switch	171
SRI	-	series Resonant Inverter	68
SW	-	power semiconductor switch in general	153
t_{ceil}	[ns]	time within a pulse when high constant voltage is applied to DBD	56
T_d	[μs]	period of a damped resonant circuit	71
$t_{discharge}$	[ns]	time at which DBD discharge occurs	47
ΔT_{device}	[K]	temperature increase of a power dissipating device	88
TDR	-	time-domain reflectometry	85
t_{fall}	[ns]	fall time of pulse (90-10% of $v_{DBD,pk}$)	56
t_{idle}	[μs]	idle time between the pulses, low voltage applied to DBD	56
Tr	-	transformer	104
T_{rep}	[μs]	pulse Repetition time	43
Tr_{ideal}	-	transformer with ideal properties	104
t_{rr}	[ns]	reverse recovery time of diodes	90
$t_{precharge}$	[μs]	time required for inductor pre-charge	155
t_{pulse}	[ns-μs]	duration of pulse	56
t_{recov}	[μs]	time required for energy recovery	155

t_{rise}	[ns]	rise time of pulse (10-90% of $v_{DBD,pk}$)	56
USP	-	universal sinusoidal pulse (topology)	170
USPSP	-	universal single period sinusoidal pulse (topology)	177
v_b	[V]	voltage across barrier of DBD	29
v_C	[V]	voltage across capacitor C	73
V_C	[cm^3]	magnetic core volume	100
v_{CDC}	[V]	voltage across DC-blocking capacitor C_{DC}	124
$V_{C,i}$	[V]	initial voltage across capacitor C	73
$V_{C,pk}$	[V]	peak voltage across capacitor C	71
v_{DBD}	[kV]	voltage across DBD	29
$v_{DS,HS}$	[V]	drain-source voltage, exemplarily of HS	69
$v_{GS,HS}$	[V]	gate-source voltage, exemplarily of HS	69
$v_{DBD(')}$	[kV]	voltage across DBD lamp (primary transferred)	29
$v_{DBD,ign}$	[kV]	outer DBD ignition voltage	34
$v_{DBD,init}$	[kV]	initial DBD voltage, related to time-domain	148
$v_{DBD,int}$	[V]	internal DBD voltage	45
$V_{DBD,pk}$	[kV]	maximum voltage across DBD (during pulse)	47, 100
$V_{DS,HS}$	[V]	drain to source voltage of high-side switch HS	69
$V_{DS,LS}$	[V]	drain to source voltage of low-side switch LS	69
v_g	[kV]	voltage across gap of DBD	29
$V_{g,ign}$	[kV]	voltage across gap of DBD at time of ignition	47
$V_{g,ext}$	[kV]	plasma extinguishing voltage	48
V_{GS}	[V]	gate-source voltage	90
v_{LS}	[V]	voltage across drain-source of low-side switch LS	117
v_{MS}	[kV]	voltage across matrix switch	163
V_{pk}	[kV]	peak voltage of L **or** C_{DBD} in a resonant circuit	71
V_S	[V]	supply voltage	69
vsef	-	factor determining the capacitive plasma energy support	42
v_{SW}	[V]	voltage across power semiconductor switch in general	153
ω_0	[1/µs]	natural angular resonance frequency of a resonant circuit	70
ω_d	[1/µs]	angular resonance frequency of a damped resonant circuit	70
$\varpi_{g,ign}$	[mJ/m]	electrical energy density of the gap	48
w	[mm]	total winding window width	107
w_p	[mm]	width of primary winding	107

w_s	[mm]	width of secondary winding	107
z	-	air gap factor	99
Z_0	[Ω]	characteristic impedance of a resonant circuit	71
ZCS	-	zero current switching (soft switching)	70
ZD	-	zener diode	95
$Z_{P(S)}$	[Ω]	characteristic impedance of a parallel (series) resonant circuit	86
Z_{excite}	[kΩ – Ω]	variable impedance of the plasma ignited within the DBD's gap	29
ZVS	-	zero voltage switching (soft switching)	70

VII. List of Publications

1. **Meisser, M., K. Messerschmidt, et al. (2012)** „MOSFET Matrix-Switch for Pulsed Operation of Plasma Optical Radiation Sources", EAPPC-BEAMS, Karlsruhe, DE.
2. **Hähre, K., M. Meisser, et al. (2012)** „Switching Speed-Control of an Optimized Capacitor-Clamped Normally-On Silicon Carbide JFET Cascode". EPE-PEMC 2012, Novi Sad, RS.
3. **Meisser, M. and K. Haehre (2012)** "Impedance characterization of high-frequency gate drive circuits for silicon RF MOSFET and Silicon-Carbide field-effect transistors". PCIM Europe. Nuremberg, DE.
4. **Meisser, M., R. Kling (2012)** "Dielektrisch behinderte Entladungs-Lampe", patent application (pending), Karlsruhe Institute of Technology, DE.
5. **Haehre, K., M. Meisser, et al. (2012)** "Characterization and comparison of commercially available Silicon Carbide (SiC) power switches." PEMD 2012. Bristol, GB.
6. **Meisser, M., K. Haehre, et al. (2012)** "Impedance characterization of high frequency power electronic circuits." PEMD 2012. Bristol, GB.
7. **Meisser, M., R. Kling, et al. (2011)** "Transformerless high voltage pulse generators for bipolar drive of Dielectric Barrier Discharges." PCIM 2011. Nuremberg, DE.
8. **Meisser, M., R. Kling, et al. (2011)** "Universal resonant topology for high frequency pulsed operation of Dielectric Barrier Discharge light sources." APEC 2011. Fort Worth, TX, USA.
9. **Meisser, M., M. Paravia, et al. (2010)** "Resonance behaviour of a pulsed electronic control gear for Dielectric Barrier Discharges." PEMD. Brighton, UK.
10. **Meisser, M., M. Paravia, et al. (2010)** "Single period sinusoidal pulse generator for efficient drive of Dielectric Barrier Discharges." PCIM. Nuremberg, DE.
11. **Meisser, M. (2010).** EP000002389047A1: „Schaltungsanordnung und Verfahren für den Betrieb einer kapazitiven Last.", Karlsruhe Institute of Technology, DE.
12. **Meisser, M., M. Paravia, et al. (2010)** "Driving of dielectric barrier discharge lamps with a pulsed transformer-less full-bridge topology." LS12. Eidhoven, NL.
13. **Paravia, M., M. Meisser, et al. (2009)** "Plasma efficiency and losses for pulsed Xe Excimer DBDs at high power densities." GEC. Sagatoga Springs, USA.
14. **Paravia, M., M. Meisser, et al. (2008)** "Arguments for increased efficiency of a Xe excimer DBD by pulsed instead of sinusoidal excitation", GEC. Dallas, TX, USA.
15. **Meisser, M. (2005).** DE102005025127A1: „Pulsweitenmodulation mit Gleichspannungs-überlagerung", DE.

VIII. Curriculum Vitae

Personal Data:

Name:	Michael Meisser (Meißer)
Date of Birth:	02.07.1980 in Schwerin, Germany
Nationality:	German
Networks:	LinkedIn  

Personal Profile:

Michael is graduate of the University of Rostock, Germany, and finished his doctorate at the Light Technology Institute (LTI) of the Karlsruhe Institute of Technology (KIT) in 2013. At the LTI, he was involved in multiple industrial projects and already gained project-lead experience. Michael is very keen to search for solutions of complex problems and is capable of familiarising with special topics within a short time frame. He is interested in new technologies, new approaches and is open-minded. Michael has very good knowledge in written and spoken English and visits international conferences and trade fairs on a regular basis. Dealing with people and spending his free-time in nature is what he enjoys. Since an age of 14, he is interested in electric and electronic circuitry.

Professional Profile:

Michael's work and research interest is mainly in the field of power electronic circuit development, especially for pulsed power applications. His research at the LTI was focussed on efficient ways of generating visible and ultra-violet optical radiation by the optimal drive of gas discharge lamps. Moreover, the characterisation and the beneficial operation of modern silicon-carbide power semiconductors, which play a major role in future power electronic concepts, was part of his studies.

Education:

11/2007 – 07/2013	PhD student: LTI of the KIT, scientific research and development in the field of pulsed power, power semiconductor switches and device characterisation.
10/2001 – 02/2007	Student: University of Rostock, Faculty of Electrical Engineering, Germany, specialisation in device systems and circuitry, final mark 1,7.
1997 – 2000	Equivalent A-level: Fachgymnasium Schwerin, Germany main subject: electrical engineering, final mark 1,3.

EXPERIENCE:

11/2007 –03/2013	Scientific co-worker: Light Technology Institute (LTI) of the Karlsruhe Institute of Technology (KIT), project work and management of industrial projects, self-reliant work on scientific challenges, scientific publication, student supervision, responsible for the power electronics laboratory and LTI's total quality management.
04/2012 – 06/2012	Visiting academic: University of Warwick, PEATER laboratory, scholarship of the Karlsruhe House of Young Scientists, design and manufacture of power electronic semiconductor device modules (power electronic packaging).
04/2007 – 06/2007	Internship: Technical University of Graz, Austria, Institute of Building Physics: manufacture, test, maintenance and repair of measurement systems.
06/2006 – 02/2007	Diploma student: Argus Electronic, topic: „Development of a measurement device for the investigation of the status of wood poles" (ultrasonic measurement), mark 1,3.
09/2005 - 02/2006	Internship: National Technical University of Athens (NTUA), Greece: implementation of Fuzzy-Logic-controllers in hardware (VHDL). Deepened English knowledge and experienced foreign cultures.
08/2004 – 03/2007	Treasurer, project manager: Electrical Engineering by Students e.V., student's company with focus in electronic product development.
03/2003 – 04/2003	Internship: NORDEX Energy, Rostock, Germany.
03/2002 – 04/2002	Internship: STN Schiffselektrik, Elmenhorst, Germany.

QUALIFICATION:

2001 – 2009	Correspondence course at Fernuni Hagen, degree: Patent Engineer, course in technical English (UNICERT III-degree).
2007 – 2013	Attended numerous courses in the fields of time management, professional creativity methods.
Permanently	Electronic CAD & simulation (EAGLE, Allegro, Matlab, COMSOL), Microsoft Office